Advances in Personalized Nanotherapeutics

Ajeet Kaushik • Rahul Dev Jayant
Madhavan Nair
Editors

Advances in Personalized Nanotherapeutics

Springer

Editors
Ajeet Kaushik
Center for Personalized Nanomedicine
Institute of Neuro-Immune Pharmacology
Department of Immunology
Herbert Wertheim College of Medicine
Florida International University
Miami, FL, USA

Rahul Dev Jayant
Center for Personalized Nanomedicine
Institute of Neuro-Immune Pharmacology
Department of Immunology
Herbert Wertheim College of Medicine
Florida International University
Miami, FL, USA

Madhavan Nair
Center for Personalized Nanomedicine
Institute of Neuro-Immune Pharmacology
Department of Immunology
Herbert Wertheim College of Medicine
Florida International University
Miami, FL, USA

ISBN 978-3-319-63632-0 ISBN 978-3-319-63633-7 (eBook)
https://doi.org/10.1007/978-3-319-63633-7

Library of Congress Control Number: 2017958816

Printed on acid-free paper

This Springer imprint is published by Springer Nature
The registered company is Springer International Publishing AG
The registered company address is: Gewerbestrasse 11, 6330 Cham, Switzerland

Endorsed by the Society of Personalized Nanomedicine

http://www.s-pnm.org/our-society.html

SPNM
Society of Personalized Nanomedicine

The Society for Personalized NanoMedicine (SPNM) aims to tailor medical intervention to patient- and disease-specific needs. The SPNM's mission is to promote research, serve as a source of information on current applications of nanotechnology, and foster the exchange of information and ideas on personalized nanomedicine. Our vision and goals are to merge interdisciplinary research in order to increase our understanding of current applications of nanotechnology. These applications include reconstructive surgery; targeted therapy; the latest research on nanodevices, drug development, and drug delivery; and the use of microelectronics and high-precision lithography for the production of nanocomposites for personalized medical use. The SPNM also promotes translational research that focuses on the interactions between the human immune system, substance abuse, HIV, and cancer, in order to create a solid ground for the development and application of groundbreaking medical devices and systems for superior diagnosis and treatment.

Madhavan Nair, Ph.D.
Founder and President, Society for Personalized Nanomedicine
Distinguished Professor and Chair, Department of Immunology
Director, Institute of NeuroImmune Pharmacology
Herbert Wertheim College of Medicine
Associate Dean of Bio-Medical Research
Associate Vice-President for NanoMedicine
Florida International University
Miami, FL, USA

Preface

Personalized health care management and optimization of the treatment of disease is crucial for improving the quality of health. Significant efforts have been made to design and develop novel nanotherapeutics strategies for the proficient monitoring and treatment of disease in a personalized manner. As per the state of the art, there are various strategies that involve the development of novel nanomaterials; novel drug delivery systems; the discovering of novel therapeutic agents; the integration of devices for better biosensing technology; and new therapeutic agents for the development of personalized nanomedicine to combat targeted diseases with no side effects. Besides nano-drug delivery, attention has also been focused on describing nano-enabled sensors, miniaturizing sensing systems, the interfacing of sensing components, and developing smart portable systems for point-of-care (POC) applications to detect biomarkers at very low levels in order to monitor the progression of targeted diseases. Such systems have also been used to assess the therapeutic efficacy of medicines that are specifically prescribed for the targeted diseases.

This book describes the fundamentals of nanomedicine; personalized therapeutics; novel nanomaterials for drug delivery; the role of nanotechnology in investigating therapeutic approaches; targeted CNS drug delivery; stimuli-responsive drug release; nano-enabled sensing systems for health care; and disease management. The future prospects of personalized nanotherapeutics and related challenges – with possible solutions – are also discussed. The book can be the source for new ideas to design and develop novel biomaterials, novel nano-formulations, targeted delivery, translational medicine, the scaling up of nanomedicine to a clinical phase, POC-sensing systems for rapid diagnostics, and the promotion of nano-pharmacology for next-generation personalized medicine.

This book will also be very useful for helping young scholars understand the exploration of state-of-the-art nanotechnology for personalized health care; it will also help researchers design their future investigations towards developing effective personalized nanomedicine and diagnostic healthcare systems. Numerous studies have reported on the design and development of nanomedicines with higher efficacy, but unfortunately such products are in the laboratory research phase only and need to be thoroughly tested, using pre-clinical or human models. Our book can be a call for experts to explore multidisciplinary research for developing novel and effective approaches to exploring smart, efficient nanocarriers for site-specific, on-demand controlled drug delivery to combat targeted diseases, and smart sensing systems to detect targeted biomarkers at the fM level, for complete personalized healthcare.

Miami, FL, USA

Ajeet Kaushik
Rahul Dev Jayant
Madhavan Nair

Contents

Contributors

Sharif Ahmad Department of Chemistry, Materials Research Laboratory, New Delhi, India

Hamed Arami Molecular Imaging Program at Stanford (MIPS), The James H Clark Center, Stanford University, Stanford, CA, USA

Department of Radiology, Stanford University School of Medicine, Stanford, CA, USA

Venkata Atluri Department of Immunology, Institute of Neuroimmune Pharmacology, Herbert Wertheim College of Medicine, Florida International University, Miami, FL, USA

Vinay Bhardwaj Department of Biomedical Engineering, Rutgers, The State University of New Jersey, Piscataway Township, NJ, USA

Jaydeep Bhattacharya School of Biotechnology, Jawaharlal Nehru University, New Delhi, India

Rashmi Chaudhari Department of Biosciences and Bioengineering, Indian Institute of Technology, Bombay, Mumbai, India

Ravi Doddapaneni Department of Ophthalmology, Bascom Palmer Eye Institute, University of Miami Miller School of Medicine, Miami, FL, USA

Anujit Ghosal Department of Chemistry, School of Basic and Applied Sciences, Galgotias University, Gautam Buddh Nagar, Uttar Pradesh, India

School of Biotechnology, Jawaharlal Nehru University, New Delhi, India

Priyanka Giri Department of Pharmacology and Toxicology, National Institute of Pharmaceutical Education and Research Hyderabad (NIPER Hyderabad), Balanagar, Telangana, India

Rahul Dev Jayant Center for Personalized Nanomedicine, Institute of Neuro immune Pharmacology, Department of Immunology, Herbert Wertheim College of Medicine, Florida International University, Miami, FL, USA

Abhijeet Joshi Centre for Biosciences and Bio-medical Engineering, Indian Institute of Technology Indore, Indore, Madhya Pradesh, India

Jyothirmai Kaligatla Department of Pharmacology and Toxicology, National Institute of Pharmaceutical Education and Research Hyderabad (NIPER Hyderabad), Balanagar, Telangana, India

Anil Kumar Kalvala Department of Pharmacology and Toxicology, National Institute of Pharmaceutical Education and Research Hyderabad (NIPER Hyderabad), Balanagar, Telangana, India

Krishna Kanhaiya Computational Biomodeling Laboratory, Turku Centre for Computer Science, Åbo Akademi University, Turku, Finland

Babak Kateb National Center for NanoBioElectronics, West Hollywood, CA, USA

California Neurosurgical Institute, Los Angeles, CA, USA

Brain Mapping Foundation, West Hollywood, CA, USA

Society for Brain Mapping and Therapeutics, West Hollywood, CA, USA

Ajeet Kaushik Center for Personalized Nanomedicine, Institute of Neuro immune Pharmacology, Department of Immunology, Herbert Wertheim College of Medicine, Florida International University, Miami, FL, USA

Wahid Khan Department of Pharmaceutics, National Institute of Pharmaceutical Education and Research Hyderabad (NIPER Hyderabad), Balanagar, Telangana, India

Ashutosh Kumar Department of Pharmacology and Toxicology, National Institute of Pharmaceutical Education and Research Hyderabad (NIPER Hyderabad), Balanagar, Telangana, India

Zimple Matharu Department of Electrical and Computer Engineering, University of California–Davis, Davis, CA, USA

Chandini C. Mohan Center for Soft and Living Matter, Institute for Basic Science (IBS), Ulsan National Institute of Science and Technology, Ulsan, Republic of Korea

Madhavan Nair Center for Personalized Nanomedicine, Institute of Neuro immune Pharmacology, Department of Immunology, Herbert Wertheim College of Medicine, Florida International University, Miami, FL, USA

Roozbeh Nikkhah-Moshaie Department of Immunology, Herbert Wertheim College of Medicine, Miami, FL, USA

Center for Personalized Nanomedicine, Florida International University, Miami, FL, USA

Eliset Perez Department of Immunology, Institute of Neuroimmune Pharmacology, Herbert Wertheim College of Medicine, Florida International University, Miami, FL, USA

Ozge Polat Department of Electrical and Computer Engineering, University of California–Davis, Davis, CA, USA

PSS Rao Department of Pharmaceutical Sciences, College of Pharmacy, The University of Findlay, Findlay, OH, USA

Abhijit Chandra Roy Soft Matter Laboratory, Department of Chemical Engineering, Indian Institute of Technology, Kanpur, Uttar Pradesh, India

Krati Sharma Scientific Technician-2, Fox Chase Cancer Center, Philadelphia, PA, USA

Eram Sharmin Department of Pharmaceutical Chemistry, College of Pharmacy, Riyadh, Kingdom of Saudi Arabia

Renu Singh Department of Bioproducts and Biosystems Engineering, University of Minnesota, Minneapolis, MN, USA

Christopher RT Stang Department of Pharmaceutical Sciences, College of Pharmacy, The University of Findlay, Findlay, OH, USA

Yasushi Takemura Department of Electrical and Computer Engineering, Yokohama National University, Yokohama, Japan

Shivani Tiwari Department of Chemistry, School of Basic and Applied Sciences, Galgotias University, Gautam Buddh Nagar, Uttar Pradesh, India

Asahi Tomitaka Center for Personalized Nanomedicine, Institute of Neuro immune Pharmacology, Department of Immunology, Herbert Wertheim College of Medicine, Florida International University, Miami, FL, USA

Arti Vashist Center for Personalized Nanomedicine, Institute of Neuro immune Pharmacology, Department of Immunology, Herbert Wertheim College of Medicine, Florida International University, Miami, FL, USA

Atul Vashist Department of Biotechnology, All India Institute Medical Sciences, New Delhi, India

Chapter 1
Nanomedicine

Vinay Bhardwaj and Roozbeh Nikkhah-Moshaie

Abstract In this chapter, we start with the origin of nanotechnology and nanomedicine and the definition of nanomedicine in the USA and Europe. The size range of nanomedicine is discussed according to different definitions and regulations, and development of nanomedicine from lab to market is reviewed. Nanomaterials for nanomedicine is divided into two categories of polymer-based and metal-based. A brief explanation of most important nanomaterials including liposome, carbohydrates, gold, silver, and iron oxide in drug delivery and release, sensing and imaging are given. Nanomedicines already showed their potential and are promising to cure cancer, neurodegenerative disorders including NeuroHIV, Alzheimer's, and Parkinson's.

Keywords Nanotechnology • Nanomedicine • Nanomaterials • Nanocarriers • Drug Delivery

1 Genesis

The genesis of nanotechnology and nanomedicine can be dated back to a popular talk "There's plenty of room at the bottom" in December 29, 1959 at California Institute of Technology by late Dr. Richard P. Feynman, a Nobel laureate in physics [1]. Although he did not refer to the words "nano, nanotechnology or nanomedicine", he prophetically envisioned the unavoidable revolution in development of tiny machines and robots for biomedical applications. In an essay to his friend and graduate student he presented the first nanomedicine proposal, he wrote, it would be interesting if one could "swallow the surgeon". You put the surgeon, those tiny

V. Bhardwaj (✉)
Department of Biomedical Engineering, Rutgers, The State University of New Jersey, Piscataway Township, NJ, USA
e-mail: bhard.vinay@gmail.com

R. Nikkhah-Moshaie
Department of Immunology, Herbert Wertheim College of Medicine, St, Miami, FL, USA

Center for Personalized Nanomedicine, Florida International University, St, Miami, FL, USA

A. Kaushik et al. (eds.), *Advances in Personalized Nanotherapeutics*,
https://doi.org/10.1007/978-3-319-63633-7_1

robots inside blood vessel and the surgeon will move around inside your heart with little knife to slice defected heart valve. The concept of nanodevices, nanoknives, nanoscissors, which we still see on many cover pages of the books and magazines was in fact his original idea. Besides those two small inventions to win $1000 prize money for Dr. Feynman's challenge to the public, building of a tiny motor by a craftsman and writing by a Stanford graduate student on a 1/25,000 smaller surface, there were no noticeable breakthroughs to make Feynmans dream true until 1974. Late Dr. Norio Taniguchi, a professor at Tokyo University of Science coined the term "Nanotechnology" to describe control of semiconductor process on the order of nanometer [2]. Later, Dr. Kim Eric Drexler, an American Engineer at Massachusetts Institute of Technology proposed the idea of molecular nanotechnology in 1986 in his book "Engines of Creation: The coming era of nanotechnology". There is a controversy on the ontological status of the founding father of nanotechnology among Drs. Feynman, Taniguchi and Drexler. Authors of this chapter agree with a fairly recent sentimental statement by Dr. Richard A. L. Jones in 2006 in context to the creator of Nanotechnology. Though very unlikely and hard for many people to accept, US president Bill Clinton can be considered creator of nanotechnology as he is the first person to believe in potential of nanotechnology and nanomedicine. He launched $475 million National Nanotechnology Initiative (NNI) in 2000. The potential of nanotechnology in medicine "nanomedicine" was also realized at the beginning of this first initiative. US National Institutes of Health (NIH) launched 4-year program for nanoscience and nanotechnology in medicine in 2002 [3]. The nanotechnology funding has now grown to the proposed $1.4 billion budget for fiscal year 2017. Following US, other countries also invested heavily in nanotechnology initiatives, India's Department of Science and Technology's modest-size Nano Science and technology Initiative of a few crore rupees in 2001 has grown to 1000 crore nanomission in 2007, The European Commission's 1.3 billion euro in 2003 to current multi-billion budget, and other more than 60 countries are heavily investing in nano research and development.

2 Definition: What Can Be Considered a Nanomedicine

Although nanotechnology and nanomedicine has become widely adopted, it is surprising to know that none of the two has globally recognized definitions. And to enforce regulation, in particular safety assessment of nanomedicine, it is important to have a standard definition and classification of nanotechnoloigy and nanomedicine. Basically, every definition of nanotechnology and nanomedicine would include three elements. First, range of size, usually upper limit set to 100 nanometers (nm) diameter. Second, measurement and transformation at nanoscale, a significant progress has been achieved with the advancement of electron microscopy [4]. Third and most important motivating factor, nanoscale-specific properties and functions over micro and bulk scale. Both sides of the Atlantic Ocean have adopted different definitions. US NNI defines nanotechnology as the ability to understand, control, and manipulate matter at the level of individual atoms and molecules, as well as "supramolecular" level involving clusters of molecules (in the range of about 0.1–100 nm) in order to create materials, devices, and systems with fundamentally new properties and functions because of their small structure [5]. However, many nanotechnologies/nanomedicines including US Food and Drug Administration (FDA)-approved, contradict this definition. Unique physiochemical properties from nanomaterials with cutoff >100 nm has been observed by scientific communities, for example, as compared to 70 nm size, 150–200 nm liposomes remain longer in the bloodstream allowing better therapeutics [6], plasmon resonance in 150 nm gold nanoshells is under clinical trials for thermal therapy of cancer [7], Myocet (180 nm) and Abraxane (130 nm) [8]. Due to this conflict in size of the nanotechnology or nanomaterilas, its definition undergoes continuous negotiation and revision. Fairly recent, European Commission's adopted new definition (2011) of nanomaterial "a natural, incidental or manufactured material containing particles, in an unbound state or as an aggregate or as an agglomerate, and where, for 50% or more of the particles in the number size distribution, one or more external dimensions is in the size range 1–100 nm [9]. Any revised definition like this has significant impact on nanotech industries as they have to abide by the new regulations, inclusion of number concentration over mass concentration that too more than equals 50 %, aggregates and upper size limited to 100 nm. An example of such an impact is that many European and American companies has started reformulating their products to a size >100 nm to bypass nanomedicine safety assessment. Contrarily, many Chief Scientific officers of the companies argued to this definition to increase nano cutoff to 300 nm [10] so that they already approved nanomedicines can retain the status. An illustration, a formulation with 40% particles of 90 nm and remaining 60% particles of 100 nm will not be considered nanomedicine according to this definition. US FDA approved nanomedicines Myocet, Abraxane, and many more, in particular liposomal nanomedicine are no longer nanomedicine according to the definition adopted by European Commission. Therefore, due to this conflict in size of the nanomedicine, it is important to inform readers that in this book we did not restrict to 100 nm as the cut off size to evaluate nanomedicine. Physio-chemical characterization of nanoparticles and nanomedicine for their shape, size, bioactive conjugation and release are also of great importance.

Broadly, nanotechnology is the engineering of functional systems at the molecular scale, and nanomedicine is a subset of nanotechnology with applications in medicine. It will not be incorrect to say that there is no nanomedicine but nanotechnology in medicine. According to European Science Foundation, nanomedicine is "the science and technology of diagnosing, treating and preventing disease and traumatic injury, of relieving pain, and of preserving and improving human health, using molecular tools and molecular knowledge of the human body" [11]. According to United States' National Institutes of Health, nanomedicine is an offshoot of nanotechnology, which refers to highly specific medical interventions at the molecular scale for curing disease or repairing damaged tissues, such as bone, muscle, or nerve.

3 Nanomedicine: Development from Bench-Side to Market

There are usually six developmental stages of a nanomedicine to reach market or patient (Fig. 1.1). (1) Synthesis and physiochemical characterization of the nanomaterial or nanoparticle, and (2) nanoformulation which is functionalization of nanoparticles for specific biomedical applications, followed by safety and performance testing (3) outside living body (in vitro usually at cellular level) or ex vivo, usually at tissue/organ levels (4) preclinical studies in mice, monkeys and other animal models (5) clinical trials in human, usually three phases, and finally (6) FDA approval for commercialization of the nanomedicine product. Majority of the nanomedicine proof-of-concept technologies/products die after in vitro testing either due to scarcity of capital to support preclinical and clinical testing or non-significant improvement when compared to already existing technologies/products. Most of the nanomedicine products fall under two categories, therapeutics and devices. Therapeutics has to go through a long process of FDA investigation before clearance, while 510 (k) premarket notification pathway for the medical devices makes the clearance fairly easy for the devices.

In this chapter we are giving you an overview of the nanomaterials and nanomedicine with a focus on FDA approved and clinically significant nanomedicines. Possible aspects of personalized nanomedicine including various nanomaterials

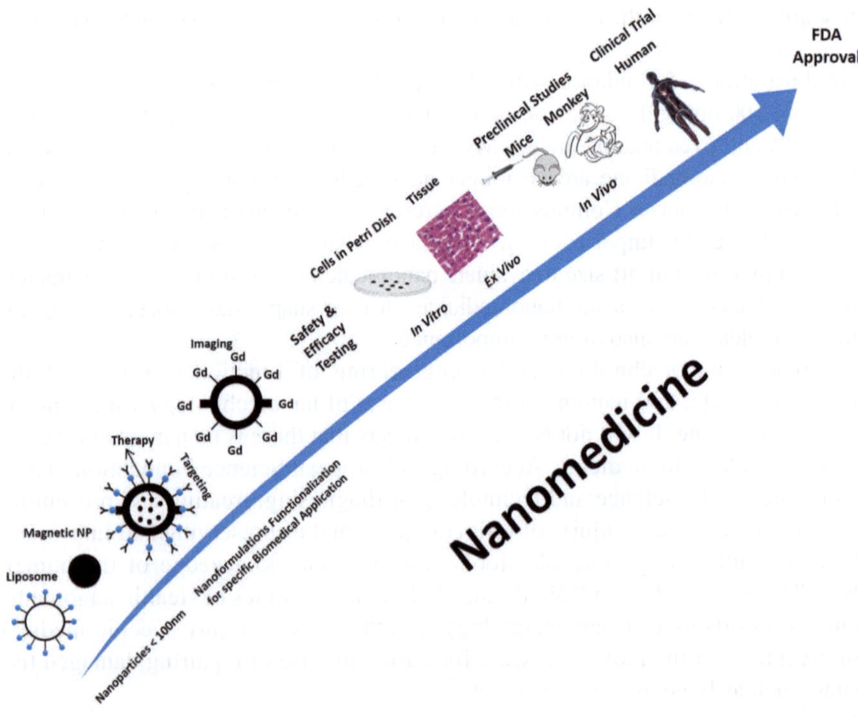

Fig. 1.1 Developmental stages of nanomedicine from lab bench to FDA approval for marketing and use in patients

with their important biomedical applications are discussed in subsequent chapters, image-guided therapy, targeted drug delivery, stimuli responsive drug delivery and controlled on-demand delivery with special emphasis on central nervous system drug delivery are described in this book.

In summary a nanoparticle can be modified for diverse medical applications. There are a few terms that are now being regularly used to further classify nanomedicine principle of design, for examples drug encapsulation (called nanoformulation) to treat cancer and neurodegenerative disorders, often called cancer nanomedicine and neuro-nanomedicine, respectively. Functionalizing nanoparticles with photosensitizer, heat generating dyes, or metal nanoparticles like gold and silver to achieve hyperethermia or photodynamic therapy, commonly called photo-nanomedicine. And regenerative nanomedicine, designing a polymer scaffold and impinging drug delivery nanocarrier, antimicrobial agents, etc., chitosan-based skin dressing with silver nanoparticles as anti-microbial agent is a typical example. Although photo-nanomedicine is discussed in this book to some extent, under section metal-based nanomedicine in this chapter, and subsequent chapters, readers interested in regenerative medicine are encouraged to read another book [12].

There are several hundreds of nanomaterials and their conjugates, for example, metallic nanomaterials, polymeric nanomaterials, metal-metal nanocomposites or hybrids, and metal-polymer nanocomposites, etc. To provide a clear understanding of the nanomaterials and the nanomedicine, we are dividing them in two broad categories as discussed below.

4 Polymer and Liposome-Based Nanomedicine

Most, if not all, polymer based nanomedicine uses either of two, (1) Liposomes and lipid based polymers, and (2) Carbohydrate and protein based polymers. Although there is a plethora of polymeric nanocarriers of medicine that are under clinical investigation, only the most successful are briefly introduced below and elaborated in subsequent chapters. For more details on polymer and liposome-based nanomedicine readers can refer to these review articles [6, 13, 14].

Since their discovery by Alec D. Bengham at University of Cambridge in 1964, liposomes have emerged to be the first and most successful nano drug delivery system or nanomedicine carrier. Liposome can simply be defined as a lipid bilayer structure or membrane around an aqueous compartment. The lipid component of the liposome can be man-made or synthetic depending on the desired properties. For example, a typical phospholipid-based liposome will have hydrophilic or aqueous core that allow encapsulation of hydrophilic drug, while the lipid membrane can encapsulate hydrophobic drug. Two biggest advantages of the liposomes are, first, a core-shell design allows better encapsulation strategy for the sustained drug release, and second, it allows encapsulation of both hydrophilic as well as hydrophobic drugs. However, these lipid based nanocarriers does not easily degrade, and toxicity is conceivable. Currently, there are approximately 12 clinically approved liposomal nanomedicine and around 80 under clinical investigation [15].

Carbohydrates, typically the polysaccharides have shown promising clinical success. Natural as well as synthetic polymers are used widely to deliver medicine. Some examples of commonly used carbohydrate polymers are chitosan, gelatin, cellulose, heparin, Polylactic acid, polyglycolic acid, and their copolymer poly(lactic-co-glycolic acid). Most widely used protein based polymer used in nano drug delivery is albumin. The inherent advantage of the carbohydrate and polymer based nanocarriers is their natural degradation by the enzymes present in the body that make these polymers and their degraded monomers highly biocompatible. However, their poor design, as compared to liposome's core-shell arrangement, is a limitation. Drug is usually conjugated to the polymer and susceptible to burst drug release. Synthesis of polymeric micelles by combining two or more polymers to form a core-shell arrangement is an alternate strategy used in the formulation of polymeric nanomedicine currently under clinical investigations.

5 Metal-Based Nanomedicine

Nanoparticles of metals, in particular noble metals, offer strong light scattering and absorption properties that holds the basis of their medicinal applications. The light scattering, for example Raman and Mie scattering is the basis of optical sensing and imaging, while light absorption to generate heat is the basis of photothermal applications. Most of the metal-based nanomedicine uses gold, silver and iron oxide. Bhardwaj et al investigated metal nanoparticles for cancer and neuro-nanomedicine, silver nanoparticles for their sensing and photo-thermal properties [16, 17], and magneto-electric nanoparticles for non-invasive drug delivery to the brain [18–20]. Gold and silver are excellent substrates for biosensor research, in particular surface-enhanced Raman spectroscopy (SERS) and surface-plasmon resonance (SPR) because of their unique localized surface plasmon resonance (LSPR) effect that arises due to the resonant oscillation of their free electrons in the presence of light [21]. While on the other hand, magnetization of iron oxide nanoparticles makes it a unique nanomaterial for magnetic-actuation for nano-electroporation, on-demand drug release, as well magnetic resonance imaging agent [19, 20], discussed in detail later in this book.

6 Nanomedicine: Progress, Prospects and Challenges

As per 2013 clinicaltrial data, there are roughly 250 nanomedicine products approved or under clinical investigation. Of 250, 54 are commercial and 46 are likely to be commercialized. Most of these nanomedicines products are either medical devices or therapeutics as categorized by U.S. Food and Drug Administration. There are 44 medical devices and 33 therapeutics so far. Among therapeutics, primarily drug delivery vehicle are liposome and polymer based nano therapeutics. A significant progress has been made towards developing nanomedicine for cancer therapeutics (Fig. 1.2). Most of the therapeutic nanomedicine have size below 350 nm (Fig. 1.3).

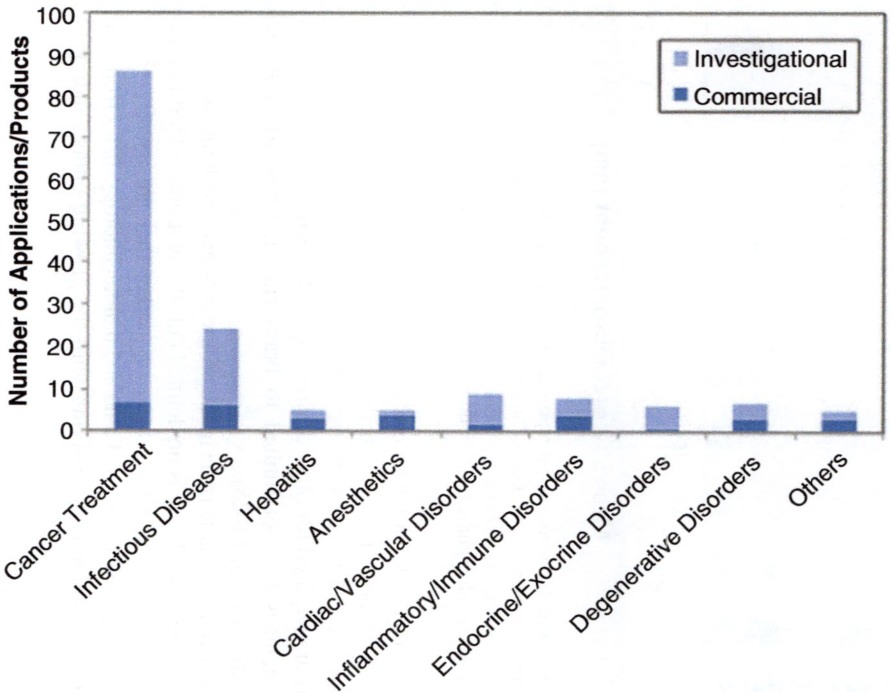

Fig. 1.2 Medical applications of commercialized and likely to be commercialized nanomedicines for therapy, as identified by year 2013. Adapted from Elsevier [22]

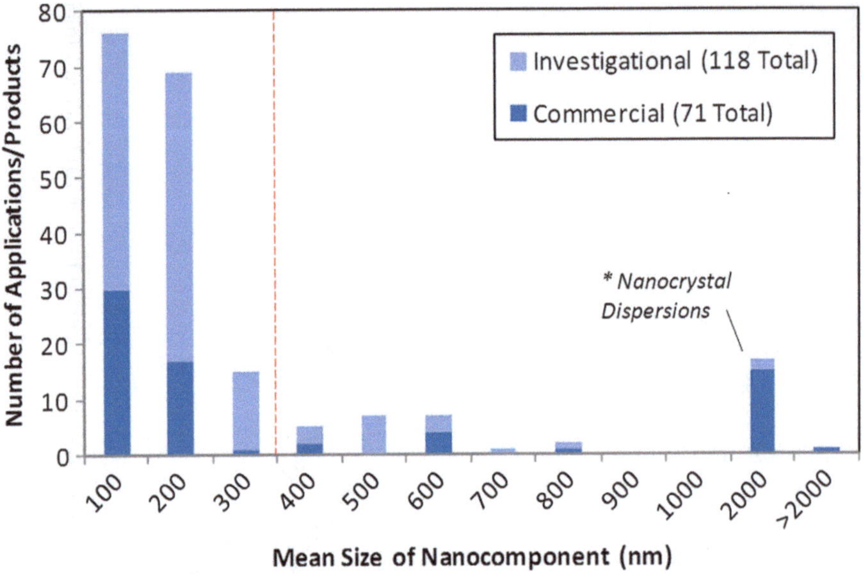

Fig. 1.3 Mean size of the nanomaterial reported in the nanomedicine approved for commercial use or likely to be commercialized. *Dotted line* indicates cutoff size used in this chapter to define nanomedicine or nanomaterials as most of them fall below this cutoff. Nanocrystal dispersion is highlighted as many products have roughly this size. Adapted from Elsevier [22]

Cancer is a major global health problem, a National Cancer Institute (NCI) initiative focused on the cancer nanomedicine. An improvement in the rate of cancer treatment has been observed with the advent of cancer nanomedicine. However, much efforts are required for other devastating diseases like neurodegenerative diseases that includes but not limited to brain cancer, neuroAIDS, Alzheimer's, Parkinson's etc. [18, 20, 23–25].

The biggest and debatable issue in the progress of nanomedicine is the size of the nanoparticles (Fig. 1.3). There is no defined cutoff any more, at least in practice, as evident by the nanomedicine products in market that are usually greater than 100 nm. The definition of nanomedicine and nanotechnology differ from country to country, and it's continuously evolving with the growing understanding of nanotechnology and nanomedicine.

7 Conclusion

Nanomaterials with size below 350 nm can be considered nanomedicine due to their effectiveness and working mechanism. Advancement in nanomedicine is occurring in both medical devices and therapeutics areas. Many nanomaterials with controlled and targeted delivery have been introduced to the market. However, much efforts are being made in research on the design and development of new nanomaterials with on-demand release and minimum toxicity. More funding is required for research, preclinical, and clinical tests in order to develop and introduce safe and effective nanomedicine to the market.

References

1. Feynman R. There's plenty of room at the bottom, American Physical Society. Pasadena: Caltech; 1959.
2. Taniguchi N. On the basic concept of nanotechnology. In: International Conference on Product Engineering. Tokyo: Japan Society of Precision Engineering; 1974. p. 18–23.
3. Nanomedicine: grounds for optimism, and a call for papers. The Lancet. 2003;362(9385):673.
4. Nikkhah-Moshaie R, Kaushik A, Jayant RD, Bhardwaj V, Nair M. TEM investigation of nanocarriers distribution in mice brain. Microsc Microanal. 2016;22(S3):1173–2.
5. Roco MC. National nanotechnology initiative: past, present, future. In: Goddard WA, Brenner DW, Lyshevski SE, Iafrate G, editors. Handbook on nanoscience, engineering and technology. 2nd ed. London: Taylor and Francis; 2007. p. 26.
6. Litzinger DC, Buiting AMJ, Rooijen NV, Huang L. Effect of liposome size on the circulation time and intraorgan distribution of amphipathic poly(ethylene glycol)-containing liposomes. Biochim Biophys Acta. 1994;1190(1):99–107.
7. Nanospectra Biosciences, Inc. Pilot study of Aurolase (tm) therapy in refractory and/or current tumors of the head and neck; 2009.
8. Pillai G. Nanomedicine for cancer therapy: an update of FDA approved and those under various stages of development. SOJ Pharm Pharm Sci. 2014;1(2):13.
9. European Commission. Commission recommendation of XXX on the definition of nanomaterial. Brussels: European Commission; 2011. p. 6.
10. European Union. Regulation (EU) No 528/2012 of the European Parliament and of the council of 22 May 2012 concerning the making available on the market and use of biocidal products, vol. L 167/1. Luxembourg: Office of the European Union; 2012.
11. European Science Foundation. Nanomedicine, An ESF—European Medical Research Councils (EMRC) Forward Look report. Strasbourg cedex, France: ESF; 2005. p. 7.
12. Steinhoff G. Regenerative medicine from protocol to patient. 2nd ed. Dordrecht: Springer; 2013.
13. Venkatraman SS, Ma LL, Natarajan JV, Chattopadhyay S. Polymer- and liposome-based nanoparticles in targeted drug delivery. Front Biosci. 2010;S2:801–14.
14. Vashist A, Kaushik A, Vashist A, Jayant RD, Tomitaka A, Ahmad S, Gupta YK, Nair M. Recent trends on hydrogel based drug delivery systems for infectious diseases. Biomater Sci. 2016;4:1535–53.
15. Bozzuto G, Molinari A. Liposomes as nanomedical devices. Int J Nanomedicine. 2015;10:975–99.
16. Bhardwaj V, Srinivasan S, McGoron AJ. Efficient intracellular delivery and improved biocompatibility of colloidal silver nanoparticles towards intracellular SERS immuno-sensing. Analyst. 2015;140(12):3929–34.
17. Srinivasan S, Bhardwaj V, Nagasetti A, Fernandez-Fernandez A, McGoron AJ. Multifunctional surface-enhanced Raman spectroscopy-detectable silver nanoparticles for combined photodynamic therapy and pH-triggered chemotherapy. J Biomed Nanotechnol. 2016;12(12):2202–19.
18. Kaushik A, Jayant RD, Nikkhah-Moshaie R, Bhardwaj V, Roy U, et al. Magnetically guided central nervous system delivery and toxicity evaluation of magneto-electric nanocarriers. Sci Rep. 2016;6:25309.

19. Kaushik A, Nikkhah-Moshaie R, Bhardwaj V, Sinha R, Alturi V, et al. Investigation of ac-magnetic field stimulated nanoelectroporation of magneto-electric nano-drug-carrier inside CNS cells. Sci Rep. 2017;7:45663.
20. Kaushik A, Jayant RD, Sagar V, Nair M. The potential of magneto-electric nanocarriers for drug delivery. Expert Opin Drug Deliv. 2014;11(10):1635–46.
21. Jain PK, Huang X, El-Sayed IH, El-Sayed MA. Noble metal at nanoscale: optical and photothermal properties and some applications in imaging, sensing, biology and medicine. Acc Chem Res. 2008;41(12):1578–86.
22. Etheridge ML, Campbell SA, Erdman AG, Haynes CL, Wolf SM, et al. The big picture on nanomedicine: the state of investigational and approved nanomedicine products. Nanomedicine. 2013;9(1):1–14.
23. Kaushik A, Jayant RD, Nair M. Advancements in nano-enabled therapeutics for neuroHIV management. Int J Nanomedicine. 2016;11:4317–25.
24. Jayant RD, Sosa D, Kaushik A, Atluri V, Vashist A, et al. Current status of non-viral gene therapy for CNS disorders. Expert Opin Drug Deliv. 2016;13(10):1433–45.
25. Nair M, Jayant RD, Kaushik A, Sagar V. Getting into the brain: potential of nanotechnology in the management of NeuroAIDS. Adv Drug Deliv Rev. 2016;103:202–17.

Chapter 2
Personalized Therapeutics: First Take Home Messages

Venkata Atluri*, Ravi Doddapaneni*, and Eliset Perez

Abstract Personalized therapeutics is the emerging field in the medicine based on the individual's unique characteristics like genetic profile/alterations, epigenetic modifications, clinical symptoms, disease biomarkers and environmental factors which play a significant role in tailoring their therapies. Although there is a significant progress in personalized therapeutics towards the treatment of few genetic disorders and life threatening diseases, there is still plenty of work to be done to make the field progress in the treatment of various diseases and make these approaches useful to patients in rural areas as well. In this chapter we have briefed few updates in the field of personalized therapeutics application in the treatment of cancer, cystic fibrosis, stroke, psychiatry and asthma.

Keywords Personalized medicine • Chronic diseases • Next generation sequencing • Therapeutics

1 Introduction

Personalized medicine is the field of healthcare that consists of treating individuals according to their unique clinical, genetic, genomic, and environmental information. Personalized Medicine Coalition defines it as "the use of new methods of molecular analysis to better manage a patient's disease or predisposition to disease". In other words, personalized medicine is about making the treatment as

*Both authors contributed equally.

V. Atluri (✉) • E. Perez
Department of Immunology, Institute of Neuroimmune Pharmacology,
Herbert Wertheim College of Medicine, Florida International University,
11200 SW 8th Street, Miami, FL 33199, USA
e-mail: dratluri@aol.com

R. Doddapaneni
Department of Ophthalmology, Bascom Palmer Eye Institute,
University of Miami Miller School of Medicine, Miami, FL 33136, USA

A. Kaushik et al. (eds.), *Advances in Personalized Nanotherapeutics*,
https://doi.org/10.1007/978-3-319-63633-7_2

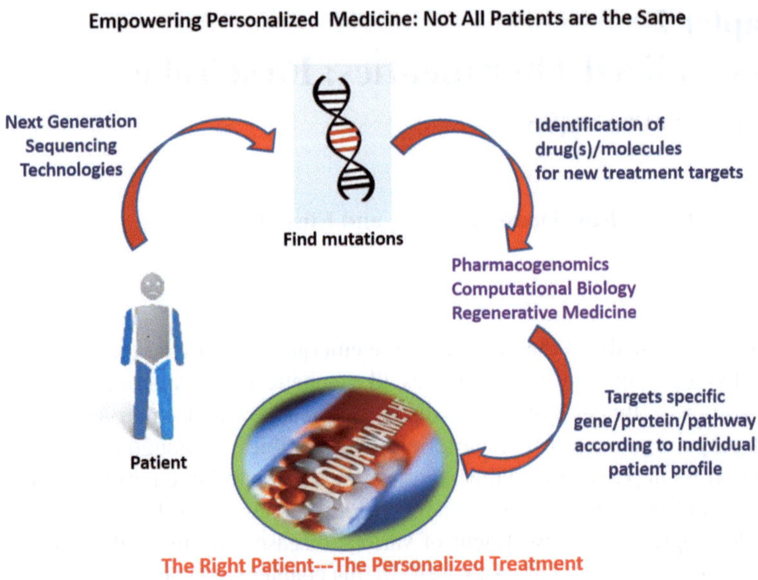

Fig. 2.1 Paving the way for personalized medicine. The promise of "personalized medicine," which is the tailoring of medical treatment to the individual characteristics, needs and preferences of each patient. The paradigmatic developments in science and technology offer new promise for developing targeted therapeutics and tools for predicting who will respond to a medical therapy or who will suffer ill effects by using next-generation technologies leads to identification of novel drug molecule(s) to target specific pathway(s)

individualized as the disease. The concept of personalized medicine dates back hundreds of years, but it was not until the turn of the twenty-first century when the Human Genome Project was successfully completed that personalized medicine became a more tangible concept. Currently, whole genome sequencing (WGS) is becoming one of the most widely used applications and is providing tremendous quantities of genome sequences relative to the past through public and private human genome sequencing projects throughout the world, which is a necessary component of personalized and precision medicine [1]. Now researchers are taking advantage of the Human Genome Project to learn how inherited differences in genes affect the body's response to medication and how these differences can be used to predict the effect of a medication on a particular individual and to prevent adverse drug reactions. Developments in the fields of genomics and other areas such as pharmacogenomics, computational biology, and regenerative medicine have played a significant role in developing tools to personalize diagnosis and treatment of diseases (Fig. 2.1).

Pharmacogenomics, which is "the study of how genes affect a person's response to drugs" is a significant area of personalized medicine where substantial progress has been made. It looks at variations in genes, such as liver enzymes, that code for proteins responsible for converting medications into their active or inactive forms. For example, the liver enzyme known as CYP2D6 acts on a quarter of all prescription drugs; it converts the painkiller codeine into its active form morphine. People who have extra copies of the CYP2D6 gene produce an excess of CYP2D6 enzyme, which causes codeine to be converted to morphine very rapidly; resulting in a drug overdose. On the other hand, some people have a different version of the CYP2D6 gene, which can result in a nonfunctional enzyme that metabolizes codeine slowly, making it ineffective at relieving pain. The Food and Drug administration has included pharmacogenomics information, such as possible side effects and differences in effectiveness for people with certain genomic variation on the labels of more than 150 medications. These advances allow doctors to use pharmacogenomics information to select the best medication and the correct dose for individual patients. The field of pharmacogenomics is currently limited, but new approaches such as the development of tailored drugs to treat diseases such Alzheimer, cancer, and HIV/AIDS are under study in clinical trials.

Personalized medicine also includes anatomically-specific devices such as the tinnitus masker, which is a personalized electronic hearing aid device that generates a masking noise, sometimes referred as white noise, for an individual suffering from hearing loss and tinnitus. According to the American Tinnitus Association tinnitus is "the perception of sound when no actual external noise is present", and even though there is no scientifically-validated cure for this health issue, the tinnitus masker has proven to be very efficient at alleviating it. Another anatomically-specific device is the Artificial Pancreas Device System, which measures patient's glucose levels and delivers patients-tailored insulin doses in people with diabetes. A different, but also very significant personalized invention in medicine was the creation of a 3D printed tracheal split. Physicians at the University of Michigan and Akron Children's Hospital utilized a computed tomography image, computer-aided design, and 3D printing to make a bioresorbable airway splint to treat an infant diagnosed with tracheobronchomalacia, which is a life-threatening condition that happens when the airways walls are so weak that they collapse during coughing or breathing [1].

Personalized medicine has also made progress for chronic diseases such as cancer, asthma, and chronic obstructive pulmonary disease. All these new discoveries open the doors to more informed medical decisions, better targeted therapies, less medicine side effects and early disease intervention. It allows people to make preventive lifestyle choices that will help counteract any biological risk they might be facing. Even potentially lifesaving medications that could be taken off the market because they pose a risk for some people, could still be available for those who could benefit from them. Personalized medicine focuses on prevention, and making a transition from treating a disease to maintaining health. Finally, it reduces healthcare cost by allowing doctors to quickly select the effective therapy for an individual patient while avoiding the costs that ineffective treatments carry. Personalized therapeutics holds the future for medicine.

2 Personalized Therapeutics in Cancer

Incidence and prevalence of cancer is of increasing concern due to the late diagnosis and poor prognosis to the treatment procedures which take weeks to months to respond with the long list of side effects to the medications. Modern Personalized medicine and therapeutics play an important role in the treatment of various types of cancer due to the altered gene expression profile in each patient's tumor cells and focuses on targeted therapy, whereas traditional personalized medicine considers patient's family history, social circumstances, environment, and lifestyle. In this targeted therapy, the information about the altered cell signaling pathway and genes/components leading to the cancer are considered. The first personalized medicine developed in the late 1970s was the selective ER modulator, tamoxifen for targeted therapy of breast cancer. Furthermore, in 20% of breast cancer patients, higher levels of HER-2 (human epidermal growth factor receptor 2) protein was observed [2–4]. Trastuzumab (herceptin), a monoclonal antibody which binds to the HER2, has been shown to selectively exert anti-tumor effects in patients with HER2-amplified breast cancer, but not in tumors with normal HER2 expression.

Role of Genetics and epigenetics; and biomarkers for early detection of colon cancer are well known. For the treatment purpose, patient's information on mutations, hypermethylation and microsatellite instability in the early polyps help to predict their response to the specific treatment regimen [5]. The treatment options for metastatic colorectal cancer (CRC) are 5-fluorouracil (5-FU), irinotecan and oxaliplatin [6]. The patient's tumors with low expression of thymidylate synthase respond well to the 5-FU treatment [5], while it is neither useful nor recommended in patients with microsatellite instability tumors [7]. Topoisomerase-1 (Topo 1) is overexpressed in nearly 50% of CRC patients and irinotecan, which is a topoisomerase-1 inhibitor reported to increase overall survival rate in these patients when compared to the low/moderate levels of Topo 1 expression [8]. Few reports suggest that irinotecan is being toxic in patients with homozygous polymorphism that reduces the UGT1A1 activity, which is essential for detoxification of irinotecan [9]. Additionally, epidermal growth factor receptor (EGFR) is another therapeutic target for cetuximab and panitumumab antibodies for the treatment of metastatic CRC. Binding of these antibodies to EGFR activates the RAS/RAF/MAPK, STAT, and PI3K/AKT signaling pathways, which together modulate cellular proliferation, adhesion, angiogenesis, migration, and survival. In 40% of CRC patients, KRAS mutations has been reported and these mutations make the tumor unresponsive to cetuximab and panitumumab treatment [10].

Genetic discoveries over the past decade have transformed the way clinicians treat and interpret disease. On the treatment side, the idea behind personalized medicine is to develop treatments that target tumor cells specifically and effectively to destroy only the cancer cells when compared to traditional chemotherapeutic approach. In the past decade and with the advent of personalized medicine, multiple advances have been made in understanding the underlying biology and molecular mechanisms of lung cancer [11]. Lung cancer is the leading cause of cancer-related deaths in the world. Non-small cell lung carcinoma (NSCLC) is a highly heterogeneous disease and the most common cause of cancer mortality worldwide with a 5-year survival rate of only 15%. Lung cancer is no longer considered a single disease entity and is now being subdivided into molecular subtypes with dedicated targeted and chemotherapeutic strategies [12]. Despite recent advances in cancer treatments, the clinical outcome among NSCLC patients is not impressive. The concept of using information from a patient's tumor profile for diagnosis and treatment decisions has revolutionized the landscape for lung cancer therapeutics. Until recently, very little was known about the genetics of this aggressive cancer. In recent years, however, lung cancer has begun to yield its secrets, and a fascinating picture is emerging of how it develops and progresses. Mutations have been found in two genes; epidermal growth factor receptor- (EGFR) and fusion genes involving the anaplastic lymphoma kinase (ALK) make them vulnerable to specific treatments. These mutations appear to be key events that provide the potential for targeted therapy and these precision therapies are lengthening progression-free survival by specifically targeting pathway(s) based on patient's profile [13]. On the other hand, new treatments for lung cancer are in clinical trials like using the patient's own immune system to block tumors [14]. The next era of personalized treatment for lung cancer will involve a comprehensive genomic characterization of adenocarcinoma, squamous-cell carcinoma, and small-cell carcinoma into various subtypes. These next generation technologies could improve early detection and diagnosis of the disease by incorporating individual molecular characteristics, which is very beneficial for joint treatment decision making in the clinics with patients and practitioners.

Adenoid cystic carcinoma (ACC) is the most common malignant epithelial cancer of the lacrimal gland (>70% of all cases) [15]. Despite a slow rate of growth, ACC's are ultimately associated with poor clinical outcome. ACC can occur in several other glandular tissues, most commonly in the salivary glands, but also in Bartholin's glands of the vagina, tracheal goblet cells, and mammary gland tissue [16]. Whole exome sequencing (WES) of ACC samples has shed light on the genetic landscape of this disease and provide evidence for Notch pathway alterations in 11 to 29% of patients [17]. The discovery of the "Philadelphia chromosome" translocation in chronic myelogenous leukemia, which creates the BCR-ABL fusion gene, ushered in an era of targeted therapy with kinase inhibitors. An important breakthrough in the understanding of the molecular pathogenesis of ACC was recently made when it was shown the ACC specific chromosomal translocations. Whole-genome sequencing (WGS) has revealed that recurrent fusion of the v-myb avian myelobastosis viral oncogene homolog (MYB) and nuclear factor I/B (NFIB) generates the MYB-NFIB transcription factor, which has been detected in a high percentage of individuals with adenoid cystic carcinoma. The MYB-NFIB fusion oncogene was identified in the majority (>85%) of adenoid cystic carcinomas (in all tissues) [18] and in 50% of lacrimal gland ACCs [19]. Therefore, the identification of recurrent fusions of the MYB-NFIB genes by next generation sequencing have begun to shed light on the molecular underpinnings, little else is known about the molecular genetics of this frequently fatal cancer. In the era of personalized medicine, we are currently on the threshold of a revolution in adenoid cystic carcinoma research, in future, it leads to identify novel therapeutic strategies using sequencing technologies.

3 Personalized Therapeutics in Cystic Fibrosis

In 2012, FDA approved a new drug Kalydeco™ (known generically as ivacaftor) for the treatment of cystic fibrosis (CF), an inherited disease that impairs the lungs and digestive system due to the specific gene mutation in a gene (CFTR-cystic fibrosis transmembrane conductance regulator) characterized by defective transport of chloride ions across several epithelial tissues [20, 21]. Although hundreds of known mutations causes the CF, Kalydeco is recommended in the patients with following CFTR gene mutations: G551D, G1244E, G1349D, G178R, G551S, S1251 N, S1255P, S549N, or S549R. The active ingredient in KALYDECO tablets and oral granules is ivacaftor, a CFTR potentiator. The CFTR protein is a chloride channel present at the surface of epithelial cells in multiple organs. Ivacaftor facilitates increased chloride transport by potentiating the channel-open probability (or gating) of the CFTR protein [22–24].

4 Personalized Therapeutics in Stroke

Recently, Dorfman et al., studied the application of personalized medicine in the chronic disease like stroke at Bridgepoint Health, Canada [25]. Acetylsalicylic acid (ASA/Aspirin) found to be ineffective in people suffering from stroke and recommended to use the Clopidogrel (antiplatelet agent) for thrombosis prevention. Clopidogrel is a pro-drug and requires the activation by cytochrome P450 CYP2C19 enzyme. In persons with mutations in CYP2C19 (inactive 2C19 enzyme) leads to less response to the Clopidogrel treatment and have greater risk for stroke/death. Therefore, FDA recommends genetic testing for choosing the better treatment regimen [11]. Furthermore, ABCB1 gene variation has also been reported to affect the clopidogrel absorption [26]. Another expensive but effective antiplatelet drug when compared to the clopidogrel is the prasugrel, which doesn't get effected by the CYP2C19 genetic variation, but has the disadvantage of higher risk of mortality due to the bleeding [27].

Warfarin is the anticoagulant and broadly used for atrial fibrillation, heart valve replacement, recent heart attack and for venous thromboembolism prophylaxis. Exact dosing of warfarin is important as it is ineffective in decreasing the risk for stroke and in higher doses it may cause intracranial bleeding. Clinical efficacy of warfarin also depends on the VKORC1 and CYP2C9 genetic variation. Dabigatran etexilate is the alternative anticoagulant for warfarin with identical efficacy without any pharmacogenetics interactions. Various other class of stroke treatment medications like angiotensin-converting enzyme (ACE) inhibitors, statins and angiotensin II receptor (AGTR2) antagonists are also effected by various metabolic genes genetic variation as well as variation in drug target genes and proteins involved in signaling cascades.

ACE inhibitors' Pharmacogenetic genes AGTR1, BDKRB1, CLCN6; NPPA

AGTR2 antagonists' Pharmacogenetic genes CYP2C9, APOB; LDLR, KNG1

Statins Pharmacogenetic genes CYP3A4, ABCB1, HTR7, HTR3B, GNB3, USP5, SLCO1B1, ABCC1, CYP2C8, CYP2C19, HCR, AGTR1, KIF6, HTR7, ITGB3

The therapeutic efficacy of statins can be measured by monitoring blood cholesterol and adverse reactions like muscle or bone pain can be eliminated by substituting with different brand with other pharmacogenetics profile.

5 Personalized Therapeutics in Psychiatry

Ozomaro et al., comprehensively reviewed the role of personalized medicine in psychiatry [28]. Polymorphism in 5′-promoter region of 5-hydroxytryptamine (5HT; serotonin) transporter (5-HTT) and 5-HT transporter-linked polymorphic region (5-HTTLPR) has been shown to influence the susceptibility for developing major depressive disorder (MDD) and response to antidepressant therapy. Many studies have linked the 5-HTTLPR long (L) allele with greater therapeutic response to selective serotonin reuptake inhibitors (SSRIs) [29–42]. Few other studies have reported that 5-HTTLPR short (S) allele is associated with poor/slower therapeutic response to SSRIs [40, 41, 43–48]. Geographic variation in 5-HTTLPR S-allele therapeutic response to SSRIs has been reported i.e. greater response [49, 50] or no relation to SSRIs [51, 52].

Geographic/ethnic/age-specific variation in relation between serotonin receptors (HTR1A and HTR2A) genetic polymorphisms and treatment response to the SSRIs has been reported. For example, in Chinese population, while few studies reported increased response to the SSRIs and HTR1A rs6295-C/C genotype association [36, 53, 54], Kato et al. [55] reported an association of increased therapeutic response to SSRIs and genotype rs6295-G/G and few studies failed to find any association [56–58]. Few studies have reported genetic variations of HTR2A (rs6311-G allele, the rs9316233-G allele, rs2224721, rs1923884, rs7997012, rs6313, rs799701, rs3125, rs1923882, and rs6314) and increased therapeutic response to SSRIs [58–62].

In the 8-week double-blind randomized control trial for the potential genetic predictors for efficacy and side effects of antidepressants paroxetine and mirtazapine, Murphy et al., reported that paroxetine treatment is associated with severe side effects in HTR2A rs6313-C/C genotype carriers while there are no side effects with the use of mirtazapine in these genetic carriers [63]. In another study, homozygous (TT) for the FK506-binding protein 5 (FKBP5) marker rs1360780 carriers reported to have faster response to SSRIs, Tricyclic antidepressants (TCAs), and mirtazapine in patients with MDD [64]. Likewise, effect of pharmacogenetics variation has been studied well in bipolar disorder therapy with respect to the lithium and found significant associations with 5-HTTLPR, CLOCK, BDNF, X-box binding protein 1 (XBP1), glycogen-synthase kinase 3 beta (GSK3B), breakpoint cluster region (BCR), cAMP responsive element binding proteins 1–3 (CREB 1, 2, and 3) and neurotrophic tyrosine kinase receptor, type 2 (NTRK2) [65–72].

6 Personalized Therapeutics in Asthma

Asthma is considered a complex genetic disease and it is a chronic disorder of the lung with variable airway obstruction, wheeze/cough and an underlying inflammatory process. There are more than 300 million subjects with doctor diagnosed current asthma worldwide, with more than 24.6 million in the United States alone [73]. For decades, first-line treatment for all but the mildest cases has included an inhaler, which patients use to deliver corticosteroid medicine into the lungs' airways. The medicines have side effects, especially for children, so physicians often initially prescribe a lower dose and increase dosage if the patient's asthma does not improve. But one-third or more of patients do not respond to treatment with any standard dosage of an inhaled corticosteroid. Recently, Genome-wide association studies have begun to identify genes underlying asthma which represent future therapeutic targets include IL1RL1, IL18R, TSLP, PDE4D, MHC II, DQ, IL-33, LRRC32, SMAD3, ORMDL3, GSDMB and IL2RB [74]. Due to these next generation sequencing technologies, in near future, asthma therapy is changing from a 'one size fits all' therapy to a 'precision medicine' model, where may prescribe the most appropriate treatment for each patient.

7 Walking Towards Personalized Therapeutics in Chronic Obstructive Pulmonary Disease

Chronic obstructive pulmonary disease (COPD) is a complex disorder and not all patients respond to all drugs. COPD is currently the fourth-leading cause of death in the USA and the world according to the WHO and unfortunately is predicted to become the third-leading cause of death worldwide by 2020 [75]. In the last 10 years, genome-wide association studies (GWAS) have demonstrated a number of regions to be strongly associated with phenotypes of COPD, including forced expiratory volume in 1 s (FEV1), forced vital capacity (FVC), FEV1/FVC and both clinically and radiologically defined emphysema [76–78]. The importance of personalized treatment in COPD is increasingly recognized and identification of phenotypes represents the first step in this process. A more precise definition of COPD phenotypes should lead to a better targeted therapeutic approach based on these phenotypes. Increased usage of next-generation sequencing will further increase our understanding of the genetic variants involved in chronic disease.

As our grasp of the complexity of chronic diseases strengthens, the ultimate goal of personalized medicine will be moving towards a more personalized therapeutic approach for effective treatment of COPD patients.

8 The Road from Next Generation Sequencing to Personalized Medicine

Currently, we are at the doorstep of yet another revolution in cancer research based around next generation DNA sequencing (NGS). NGS technologies benefits have multi-fold compared to first-generation sequencing [79]. To begin with, NGS methods do not require a bacterial cloning procedure and prepare libraries for sequencing in a cell free system. Second, NGS technologies can process millions of sequencing reactions in parallel. Third, detection of bases is performed cyclically and in parallel. Next generation sequencing could be used to personalize the monitoring of the disease by interrogating tumor-specific DNA rearrangements in the patient's plasma. The NGS or second-generation sequencing technologies have the advantage of performing enormous numbers of sequencing reactions in parallel in a time- and cost-effective manner. By sequencing both the tumor and the germline DNA, NGS may reveal the somatic genetic alterations in a cancer genome. The advantages of this approach are considerable since it could be used as an accurate non-invasive indicator of the tumor burden and provide an early indication of treatment efficacy or disease recurrence. Given the heterogeneity and complexity of various cancer treatments with respect to histology and tumor stage, genomic characterization has been developed as one of many tools which can assist physicians in this era of personalized medicine. We are currently on the threshold of a revolution in cancer research thanks to the emergence of novel technologies based on next generation sequencing (NGS). These major improvements allow scientists to process sequencing of entire genomes with low cost and in a very short period to diagnose the disease and these findings lead to close collaboration between the laboratory and the clinic. To enhance the benefits of personalized medicine, there are different hurdles to overcome such as, NGS have mainly been working on frozen tissues, which unfortunately are not available for each single patient in contrast to formalin fixed paraffin embedded (FFPE) tissues, especially in smaller hospitals. However, altogether, the promises for NGS in cancer are tremendous for increasing our understanding of the diseases, to identify new treatment targets, and to move towards genome-informed personalized medicine. These advancements are moving from the bench to the bedside at a rapid pace, with the hope of translating into improvements in clinical outcomes.

Acknowledgement This work was supported by the Herbert Wertheim College of Medicine (HWCOM) Pilot Grant (Project ID # 800008539), Florida International University, Miami, FL.

References

1. Hamburg MA. Paving the way for personalized medicine: FDA'S role in a new era of medical product development. Silver Spring, MD: FDA; 2013. [cited 2017 04/12] Available from: https://www.fda.gov/downloads/scienceresearch/specialtopics/personalizedmedicine/ucm372421.pdf
2. Drebin JA, et al. Down-modulation of an oncogene protein product and reversion of the transformed phenotype by monoclonal antibodies. Cell. 1985;41(3):695–706.
3. Vogel CL, et al. Efficacy and safety of Trastuzumab as a single agent in first-line treatment of HER2-overexpressing metastatic breast cancer. J Clin Oncol. 2002;20(3):719–26.
4. Seidman AD, et al. Randomized phase III trial of weekly compared with every-3-weeks paclitaxel for metastatic breast cancer, with Trastuzumab for all HER-2 overexpressors and random assignment to Trastuzumab or not in HER-2 nonoverexpressors: final results of cancer and leukemia group B protocol 9840. J Clin Oncol. 2008;26(10):1642–9.
5. Baehner FL, et al. Genomic signatures of cancer: basis for individualized risk assessment, selective staging and therapy. J Surg Oncol. 2011;103(6):563–73.
6. Moorcraft SY, Smyth EC, Cunningham D. The role of personalized medicine in metastatic colorectal cancer: an evolving landscape. Ther Adv Gastroenterol. 2013;6(5):381–95.
7. Ribic CM, et al. Tumor microsatellite-instability status as a predictor of benefit from fluorouracil-based adjuvant chemotherapy for colon cancer. N Engl J Med. 2003;349(3):247–57.
8. Braun MS, et al. Predictive biomarkers of chemotherapy efficacy in colorectal cancer: results from the UK MRC FOCUS trial. J Clin Oncol. 2008;26(16):2690–8.
9. Palomaki GE, et al. Can UGT1A1 genotyping reduce morbidity and mortality in patients with metastatic colorectal cancer treated with irinotecan? An evidence-based review. Genet Med. 2009;11(1):21–34.
10. Amado RG, et al. Wild-type KRAS is required for Panitumumab efficacy in patients with metastatic colorectal cancer. J Clin Oncol. 2008;26(10):1626–34.
11. Samikkannu T, Atluri VS, Nair MP. HIV and cocaine impact glial metabolism: energy sensor AMP-activated protein kinase role in mitochondrial biogenesis and epigenetic remodeling. Sci Rep. 2016;6:31784.
12. Jemal A, et al. Global cancer statistics. CA Cancer J Clin. 2011;61(2):69–90.
13. Hensing T, et al. A personalized treatment for lung cancer: molecular pathways, targeted therapies, and genomic characterization. Adv Exp Med Biol. 2014;799:85–117.
14. Farkona S, Diamandis EP, Blasutig IM. Cancer immunotherapy: the beginning of the end of cancer? BMC Med. 2016;14:73.
15. Wright JE, Rose GE, Garner A. Primary malignant neoplasms of the lacrimal gland. Br J Ophthalmol. 1992;76(7):401–7.
16. Khan AJ, et al. Adenoid cystic carcinoma: a retrospective clinical review. Int J Cancer. 2001;96(3):149–58.
17. Stephens PJ, et al. Whole exome sequencing of adenoid cystic carcinoma. J Clin Invest. 2013;123(7):2965–8.
18. Ho AS, et al. The mutational landscape of adenoid cystic carcinoma. Nat Genet. 2013;45(7):791–8.
19. von Holstein SL, et al. Adenoid cystic carcinoma of the lacrimal gland: MYB gene activation, genomic imbalances, and clinical characteristics. Ophthalmology. 2013;120(10):2130–8.
20. Bompadre SG, Li M, Hwang TC. Mechanism of G551D-CFTR (cystic fibrosis transmembrane conductance regulator) potentiation by a high affinity ATP analog. J Biol Chem. 2008;283(9):5364–9.
21. Rachul C, Toews M, Caulfield T. Controversies with Kalydeco: newspaper coverage in Canada and the United States of the cystic fibrosis "wonder drug". J Cyst Fibros. 2016;15(5):624–9.
22. Eckford PD, et al. Cystic fibrosis transmembrane conductance regulator (CFTR) potentiator VX-770 (ivacaftor) opens the defective channel gate of mutant CFTR in a phosphorylation-dependent but ATP-independent manner. J Biol Chem. 2012;287(44):36639–49.

23. Van Goor F, et al. Rescue of CF airway epithelial cell function in vitro by a CFTR potentiator, VX-770. Proc Natl Acad Sci. 2009;106(44):18825–30.
24. Sloane PA, Rowe SM. Cystic fibrosis transmembrane conductance regulator protein repair as a therapeutic strategy in cystic fibrosis. Curr Opin Pulm Med. 2010;16(6):591–7.
25. Dorfman R, et al. Application of personalized medicine to chronic disease: a feasibility assessment. Clin Transl Med. 2013;2(1):16.
26. Mega JL, et al. Genetic variants in ABCB1 and CYP2C19 and cardiovascular outcomes after treatment with clopidogrel and prasugrel in the TRITON-TIMI 38 trial: a pharmacogenetic analysis. Lancet. 2010;376(9749):1312–9.
27. Wiviott SD, et al. Prasugrel versus clopidogrel in patients with acute coronary syndromes. N Engl J Med. 2007;357(20):2001–15.
28. Ozomaro U, Wahlestedt C, Nemeroff CB. Personalized medicine in psychiatry: problems and promises. BMC Med. 2013;11:132.
29. Yu A-M, et al. Regeneration of serotonin from 5-methoxytryptamine by polymorphic human CYP2D6. Pharmacogenetics. 2003;13:307.
30. Mrazek DA, et al. SLC6A4 variation and citalopram response. Am J Med Genet B Neuropsychiatr Genet. 2009;150B:341.
31. Min W, et al. Monoamine transporter gene polymorphisms affect susceptibility to depression and predict antidepressant response. Psychopharmacology. 2009;205:409.
32. Huezo-Diaz P, et al. Moderation of antidepressant response by the serotonin transporter gene. Br J Psychiatry. 2009;195:30.
33. Bozina N, et al. Association study of paroxetine therapeutic response with SERT gene polymorphisms in patients with major depressive disorder. World J Biol Psychiatry. 2008;9:190.
34. Smeraldi E, et al. Serotonin transporter gene-linked polymorphic region: possible pharmacogenetic implications of rare variants. Psychiatr Genet. 2006;16:153.
35. Smeraldi E, et al. Polymorphism within the promoter of the serotonin transporter gene and antidepressant efficacy of fluvoxamine. Mol Psychiatry. 1998;3:508.
36. Hong C-J, et al. Response to fluoxetine and serotonin 1A receptor (C-1019G) polymorphism in Taiwan Chinese major depressive disorder. Pharmacogenomics J. 2006;6:27.
37. Kato M, et al. Effects of the serotonin type 2A, 3A and 3B receptor and the serotonin transporter genes on paroxetine and fluvoxamine efficacy and adverse drug reactions in depressed Japanese patients. Neuropsychobiology. 2006;53:186.
38. Kato M, et al. Controlled clinical comparison of paroxetine and fluvoxamine considering the serotonin transporter promoter polymorphism. Int Clin Psychopharmacol. 2005;20:151.
39. Rausch JL, et al. Initial conditions of serotonin transporter kinetics and genotype: influence on SSRI treatment trial outcome. Biol Psychiatry. 2002;51:723.
40. Zanardi R, et al. Factors affecting fluvoxamine antidepressant activity: influence of pindolol and 5-HTTLPR in delusional and nondelusional depression. Biol Psychiatry. 2001;50:323.
41. Serretti A, et al. Further evidence of a combined effect of SERTPR and TPH on SSRIs response in mood disorders. Am J Med Genet B Neuropsychiatr Genet. 2004;129B:36.
42. Murphy GM, et al. Effects of the serotonin transporter gene promoter polymorphism on mirtazapine and paroxetine efficacy and adverse events in geriatric major depression. Arch Gen Psychiatry. 2004;61:1163.
43. Smits K, et al. Serotonin transporter polymorphisms and the occurrence of adverse events during treatment with selective serotonin reuptake inhibitors. Int Clin Psychopharmacol. 2007;22:137.
44. Smits KM, et al. The influence of 5-HTTLPR and STin2 polymorphisms in the serotonin transporter gene on treatment effect of selective serotonin reuptake inhibitors in depressive patients. Psychiatr Genet. 2008;18:184.
45. Durham LK, et al. The serotonin transporter polymorphism, 5HTTLPR, is associated with a faster response time to sertraline in an elderly population with major depressive disorder. Psychopharmacology. 2004;174:525.
46. Arias B, et al. 5-HTTLPR polymorphism of the serotonin transporter gene predicts non-remission in major depression patients treated with citalopram in a 12-weeks follow up study. J Clin Psychopharmacol. 2003;23:563.

47. Joyce PR, et al. Age-dependent antidepressant pharmacogenomics: polymorphisms of the serotonin transporter and G protein beta3 subunit as predictors of response to fluoxetine and nortriptyline. Int J Neuropsychopharmacol. 2003;6:339.
48. Pollock BG, et al. Allelic variation in the serotonin transporter promoter affects onset of paroxetine treatment response in late-life depression. Neuropsychopharmacology. 2000;23:587.
49. Kim H, et al. Monoamine transporter gene polymorphisms and antidepressant response in koreans with late-life depression. JAMA. 2006;296:1609.
50. Yoshida K, et al. Prediction of antidepressant response to milnacipran by norepinephrine transporter gene polymorphisms. Am J Psychiatry. 2004;161:1575.
51. Yoshimura R, et al. Rapid response to paroxetine is associated with plasma paroxetine levels at 4 but not 8 weeks of treatment, and is independent of serotonin transporter promoter polymorphism in Japanese depressed patients. Hum Psychopharmacol. 2009;24:489.
52. Ng CH, et al. Serotonin transporter polymorphisms and clinical response to sertraline across ethnicities. Prog Neuro-Psychopharmacol Biol Psychiatry. 2006;30:953.
53. Yu YW-Y, et al. Association study of two serotonin 1A receptor gene polymorphisms and fluoxetine treatment response in Chinese major depressive disorders. Eur Neuropsychopharmacol. 2006;16:498.
54. Serretti A, et al. The C(−1019)G polymorphism of the 5-HT1A gene promoter and antidepressant response in mood disorders: preliminary findings. Int J Neuropsychopharmacol. 2004;7:453.
55. Kato M, et al. Effect of 5-HT1A gene polymorphisms on antidepressant response in major depressive disorder. Am J Med Genet B Neuropsychiatr Genet. 2009;150B:115.
56. Levin GM, et al. Assessment of human serotonin 1A receptor polymorphisms and SSRI responsiveness. Mol Diagn Ther. 2007;11:155.
57. Suzuki Y, Sawamura K, Someya T. Polymorphisms in the 5-hydroxytryptamine 2A receptor and CytochromeP4502D6 genes synergistically predict fluvoxamine-induced side effects in Japanese depressed patients. Neuropsychopharmacology. 2006;31:825.
58. Peters EJ, et al. Investigation of serotonin-related genes in antidepressant response. Mol Psychiatry. 2004;9:879.
59. Uher R, et al. Genetic predictors of response to antidepressants in the GENDEP project. Pharmacogenomics J. 2009;9:225.
60. Peters EJ, et al. Resequencing of serotonin-related genes and association of tagging SNPs to citalopram response. Pharmacogenet Genomics. 2009;19:1.
61. Wilkie MJV, et al. Polymorphisms in the SLC6A4 and HTR2A genes influence treatment outcome following antidepressant therapy. Pharmacogenomics J. 2009;9:61.
62. McMahon FJ, et al. Variation in the gene encoding the serotonin 2A receptor is associated with outcome of antidepressant treatment. Am J Hum Genet. 2006;78:804.
63. Murphy GM, et al. Pharmacogenetics of antidepressant medication intolerance. Am J Psychiatry. 2003;160:1830.
64. Binder EB, et al. Polymorphisms in FKBP5 are associated with increased recurrence of depressive episodes and rapid response to antidepressant treatment. Nat Genet. 2004;36:1319.
65. Rybakowski JK, et al. Prophylactic effect of lithium in bipolar affective illness may be related to serotonin transporter genotype. Pharmacol Rep. 2005;57:124.
66. Masui T, et al. A possible association between the −116C/G single nucleotide polymorphism of the XBP1 gene and lithium prophylaxis in bipolar disorder. Int J Neuropsychopharmacol. 2006;9:83.
67. Benedetti F, et al. Actimetric evidence that CLOCK 3111 T/C SNP influences sleep and activity patterns in patients affected by bipolar depression. Am J Med Genet B Neuropsychiatr Genet. 2007;144B:631.
68. Rybakowski JK, et al. Response to lithium prophylaxis: interaction between serotonin transporter and BDNF genes. Am J Med Genet B Neuropsychiatr Genet. 2007;144B:820.
69. Adli M, et al. Response to lithium augmentation in depression is associated with the glycogen synthase kinase 3-beta -50T/C single nucleotide polymorphism. Biol Psychiatry. 2007;62:1295.

70. Masui T, et al. A possible association between missense polymorphism of the breakpoint cluster region gene and lithium prophylaxis in bipolar disorder. Prog Neuro-Psychopharmacol Biol Psychiatry. 2008;32:204.
71. Mamdani F, et al. Lithium response and genetic variation in the CREB family of genes. Am J Med Genet B Neuropsychiatr Genet. 2008;147B:500.
72. Dmitrzak-Weglarz M, et al. Association studies of the BDNF and the NTRK2 gene polymorphisms with prophylactic lithium response in bipolar patients. Pharmacogenomics. 2008;9:1595.
73. Akinbami OJ, Moorman JE, Liu X. Asthma prevalence, health care use, and mortality: United States, 2005–2009. Washington, DC: US Department of Health and Human Services, Centers for Disease Control and Prevention, National Center for Health Statistics; 2011.
74. Moffatt MF, et al. A large-scale, consortium-based genomewide association study of asthma. N Engl J Med. 2010;363(13):1211–21.
75. Agusti A, MacNee W. The COPD control panel: towards personalised medicine in COPD. Thorax. 2013;68(7):687–90.
76. Repapi E, et al. Genome-wide association study identifies five loci associated with lung function. Nat Genet. 2010;42(1):36–44.
77. Manichaikul A, et al. Genome-wide study of percent emphysema on computed tomography in the general population. The multi-ethnic study of atherosclerosis lung/SNP health association resource study. Am J Respir Crit Care Med. 2014;189(4):408–18.
78. Pillai SG, et al. A genome-wide association study in chronic obstructive pulmonary disease (COPD): identification of two major susceptibility loci. PLoS Genet. 2009;5(3):e1000421.
79. Meldrum C, Doyle MA, Tothill RW. Next-generation sequencing for cancer diagnostics: a practical perspective. Clin Biochem Rev. 2011;32(4):177–95.

Chapter 3
Nanotechnology for Therapeutics

Anujit Ghosal, Arti Vashist, Shivani Tiwari, Eram Sharmin, Sharif Ahmad, and Jaydeep Bhattacharya

Abstract Tremendous growth in the field of pharmacology and therapeutics has been observed due to revolutionised development of novel drug delivery systems predominantly based on "Nanotechnology". Treatment of wide varieties of diseases is made possible by miniaturisation of drug delivery systems. Nanotechnology delivers a unique approach, which promises higher drug efficacy, targeted drug delivery, on demand delivery, biocompatibility, etc. The importance of nanotechnology can be visualised by its ability of addressing several problems in central areas of biomedical, chemical, mechanical and electronics. Here, we discuss how nano-therapeutics can be fruitful for the treatment of brain diseases such as human immunodeficiency virus (HIV), Parkinson's, cancer, Alzheimer and their drug delivery mechanism. In this regard, the challenges involved and required future developments in drug delivery systems becomes few important topics to be worked on for expanding the utilization of nano-therapeutics.

A. Ghosal (✉)
Department of Chemistry, School of Basic and Applied Sciences,
Galgotias University, Gautam Buddh Nagar, Uttar Pradesh, India

School of Biotechnology, Jawaharlal Nehru University, New Delhi, India
e-mail: anuj.ghosal@gmail.com

A. Vashist
Center for Personalized Nanomedicine, Institute of NeuroImmune Pharmacology,
Department of Immunology, Herbert Wertheim College of Medicine,
Florida International University, Miami, 33199 FL, USA

S. Tiwari
Department of Chemistry, School of Basic and Applied Sciences,
Galgotias University, Gautam Buddh Nagar, Uttar Pradesh, India

E. Sharmin
Department of Pharmaceutical Chemistry, College of Pharmacy,
Riyadh, Kingdom of Saudi Arabia

S. Ahmad (✉)
Department of Chemistry, Materials Research Laboratory, Jamia Millia Islamia, New Delhi, India
e-mail: sharifahmad_jmi@yahoo.co.in

J. Bhattacharya
School of Biotechnology, Jawaharlal Nehru University, New Delhi, India

A. Kaushik et al. (eds.), *Advances in Personalized Nanotherapeutics*,
https://doi.org/10.1007/978-3-319-63633-7_3

Graphical Abstract

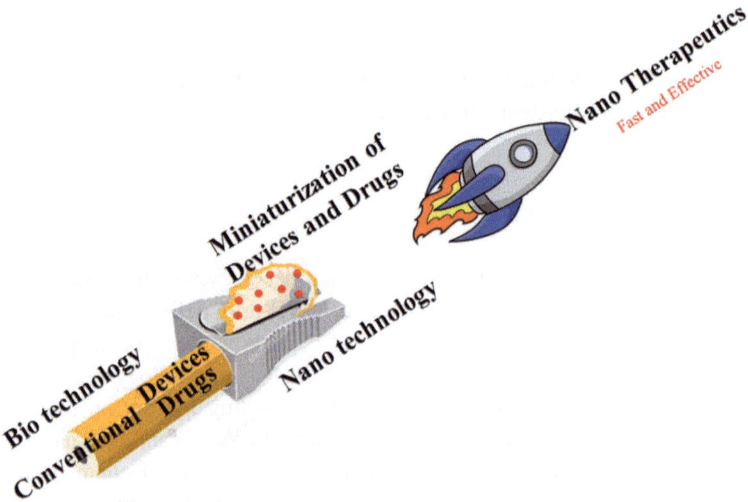

The transition of conventional theruputics to highly effective nanotherapeutics

Keywords Nanotechnology • Nano neurotherapeutics • Nanodevices • Nanobots • Drug deliveery systems

1 Introduction

Along with the development of human race their requirements for food resources, bio medical challenges, arsenals, and others have increased tremendously. These challenges can only be satisfied by the help of scientific research and development. The disastrous change in environments, excessive working culture, use of dietary supplements and bad habits have affected adversely, producing many carcinogenic and bio hazardous diseases. Maintaining proper health conditions, knowledge of any change in biological systems, and instant or long term treatment of any fatal diseases have become priority using smart medicines. Even the efficacy of many drugs have been seen to be based on the delivery system, mechanism (release as well as working) and rate of its release [1–3].

The development of material for bio medical applications with multifunctional properties like hydrophobicity, porosity, biocompatibility, aqua stability, swelling and stimuli responsiveness have been the focus of research [4–6]. In this regards, "Nanotechnology" is a highly rewarding and challenging field, which pave ways for growth and development of material with tuneable and multi-functional properties. These nanomaterials can be used for tissue engineering, stimuli responsive drug delivery system, neuro drug delivery, HIV, energy conversion, water purification, etc. [4, 7, 8]. It involves the manipulation of the matter at atomic or molecular level and study the change in basic fundamental properties of the same. This technology not only tune the properties of existing materials but also develop new advanced materials [9]. It focusses on the development of materials as well as techniques used for the same and try to imbibe the natural synthesis process i.e. through one step at a time i.e. atom by atom or at molecular level [10]. Nanotechnology has affected all the areas of research and development but have revolutionizing the field of health and medicine [11]. Several important avenues of medical sciences like sensing, therapeutics, targeted drug delivery, scaffold materials, gene therapy and imaging have thrived after the utilization of nanotechnology [12–16].

Nano materials such as metal and metal oxide nanoparticles, core shell nanoparticles, porous nanostructures, polymer coated nanomaterials and hydrogels have been tremendously used for in-vivo and in vitro drug delivery applications [17]. Gemma K. Alderton have prepared nanoporous silica-based materials as a carrier for sparingly soluble drug and then modified by using microalga *Thalassiosira pseudonana* and immunoglobulin G (IgG)-binding domain of protein G by Delalat et al. The modification lead to the targeted drug delivery to neuroblastoma xenografts tumour cells [15, 18]. Nanotechnology is even helpful in delivering drugs to brain by crossing blood brain barrier for neurological disorder like Alzheimer's, Parkinson's, fits, etc. Neurotheruputics has been achieved by polymeric nanoparticles, solid lipid nanoparticles, nanostructured lipid carriers, microemulsion, nanoemulsion, liquid crystals and others [19, 20].

Nanotechnology have been helpful in increasing the effectiveness of drugs several folds due to beneficial effects associated with nano size along and so far is a successfully solution for the targeted drug delivery purposes, cell or tissue repair at cellular lever to proteins, stimuli responsiveness it may be to pH, osmotic pressure, temperature gradient, and in many instances have shown the tendency to cross blood brain barrier and early diagnostics abilities [21]. The biocompatibility, multi functionality and effectiveness of nano sized drugs and nano drug carriers can be improved by providing an jacket of polymeric or colloidal material which may or may not be self-healing in nature [22, 23]. These smart drug carriers can also act as on demand drug delivery agents. Dihydroxyphenylalanine, polyethylene glycol, mussel adhesive proteins, polymethylmethacrylate coated or surface engineered nano particles have been used as drug carriers. This chapter incorporates the uses of nanotechnologies, various strategies, and applications nanotechnology in therapeutics.

2 Nanodevices in Therapeutics

The tailoring of material at atomic scale gives the opportunity for studying the nano effect on existing characteristic properties of material (bulk) and generation of new unique properties The interdisciplinary nature of nanotechnology involving physical, chemical, biological, medicinal, engineering, and other branches of sciences makes it complex as well as very innovating [24, 25]. Most importantly, the unpredictable field of nano biotechnology, which covers the effect of nano sized materials in biology or biological processes. The role of nanotechnology in therapeutics is becoming very important due to merging of biotechnology and nanotechnology, which are the most themed research areas.

The molecular or cellular level (DNA) actions are known to derive various macroscopic behaviours in living organisms. Studies at such depth i.e. about the correlations in molecular level with macroscopic behavioural actions of living organisms cannot be studied or analysed through conventional technologies and can only achieved by interdisciplinary work culture. Further, the impact of nanotechnology in electronic, mechanics and optics have forced the study of nano effects in biological/medical sciences.

The fast growth in the sector of nanotechnology, electronics, bio medical sciences, has turned many previously imaginary objectives to reality. The development of nano/micro devices, miniaturized structures, micro-sensors, actuators, transducers and microelectronics, Micro-Electro-Mechanical Systems (MEMS) , Nano-Electro-Mechanical Systems (NEMS) are few examples of such achievements [26–30]. MEMS/NEMS-based micro pumps used for drug delivery, sensors and micro analysis in biological systems. Among these actuator micropumps based on electrochemical, electrowetting, anticorrosive and ion conductive polymer film (ICPF) are promising in terms of flow rates at very low applied voltage [31–35].

In a hypothetical situation after a few generations from now "Nanobots, nanoids, nanites or nanomites" can be applied in advanced medical treatments and diagnosis of diseases like diabetes, cancer, cellular, and can perform nonosurgery and genetherapy [36, 37]. The focus have been on development of these smart workers as they can work within the cell due to correlation in size with that of biological materials like proteins, lipids, DNA, tissues, etc. A particular development in drug delivery system through nanotechnology has been shown in Fig. 3.1. Its ability to interact with brain, cells, DNA, ulcers within and outside the body makes the nanomaterial's unique and potential candidate for their endless uses in biomedical industries.

As per the literature various nano forms or nano-scale materials (Fig. 3.2) have been progressively used in improving imaging, nano-sensing, cardiovascular, nervous disorders, targeted delivery of materials across gut, gene delivery, preparation of artificial implants, cross blood brain barrier and tune bio characteristics of biomolecules, and thus, have become a tool to explore dark avenues in medical sciences [38–43].

3 NanoMedicines

One of the important parameters in therapeutics is drug deliver that to the effective or concerned area. Nano medicine are one of the vibrantly growing part in therapeutics, which are engineered to be highly specific, targeted for better medical results, improved sensing, imaging and treatment of diseases [20, 44, 45]. The huge inflow of nano medicines in commercial market around the globe puts the future scenario

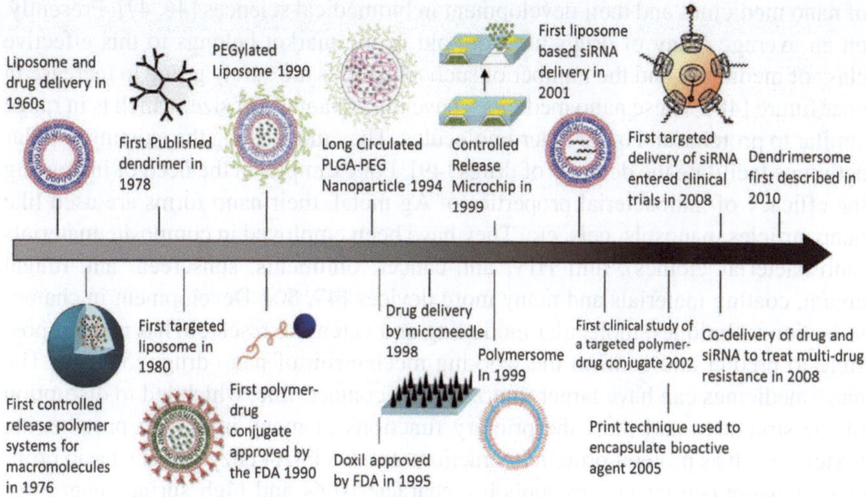

Fig. 3.1 Timeline of nanotechnology-based drug delivery. Here, we highlight some delivery systems that serve as important milestones throughout the history of drug delivery. "Reprinted with permission from Reference [13] Copyright (2010) American Chemical Society"

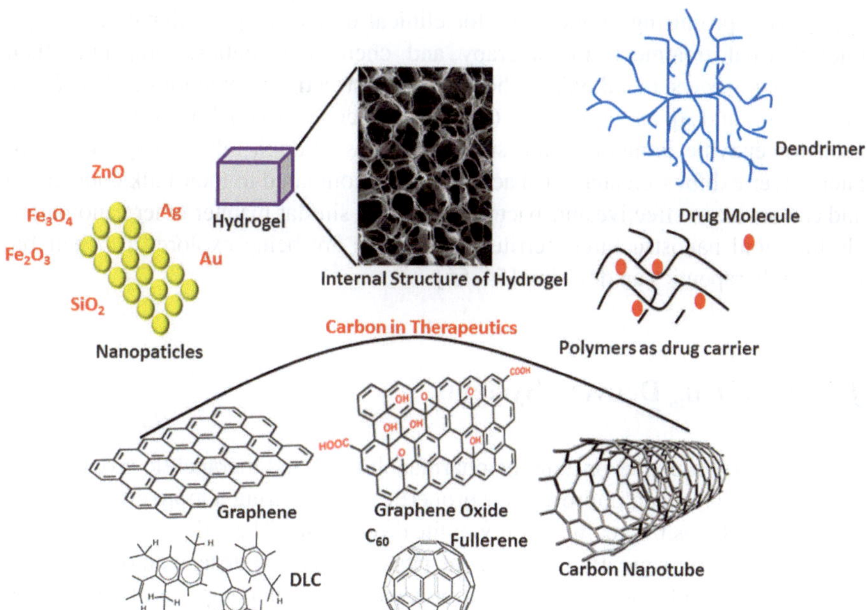

Fig. 3.2 Structures of few potential nanodrug carriers

of nano medicines and their development in biomedical sciences [46, 47]. Presently, on an average many of drugs used or sold in the market belongs to this effective class of medicines and the number of such medicines are surely going to increase in near future [48]. These nano medicines have the advantage of size, which is in range similar to proteins and other macromolecules. They move along the existing cellular activities facilitate the delivery of drugs [49]. For example, in the need of increasing the efficacy of antibacterial properties of Ag metal, their nano forms are used like nanoparticles, nanosols, gels, etc. They have been employed in composite materials (anti-bacterial clothes), anti HIV, anti-cancer, ointments, sunscreen, anti-fungal cream, coating materials and many more devices [47, 50]. Development in characterisation techniques, molecular modelling and extensive research, has made it possible to predict and confirm the working mechanism of nano drugs [51, 52]. The nano medicines can have larger interaction or contact time, which led to disruption of the structures and thus, the primary functions of many important proteins i.e. toxicity, such as protein–protein interactions and CYP3A4 enzyme. The basic property of nanomaterial i.e. hydrophobic characteristics and high surface energy of these materials along with aromatic residues are seem to be the reason for such cell disruption via., π–π stacking interactions, modulation of oxidative stress, anti-angiogenesis, and immune stimulatory activity [51, 53]. However, the gadolinium metallofullerenes ($Gd@C_{82}(OH)_{22}$), functionalized fullerenes are found to inhibit tumor growth, metastases, magnetic resonance imaging (MRI) contrast-enhancing agent, have promising applications for clinical diagnosis, photodynamic therapy, photothermal treatment, radiotherapy and chemotherapeutics. Graphene, their oxides, nanosheets, metal oxides show strong destructive interactions to *E. coli* cell membranes and Aβ amyloid fibrils (anti Alzheimer's disease), however, can also be used for enzyme immobilisation such as lipases through lid opening. Similarly, nano silver exhibits greater anti-bacterial effect compared to their bulk counterpart and can act as an effective anti-bacterial agent. In similar manner other nanomaterials like gold nanostructures, ferrites, zinc oxide are being explored for their bio medical therapeutics an devicing [54–56].

4 Nano Drug Delivery Systems

A drug carrier can alter or tune the pharmacological and therapeutic properties of many drugs. They can improve their properties by acquiring a more effective drug delivery mechanism, specific affinity of the drug carrier. The unique characteristics of the nano carriers greatly influence the performance and dosage amounts. Rapid onset of therapeutic action, improved bioavailability, biocompatibility and lower dosage requirements can be achieved due to enhanced surface active site of the carrier, increased adhesion with biological surfaces, higher resistant to precipitation and rapid dispersion abilities [57]. Engineered nanostructures can be utilised for loading and delivery of therapeutic drugs, DNA, proteins and other macromolecules to the targeted cells and tissues [58]. This in turn enhances the therapeutic

Table 3.1 Nanoscale application in medicine

Drug delivery	Nanoscale delivery vehicles can (1) enhance the therapeutic efficacy and minimize adversities associated with available drugs; (2) enable new classes of therapeutics; and (3) encourage the re-investigation of pharmaceutically suboptimal but biologically active new molecular entities that were previously considered undevelopable.
In vitro diagnostics	Nanotechnology based sensore (e.g., nanowires, nanotubes, nanoparticles, cantilevers, and micro/nanoarrays) can enable fast and high throughput detection of disease biomarkers with higher sensitivity and lower sample consumption. Nanotechnology also offers hope for the early detection of viruses, bacteria and circulating tumor cells, as well as for single cell analysis.
In vivo imaging	Targeted imaging nanoprobes (e.g., magnetic nanoparticles, quantum dots, and carbon nanotubes) could provide a faster, less invasive, and more accurate way to diagnose diseases (e.g., cancer) at their earliest stages and monitor disease progression. Some other possible opportunities include reporting in vivo efficacy of therapeutics and tracking nanocarrier biodistribution in the body. In addition, imaging nanoprobes could help surgeons to locate tumors and their margins, identify important adjacent structures, and map sentinel lymph nodes.
Therapy techniques	Certain nanomaterials have unique therapeutic properties that differ from conventional drugs, and can, therefore, be directly used to treat diseases. For example, hafnium oxide and gold based nanoparticles can greatly enhance X—Ray therapy, gold nanoshells/nanorods, carbon nanotubes, magnetic nanoparticles can induce hypothermia to kill cancer cells; and nanocrystalline silver is being used as an antimicrobial agent.
Biomaterials	Biocompatible nanomaterials that have optimal mechanical properties can be used as medical implants, such as dental restoratives and bone substitutes (also categorized as hard-tissue engineering). Nanocoatings or nanostructures surfaces can also improve the biocompatibility and adhesion of biomaterials.
Tissue engineering	Nanotechnology can enable the design and fabrication of biocompatible scaffolds at the nanoscale and control the spatiotemporal release of biological factors, resembling native extracellular matrix, to direct cell behaviors, and eventually lead to the creation of implantable tissues.

Reprinted with permission from Reference [13]. Copyright (2010) American Chemical Society

benefit at lower dosage, cost and side effects of many drugs. Various utilities of nanotechnology in biomedical application are provided in Table 3.1. Above all the properties of the nanocarriers can be tuned as per the requirement by manipulating the size, morphology and surface functionalities of the carriers which in turn can initiate both passive and active drug targeting is possible [59]. Surface functionalization of such nano carriers using antibodies or aptamers can make them site-specific. Inherit properties of few nano carriers like magnetisation can be utilised for guided delivery of drug at specific areas [60]. Phosphonate, polyethylene glycol (PEG), polyethylenimine–polyethylene glycol (PEI–PEG), Tetraethylorthosilicate (TEOS), vegetable oils, polyols, polyelectrolyte, poly(lactic-co-glycolic acid), chitosan coating, etc. have been used for surface functionalization's and increasing the drug loading capacities of the carriers [61–63]. Controlled and sustained release of drug can also be achieved by these nanoparticles which in turn improve therapeutic efficacy and reduce side effects. They can further enhance the stability of drugs in storage or unused conditions, rendering them suitable for target drug delivery system, decreasing the toxicity or side effect of drugs, and changes drug delivery mechanism. A wide variety of therapeutic drug nano-carriers are being prepared and extensively investigated for their efficacy in this area [64, 65]. The drugs can be loaded or immobilised on to nanostructure via., encapsulation, absorption, adsorption, dispersion, etc. [66, 67]

Now a days, biopharmaceutical companies, researchers and scientists have focussed on the development of substances which are having both biodefense and medical applications. The main objective of research is to find new, faster and more effective countermeasures against these damages. This possibly can be achieved by developing multifunction drugs (say nano drugs), innovative drug delivery mechanisms (nano droids, nano gels, etc.) and systems which can protect and treat against genetically engineered infectious diseases. Such therapeutic technologies can be very effective towards chemical, biological, radiological (nuclear) attacks, and outbreaks of pathogens or any toxicants. However, the costs, efficacy, toxicity, dose and specificity of these new therapies are expected to be in correct order. In order to meet these demands, many well established and efficiently drugs are being delivered using new drug delivery systems [68, 69]. The optimization of these advanced therapeutics may result in addition of value added advancements in human health through their use in biosensors, CNS repairing, biodefense, wound healing, oncology, orthopaedics, bone graft filler, intracellular drug delivery, gene and cancer therapies.

5 Advantages of Modern Drug Delivery System over Traditional

Presently, most of the drugs and their delivery follows traditional administration methods i.e. via. Mouth. So, the drugs has to travel an unrequired path to reach the affected area of the body or brain say from mouth to the dopamine receptor. These drugs has to go through many biological cycles, different area of gastric pH in stomach, digestive vials, proteins, amino acids and many others. Which resulted in increment of the required dosage amount of drug, their side effects, and decrease in the bioavailability and may have bio compatibility issues due to difference in interaction levels with different passing phases. Neurotransmitters drugs like levodopa, or steroids like betamethasone, deflazacort, dexamethasone, hydrocortisone, methylprednisolone, prednisolone and fludrocortisone on oral delivery lead to many side effects such as outweigh, osteoporosis, and suppression of the immune system thus, modifies the drugs actual action. This problem can be tackled by using other routes of administration and systems such as intranasal, infusion pumps, skin patches rectal, sublingual or pulmonary, etc. However, for neuro drug delivery purposes and to have certain control over dopaminergic stimulation, none of the system work very efficiently.

Research has shown that through infusion pumps containing drugs in nano range can be delivered at subcutaneous and intraduodenal level directly which avoid the oscillations of both plasma and striatal levels. This control or drug administering ability maintains the physiological stimulation of dopamine receptors, controls motor fluctuations and dyskinesias. Another effective delivery system is transdermal delivery by patches containing lisuride or rotigotine [70, 71]. Continuous research is on to device new and more effective methodologies to deliver the neurodrug effectively. The typical delivery mechanism involves release of drug from loaded patches due to ionic gradients to the silicon matrix and then to the stratum corneum to reach the epidermis. Sustained delivery of drugs to different dopamine receptors in required amount is the key to control any such neurological disorder. Now, nasal route delivery of drug has been focussed as this can circumvent the blood–brain barrier which along with many other enzymes which restricts the entry of drugs and many neurotherapeutics for maintaining internal milieu of the brain. This is a non-invasive route of delivery which avoid hepatic first-pass metabolism, practical, and safe. Few other delivery pathways involves olfactory and/or trigeminal nerve pathways, but their existence in humans have not been confirmed [72, 73].

6 Nano-Neurotherapeutics

The term "Neurotherapeutics" is associated with the therapeutic treatments for Neurological disorders, which are caused due to imbalance in central nervous system (CNS). CNS is the CPU of any living organism, i.e. it controls most of the functions in body and mind. It is consisted of brain and the spinal cord. The brain is the interpreter of the signals, controlling the functioning of internal organs and responses for external situations. Whereas, the spinal cord channelizes the communication in between brain and body. Any disturbance in their coordination lead to generation of diseases of the brain, spine and nerves, overall, accounting for more than 500 diseases of the nervous system. Discrepancies in proper functioning of CNS may cause brain tumors, epilepsy, Parkinson's disease, Alzimer's disease, dementia and strokes along with impairment of proper functioning required for sustain healthy life. Compared to many organs and tissues within the body which have a self-recovery or self-repairing functioning, CNS once damaged do not self-repair because of inability of CNS to create new cells and also cannot recreate the connections in CNS due to complexity in CNS involving generation of Neurons, millions of connections, synapses, neurotransmitters, etc. In this time of development, new drug, delivery systems, computational models are constantly helping in designing new and effective remedies for such neuro diseases [74].

7 Nano-Neurotherapeutics for Parkinson's and Alzheimer Disease

Neuro disorders particularly Parkinson's and Alzheimer disease are complex and heterogeneous disorders, including etiologic, symptomatic, therapeutic and evolutional changes affecting human brain and compensatory mechanisms [75]. It is a neurophysiological disturbances generated due to decrease in the level of dopamine in the striatum and in other nuclei, which in response increases firing rate of neurons with abnormal firing pattern in the subthalamic nucleus and internal segment of the globus pallidus [76]. The extent of disease and its anatomical effects on different cells depends on genetics of patient. Different nucleus of brain at a molecular level get genetically modified due to the created environmental resulting in mis-folding of proteins with consequences at the cellular level dysfunction of the ubiquitin proteasome system, mitochondria and protein aggregation. The apoptotic death of dopaminergic cells triggers the production of biochemical and neurophysiological disturbances in the striatum and in other nuclei. This increases firing rate with abnormal firing pattern in the subthalamic nucleus and internal segment of the globus pallidus. Parkinson's disease is dynamic, especially when treated, with signalling pathways and neurotransmission systems in constant change.

Several needs of patients with Parkinson's disease are not fully understood, so new diagnostic and therapeutic tools are required every now and then for the identification and solution of any particular problem. For example, need of reliable biomarker which can diagnosis presymptom of Parkinson's disease and mechanisms of cell death to start neuroprotective treatments which can stop progression of the disease. Figure 3.3 show the scientific approaches which are important for effective treatment of Neuro degenerative diseases.

Nanotechnology can be helpful in determining the actual cause of initiation of Parkinson's disease. It can be helpful for regeneration of damaged parts in the brain and provides an effective as well as a safer route of drug delivery to the brain along with efficient cell generation. All this can increase the survival of implanted cells if any, and improves the targeted delivery of molecules to affected areas of the brain. New types of diagnostic and therapeutic tools includes engineered nano probes/materials at least in one dimension having different functionalities with possibilities for constant manipulation and generation of new functional properties. Nanotechnology have the ability to cease a neuropathological state, provide neuroprotection and can delivery drugs and small molecules across the blood brain barrier, which can help in functional regeneration of damaged neurons [4, 77]. Nanotechnology could have a revolutionary impact on diagnosis and therapy of neurodegenerative diseases At nanoscale, the efficacy of material increases in terms of target cells/tissues stimulation cellular level interaction within biological systems, ability to cross certain blood barriers and minimizing undesirable effects of certain drug molecules. Despite these inherent complexities, nanoscience could decisively contribute to the achievement of some major research objectives. Besides

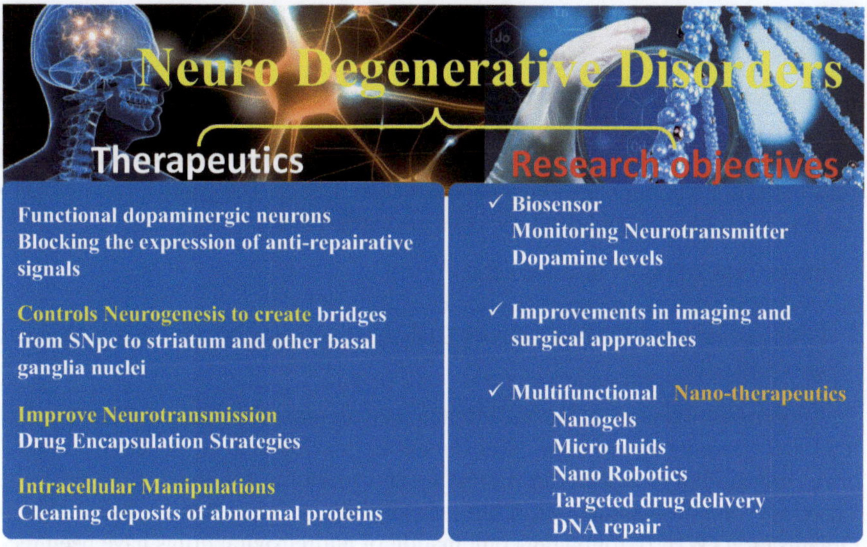

Fig. 3.3 Different scientific approaches for effective tackling of neuro disorders (Parkinson's, Alzheimers, etc)

diagnostic purposes, nanotechnology could provide devices to limit and reverse neuropathological disease states, to support and promote functional regeneration of damaged neurons, to provide neuro protection and to facilitate the delivery of drugs and small molecules across the blood brain barrier [29].

Treatment of Parkinson's disease involves use of more than 12 or more drugs or pills every day for different behavioural and neurotransmitter systems. To simplify this cumbersome and hectic treatment, the main aim is concluded to reduce this number of tabs to one or two capsules, inbuilt with all the necessary drugs in the precise doses, to be taken once a day. For example, treatment other brain pathologies, such as depression and schizophrenia, has been done using single drug, known as "dirty drugs" (Clozapine) having combined actions equivalent to the use of multi drugs. These magic tabs having the influence of nanotechnology can achieve a complex action with a single tablet by improving motor, emotional and cognitive symptoms without any complications of dyskinesias or psychiatric. Amantadine, Budipine are antiparkinsonian drugs with multiple actions but are having issues like their efficacy, compositions, unfavourable benefit and side effect [78, 79]. The efficacy, delivery mechanism and their effects can be controlled by nanotechnology as it has been used as a tools to selectively work with complex biological systems and can simultaneously modulate/respond to the system, normalize the concentration and the activity of neurotransmitters in the most physiologically possible fashion as compared to pure pharmacological approach [72].

8 Nanotherapeutic Formulation for HIV Treatment

Antiretroviral drugs which are generally used for the treatment of HIV, do not have the ability to kill the virus. However, when administered properly along with other responsive drugs they can check the growth of the virus. HIV primarily infects vital cells in the human immune system such as helper T cells. The lower levels of CD4+ T cells, macrophages, and dendritic cells are caused due to

1. direct viral killing of infected cells;
2. increased rates of apoptosis in infected cells; and
3. killing of infected CD4+ T cells by CD8 cytotoxic lymphocytes that recognizes infected cells.

Antiretroviral treatment lets the immune system recover by suppressing the viral load and prevents its passage, viral mutation, development of resistance against drug, substantive reductions in morbidity and mortality for infected individuals. There are lots of combinations of antiretroviral drugs but as per the World Health Organisation (WHO), following combination of antiretroviral drugs have been recommended for people starting HIV treatment [80].

- TDF (tenofovir) please add AZTTP anti-HIV drug also
- either 3TC (lamivudine) or FTC (emtricitabine)
- and EFV (efavirenz)

However, the traditional oral delivery of antiretroviral drugs has a number of limitations like high drug dose, side effects, percent efficacy, solubility and stability in aqueous system. The liquid formulation of many drugs are not available for administration purposes like Efavirenz. Further, very few of these drugs can be administered intravenously to increase the effectiveness and decrease the drug dosage. Nanomedicines formulations and nano drug carriers are the probable solution for these problems. The adherence of antiretroviral drugs is a very important factor, as after proper optimisation such nanomedicines can be utilised for lifelong daily dosing purpose. The nanostructure of optimised size which can penetrate into cell and tissue containing crystalline drug particles can be produce to target immune cells for sustained release of antiretroviral drugs. Diego A Chiappetta et al., improved the pediatric HIV pharmacotherapy by increase the bioavailability, aqueous solubility and improving the drug administration. The pharmacokinetic studies indicates that the incorporation of Efavirenz in the core of linear and branched poly(ethylene oxide)–poly(propylene oxide) block copolymer micelles increases the bioavailability of same [81]. In comparison to the traditional methodologies of treatment, nanotheruputic improves the efficiency of the same drug by improving the penetration ability in infected target tissues and cells. The nano transport system facilitate the delivery across the blood–brain barrier, and efficient internalization in cells by nonspecific or receptor-mediated endocytosis. As many metallic and non-metallic nanoparticles like silver, gold or ferrite nanoparticles are known for their ability to virus targeting, bio sensing [56] and formulations [82] which improves the diagnostic or therapeutic ability manyfolds [83–85].

9 Challenges and Future Prospects

The main challenges associated with nanotechnology applications in therapeutics is the deficiency of the complete knowledge in what type of bio response would be trigger due to interaction particular nanoparticles, nano drug, or carriers to biological system. The development of newer and effective delivery systems have led to the research in nano therapeutics, where a lot more distance has to be covered, particularly in the area of nanobots, nano swimmers, etc. Further, failing in regeneration of efficiency of *in vitro* results into *in vivo* results due to certain technical reasons like 2D and 3D cell culturing, limited cell imaging using UV or TEM analysis. Scaling up and regeneration of the custom tailoring of surface linkages over the drug and carrier is a challenging task which is indeed a multidisciplinary collaboration work among synthetic chemists, biologists, and biomedical engineers [86]. The complete knowledge of pharmacokinetics during metabolic stability, clearance rate, premature drug release and trouble in penetration of barriers like BBB, tissues with in the human bodies is most important aspect.

On the other hand, nano neuro therapeutics is dealing with numerous related complexity of the diseases. Nanotechnology allows an intervention at a molecular or desired cellular signalling pathway which is highly unspecific in nature. Further development of innovative technologies that are able to multitask such ability to perform diverse cellular and physiological functions, targeting multiple receptors or ligands is still in progress. Inability to exactly understand the numerous interdependent molecular and biochemical events which are linked can be disrupted due to predict the clinical expressions resulted from interactions of numerous neurons and interdependent cell groups. Finally, long-term effects of nanotechnologies are unknown.

Spreading education for diseases, acceptance of the illness, and public participation in awareness and research purposes will improve the understanding as well as the management of neuro deficiencies and their treatments. Active and passive targeting approaches for delivery of surface-engineered nanocarrier systems would bring desirable effects and put forward further progress in this field. Scientist and researchers are optimistic about nanotechnology-based nose to brain drug delivery, which would eventually become an important consent to the therapies offered to the Parkinson's, HIV, Alzheimer's patients in near future.

10 Conclusion

The present treatments which involve brain mapping, cyber knife, deep brain stimulation, gamma knife exhibit many limitations including high cost, risky and sometimes result into adverse health effects. Nanotherapeutics offer better alternative to other existing treatments, provided proper implantation or delivery mechanism is followed. Nano medicines or nanotherapeutics can involve monitoring, repair, construction, and control of human biological systems at the molecular level, using engineered nano devices and nanostructures. They have been utilised in techniques for early diagnosis and effective treatment of central nervous disorders like Alzheimer's, Parkinson's, fits and others.

Acknowledgements The authors appreciate the financial support received from the Council of Scientific and Industrial research (CSIR), New Delhi, India for this work. I am also thankful for the financial support given by Department of Chemistry, School of Basic and Applied Sciences, Galgotias University, Gautam Buddh Nagar, Uttar Pradesh, India. Dr. Anujit Ghosal also expresses his sincere gratitude to the Government of India, Science and Engineering Research Board (SERB) (NPDF/2016/003866).

References

1. Kabu S, Gao Y, Kwon BK, Labhasetwar V. J Control Release. 2015;219:141–54.
2. Kamaly N, Yameen B, Wu J, Farokhzad OC. Chem Rev. 2016;116:2602–63.
3. Oberoi RK, Parrish KE, Sio TT, Mittapalli RK, Elmquist WF, Sarkaria JN. Neuro-Oncol. 2016;18:27–36.
4. Vashist A, Kaushik A, Vashist A, Jayant RD, Tomitaka A, Ahmad S, Gupta Y, Nair M. Biomater Sci. 2016;4:1535.
5. Nair M, Jayant RD, Kaushik A, Sagar V. Adv Drug Deliv Rev. 2016;103:202.
6. Rao V. Transl Biomed. 2016;7:2.
7. Peng S, Jin G, Li L, Li K, Srinivasan M, Ramakrishna S, Chen J. Chem Soc Rev. 2016;45:1225–41.
8. Srikanth M, Kessler JA. Nat Rev Neurol. 2012;8:307–18.
9. Jung YC, Bhushan B. Nanotechnology. 2006;17:4970.
10. Sharmin E, Akram D, Vashist A, Wani M, Ahmad A, Zafar F, Ahmad S. Chemistry of phytopotentials: Health, energy and environmental perspectives. Berlin: Springer; 2012. p. 223–7.
11. Park K. J Control Release. 2007;120:1.
12. Seeman NC. Chem Biol. 2003;10:1151–9.
13. Shi J, Votruba AR, Farokhzad OC, Langer R. Nano Lett. 2010;10:3223–30.
14. Koo OM, Rubinstein I, Onyuksel H. Nanomedicine. 2005;1:193–212.
15. Alderton GK. Nat Rev Cancer. 2016;16:5–5.
16. DeVience SJ, Pham LM, Lovchinsky I, Sushkov AO, Bar-Gill N, Belthangady C, Casola F, Corbett M, Zhang H, Lukin M. Nat Nanotechnol. 2015;10:129–34.
17. Vashist A, Shahabuddin S, Gupta YK, Ahmad S. J Mater Chem B. 2013;1:168–78.
18. Delalat B, Sheppard VC, Ghaemi SR, Rao S, Prestidge CA, McPhee G, Rogers M-L, Donoghue JF, Pillay V, Johns TG. Nat Commun. 2015;6:8791.
19. Fonseca-Santos B, Gremião MPD, Chorilli M. Int J Nanomedicine. 2015;10:4981.
20. Aparicio-Blanco J, Martín-Sabroso C, Torres-Suárez A-I. Biomaterials. 2016;103:229–55.
21. Sharma R, Agrawal U, Mody N, Vyas SP. Biotechnol Adv. 2015;33:64–79.
22. Merino S, Martín C, Kostarelos K, Prato M, Vázquez E. ACS Nano. 2015;9:4686–97.
23. Kaushik NK, Kaushik N, Pardeshi S, Sharma JG, Lee SH, Choi EH. Mar Drugs. 2015;13:6792–817.
24. Ghosal A, Shah J, Kotnala RK, Ahmad S. J Mater Chem A. 2013;1:12868–78.
25. Khan R, Kaushik A. Nanobiotechnology for sensing applications: from lab to field. Oakville, ON: Apple Academic; 2016. p. 265–86.
26. Vasudev A, Kaushik A, Tomizawa Y, Norena N, Bhansali S. Sensors Actuators B Chem. 2013;182:139–46.
27. Leondes CT. Mems/Nems: (1) Handbook techniques and applications design methods, (2) Fabrication techniques, (3) manufacturing methods, (4) Sensors and actuators, (5) Medical applications and MOEMS. New York: Springer Science & Business Media; 2007.

28. Manickam P, Kaushik A, Karunakaran C, Bhansali S. Biosens Bioelectron. 2017;87:654–68.
29. Vashist A, Ahmad S. Curr Pharm Biotechnol. 2015;16:606–20.
30. Vashist A, Vashist A, Gupta Y, Ahmad S. J Mater Chem B. 2014;2:147–66.
31. Korotcenkov G, Cho B. Porous silicon: from formation to application: biomedical and sensor applications. Biomed Sens Appl. 2016;2, 299
32. Vashist A, Ghosal A, Gupta YK, Ahmad S, Nair M. Nanocomposite hydrogels for neuro drug delivery. J Neuroimmune Pharmacol. 2016;11:s48–s49. New York: Springer.
33. Rawat NK, Ghosal A, Ahmad S. RSC Adv. 2014;4:50594–605.
34. Ghosal A, Rahman OU, Ahmad S. High-performance soya polyurethane networked silica hybrid nanocomposite coatings. Ind Eng Chem Research. 2015;54:12770–87.
35. Sharmin E, Akram D, Ghosal A, ur Rahman O, Zafar F, Ahmad S. Prog Org Coat. 2011;72:469–72.
36. Muthukumaran G, Ramachandraiah U, Samuel D. In Advanced Materials Research, 2015;1086:61–7.
37. Vashist A, Ahmad S. Orient J Chem. 2013;29:861–70.
38. Masoomi MY, Morsali A. Coord Chem Rev. 2012;256:2921–43.
39. Ariga K, Ito H, Hill JP, Tsukube H. Chem Soc Rev. 2012;41:5800–35.
40. Delgado JL, Herranz MÁ, Martin N. J Mater Chem. 2008;18:1417–26.
41. Hwang SR, Kim K. Arch Pharm Res. 2014;37:24–30.
42. Abdelmohsen LK, Peng F, Tu Y, Wilson DA. J Mater Chem B. 2014;2:2395–408.
43. Brzeziński M, Biela T. Polym Int. 2015;64:1667–75.
44. Hartley JM, Kopeček J, Nam J, Jung S, Hwang S, Song J, Kim S, Sultana S, Sultana S, Islan GA. Smart pharmaceutical nanocarriers. Singapore: World Scientific; 2016. p. 373–413.
45. Phillips AJ. Developing a predictable regulatory path for nanomedicines by accurate and objective particle measurement. In: Braddock M, editor. Nanomedicines. Cambridge: Royal Society of Chemistry. 2016;253–80.
46. Jayant RD, Sosa D, Kaushik A, Atluri V, Vashist A, Tomitaka A, Nair M. Expert Opin Drug Deliv. 2016;13:1433.
47. Kaushik A, Jayant RD, Nair M. Int J Nanomedicine. 2016;11:4317.
48. Bawa R, Audette GF, Rubinstein I. Handbook of clinical nanomedicine: nanoparticles, imaging, therapy, and clinical applications. Boca Raton: CRC; 2016.
49. Dixit CK, Kaushik A. Nanobiotechnology for sensing applications: from lab to field. Oakville, ON: Apple Academic; 2016. p. 351–4.
50. Akhavan O, Ghaderi E. Enhancement of antibacterial properties of Ag nanorods by electric field. Sci Technol Adv Mater. 2016;10:015003.
51. Zhou R. Nanomedicine. 2016;12:468.
52. Zheng W, Wei M, Li S, Le W. Nanomedicine. 2016;11:1417–30.
53. Chen Z, Ma L, Liu Y, Chen C. Theranostics. 2012;2:238–50.
54. Gutiérrez L, Costo R, Grüttner C, Westphal F, Gehrke N, Heinke D, Fornara A, Pankhurst Q, Johansson C, Veintemillas-Verdaguer S. Dalton Trans. 2015;44:2943–52.
55. Soenen SJ, Parak WJ, Rejman J, Manshian B. Chem Rev. 2015;115:2109–35.
56. Kaushik A, Vabbina PK, Atluri V, Shah P, Vashist A, Jayant RD, Yandart A, Nair M. Biosens Bioelectron. 2016;86:426–31.
57. Such GK, Yan Y, Johnston AP, Gunawan ST, Caruso F. Adv Mater. 2015;27:2278–97.
58. Kaushik A, Tiwari S, Jayant RD, Vashist A, Nikkhah-Moshaie R, El-Hage N, Nair M. Trends Biotechnol. 2017, 35:308–17.
59. Singh R, Norret M, House MJ, Galabura Y, Bradshaw M, Ho D, Woodward RC, Pierre TGS, Luzinov I, Smith NM. Small. 2016;12:351–9.
60. Hanafi-Bojd MY, Jaafari MR, Ramezanian N, Xue M, Amin M, Shahtahmassebi N, Malaekeh-Nikouei B. Eur J Pharm Biopharm. 2015;89:248–58.
61. Loureiro JA, Gomes B, Fricker G, Coelho MAN, Rocha S, Pereira MC. Colloids Surf B Biointerfaces. 2016;145:8–13.
62. Mary TA, Shanthi K, Vimala K, Soundarapandian K. RSC Adv. 2016;6:22936–49.
63. Desai D, Prabhakar N, Mamaeva V, Karaman DŞ, Lähdeniemi IA, Sahlgren C, Rosenholm JM, Toivola DM. Int J Nanomedicine. 2016;11:299.

64. Pathak Y, Thassu D. Drug delivery nanoparticles formulation and characterization. Boca Raton, FL: CRC; 2016.
65. Vij N, Min T, Bodas M, Gorde A, Roy I. Nanomedicine. 2016;12:2415–27.
66. Siqueira JR, Caseli L, Crespilho FN, Zucolotto V, Oliveira ON. Biosens Bioelectron. 2010;25:1254–63.
67. Zamani M, Prabhakaran MP, Ramakrishna S. Int J Nanomedicine. 2013;8:2997–3017.
68. Wang Q, Cheng H, Peng H, Zhou H, Li PY, Langer R. Adv Drug Deliv Rev. 2015;91:125–40.
69. Xing J-F, Zheng M-L, Duan X-M. Chem Soc Rev. 2015;44:5031–9.
70. Woitalla D, Müller T, Benz S, Horowski R, Przuntek H. Focus on extrapyramidal dysfunction. Wien: Springer; 2004. p. 89–95.
71. Deshmukh AM, Dixit A, Kumar P, Bindra V. U.S. Patent 14/893,020. 2014 May 19. 2016.
72. Kulkarni AD, Vanjari YH, Sancheti KH, Belgamwar VS, Surana SJ, Pardeshi CV. J Drug Target. 2015;23:775–88.
73. Ali J, Ali M, Baboota S, Kaur Sahni J, Ramassamy C, Dao L. Curr Pharm Des. 2010;16:1644–53.
74. Helie S, Chakravarthy S, Moustafa AA. Front Comput Neurosci. 2015;7:174.
75. Kaushik A, Jayant RD, Tiwari S, Vashist A, Nair M. Biosens Bioelectron. 2016;80:273–87.
76. Lee JJ, Oh JS, Ham JH, Lee DH, Lee I, Sohn YH, Kim JS, Lee PH. Neurobiol Aging. 2016;38:197–204.
77. Kaushik A, Tiwari S, Jayant RD, Marty A, Nair M. Towards detection and diagnosis of Ebola virus disease at point-of-care. Biosens Bioelectron. 2016;75:254–72.
78. Giuffrida A, Martinez A. Levodopa-induced dyskinesia in Parkinson's disease. London: Springer; 2014. p. 245–64.
79. Müller T. J Parkinsonism Restless Legs Synd. 2015;5:11–7.
80. World Health Organization. Consolidated guidelines on the use of antiretroviral drugs for treating and preventing HIV infection. Geneva: WHO; 2013.
81. Chiappetta DA, Hocht C, Taira C, Sosnik A. Nanomedicine. 2010;5:11–23.
82. Dixit C, Kaushik AK. Microfluidics for biologists: fundamentals and applications. Switzerland: Springer; 2016.
83. Yadavalli T, Shukla D. Nanomedicine. 2016;13:219. doi:10.1016/j.nano.2016.08.016.
84. Chris GW, Duu-Jong L. Nanotechnology. 2016;27:365101.
85. Whiteley CG, Shing CY, Kuo CC, Lee D-J. J Taiwan Inst Chem Eng. 2016;60:83–91.
86. Wong PT, Choi SK. Chem Rev. 2015;115:3388–432.

Chapter 4
Image-Guided Therapy

Asahi Tomitaka, Hamed Arami, Yasushi Takemura, and Madhavan Nair

Abstract Image-guided therapy using nanoparticles have been used for various therapeutic applications such as drug and gene delivery, hyperthermia, and photodynamic therapy. These image-guided approaches help to achieve more efficient treatments by visualizing biodistribution of therapeutic agents and quantifying them at targeted tissues. This allows not only monitoring of their accumulation at targeted sites, but also highly localized treatments by controlled activation of therapeutic agents at the desired tissues, when needed. Here, we summarize the imaging modalities used for image-guided therapies and discuss recent advances on nanoparticle-based image-guided therapies.

Keywords Image-guided therapy • Drug delivery • Hyperthermia • Photodynamic therapy • Nanomedicine

1 Introduction

Nanomedicine has been growing rapidly over the past few decades. Introducing nanotechnology into medical sciences has broadened nanomedicine applications significantly. Various types of multifunctional nanomaterials have been developed for simultaneous therapy and medical imaging. Their small sizes (often ranging

A. Tomitaka (✉) • M. Nair
Center for Personalized Nanomedicine, Institute of Neuro immune Pharmacology, Department of Immunology, Herbert Wertheim College of Medicine, Florida International University, Miami, FL 33199, USA
e-mail: atomitak@fiu.edu

H. Arami
Molecular Imaging Program at Stanford (MIPS), The James H Clark Center, Stanford University, Stanford, CA 94305, USA

Department of Radiology, Stanford University School of Medicine, Stanford, CA 94305, USA

Y. Takemura
Department of Electrical and Computer Engineering, Yokohama National University, Yokohama 240-8501, Japan

A. Kaushik et al. (eds.), *Advances in Personalized Nanotherapeutics*,
https://doi.org/10.1007/978-3-319-63633-7_4

from 10 to 100 nm) are comparable to biological molecules such as proteins and viruses and their functionalities enable delivery of therapeutic agents and imaging contrast agents to target tissues.

Drug delivery is one of the most promising areas in nanomedicine. The conventional chemotherapy treatments are inefficient and cause toxicity to healthy tissues due to poor pharmacokinetics and inappropriate biodistribution of the drugs [1]. Delivery of drugs to specific sites and their release using nanoparticles make it possible to reduce such side-effects and treat the target site efficiently with minimal amount of drugs. Nanoparticles have been also used as platforms to carry other agents for gene therapy and photodynamic therapy (PDT). Intrinsic characteristics of the nanoparticles can be also used for treatments such as hyperthermia. In addition, nanoparticles can be used as contrast agents for diagnosis of disease at earlier stages [2].

Since the approval of the first nano-drug from the Food and Drug Administration (FDA) in 1995 [3], various nanoparticles have been developed and some of them made their way to the market. In recent years, the advancements of nanotechnology have brought new generation of multifunctional nanoparticles which integrate multiple functionalities in one nano-formulation. Emerging of multifunctional nanoparticles with both therapeutic and imaging capabilities enabled innovative image-guided approaches for various therapies (Fig. 4.1).

The most attractive part of nanoparticle-based image-guided therapies is its ability for quantitative visualization of therapeutic agents at their targeted sites after administration. This also allows evaluation of treatment efficacy. Another advantage is its ability to trigger drug release and activate therapeutic agents locally at the target sites based on this visualization.

2 Medical Imaging Modalities

Magnetic resonance imaging (MRI) is a powerful medical imaging modality which gives high spatial resolution images without radiation risk. MRI acquires signals from hydrogen protons of subject. Under external static magnetic field (B_0), the spins of hydrogen protons precess with a Larmor frequency, while they align parallel to the direction of B_0. The spins align antiparallel to the direction of B_0 when additional radio-frequency (RF) pulse with the Larmor frequency is applied. After RF pulse is removed, the protons relax through the recovery of magnetization parallel to the direction of B_0 (longitudinal or T_1 relaxation) and the decay of transverse magnetization caused by the de-phasing of the spins (transverse or T_2 relaxation) [4]. The tissue specific parameters of proton density, T_1 relaxation, and T_2 relaxation give contrast to the MR images. Due to this principle, MRI gives great spatial resolution for soft tissues which contain more water molecules. The major drawback of this technique is its relatively poor sensitivity, due to the weak signal from protons [5]. T_1 and T_2 contrast agents which shorten relaxation times have been developed to overcome its poor sensitivity and improve specificity. Gd(III) chelates are the

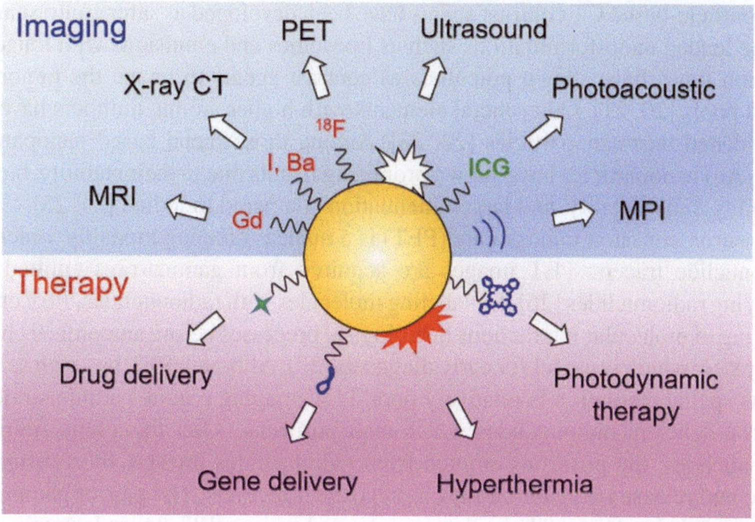

Fig. 4.1 Schematic illustration of multifunctional nanoparticles and their applications for imaging and therapies

most widely used T_1 contrast agents in clinical applications [6]. Although Gd(III) chelates have good paramagnetic property and shorten T_1 relaxation, the toxicity of gadolinium ions and their rapid renal clearance limit their use [7]. To reduce toxicity, avoid rapid clearance, and improve specificity, gadolinium has been incorporated into or conjugated on various types of nanoparticles such as liposomes [8], polymer nanoparticles [9], and micelles [10]. Ligand conjugation on these nanoparticles have been reported to improve their targeting ability [11]. Iron oxide nanoparticles have been used as T_2 contrast agents for various MR imaging applications such as brain and liver cancer diagnosis [12]. Due to their unique magnetic properties including MR imaging and magnetic targeting capability, iron oxide nanoparticles have been studied extensively as MRI contrast agents [13, 14]. T_1/T_2 dual contrast agents which contain both of T_1 and T_2 contrast agents in one formulation have been also reported [15, 16].

X-ray Computed tomography (CT) is a non-invasive imaging modality which visualizes anatomical details using the X-ray attenuation properties of the tissues. The three-dimensional (3D) images are acquired by an X-ray source and a detector which rotates around patient's body [17]. Since bones have higher attenuation compared to soft tissues, X-ray CT images provide detailed anatomical information for bones and other areas which have high attenuations. Although X-ray CT is known for its wide availability, fast scanning, and low cost, it is less efficient for imaging of soft tissues with similar attenuation properties [18]. High atomic number elements such as iodine and barium have been used as contrast agents to improve the contrast of soft tissues [19]. However, iodinated agents have short blood circulation time and barium sulfate commonly used for gastrointestinal (GI) tract imaging raises toxicity concerns [17]. To improve their circulation time and toxicity, nanoparticle-based CT contrast agents have been developed as safer contrast agents. Iodine loaded nano-formulations such as liposomes and emulsions with longer circulation times have shown potentials as contrast agents to image the tumors and blood pools [20, 21]. Other metal elements with higher atomic numbers have been also loaded into nanoparticles [22, 23]. Among those metal based nanoparticles, gold (Au) nanoparticles have shown promising results due to their stability, biocompatibility, functionality, and larger attenuation compared to iodine [24, 25].

Positron emission tomography (PET) is a nuclear imaging modality which uses radionuclide tracers. PET images are acquired from gamma-rays emitted from decaying radionuclides [26]. By labeling molecules with radionuclides, PET enables imaging of molecular interactions in biological processes before anatomical changes take place, which is useful for early diagnosis [27]. Although PET has high sensitivity, its spatial resolution is relatively poor. PET imaging requires administration of neutron-deficient radionuclides which emit positrons. After the administration of radionuclides, the positrons emitted from radionuclides travel a short distance in tissue and release two gamma-rays in opposite directions. The pair of gamma-rays are detected simultaneously by detectors around patient. 18F-fluorodeoxyglucose is the most widely used radionuclide tracer for PET imaging, due to its longer half-life compared to other radioisotopes [26]. Since PET is most effective when it is combined with anatomical imaging such as MRI and CT, radionuclides have been loaded into nanoparticles with other contrast agents [28, 29].

Ultrasound imaging is one of the most widely used imaging modalities with real-time imaging capability. Its low cost, portability, and no radiation risk are suitable for primary diagnostics. Ultrasound imaging is based on the pulse-echo principle which involves transmission of ultrasound pulse from a transducer into the body and detection of its reflected sound waves from the tissues. The image is generated from the detected echoes which depend on the different acoustic impedance of the tissues [30]. Although ultrasound imaging has high spatial resolution, its poor sensitivity and poor soft tissue contrast are drawbacks. To improve the sensitivity and the contrast in soft tissues, various gas-filled microbubbles have been developed as contrast agents. Microbubble contrast agents which utilize serum albumin and phospholipid have been used in clinical applications [31].

Advances in nanoparticle synthesis and characterizations have also provided opportunities for development of new imaging modalities such as photoacoustic tomography (PAT) [32] and magnetic particle imaging (MPI) [33]. Photoacoustic tomography (PAT) is an emerging imaging technique, which combines optical and ultrasound imaging. By combining two imaging systems, it overcomes limitations of each individual modality, such as lack of penetration depth of optical imaging or poor sensitivity of ultrasound imaging, and provides physiological information such as hemoglobin concentration and angiogenesis based on optical absorption in tissues. For the imaging, light is exposed to the subject and the absorption of light increases the temperature of the subject. This temperature rise causes thermoelastic expansion and creates acoustic waves. The images are generated by the acoustic waves detected by piezoelectric devices [34]. Since acoustic waves are scattered much less than optical waves, PAT enables high resolution imaging in deeper tissues

Table 4.1 Comparison of medical imaging systems

Table	Signal measured	Penetration	Sensitivity	Spatial resolution	Image information	Cost
MRI	Radio wave	No limit	10^{-9}–10^{-6}	50 μm	Anatomical Physiological Molecular	High
CT	X-rays	No limit	10^{-6}	50 μm	Anatomical Physiological	Medium
PET	γ-rays	No limit	10^{-15}	1–2 mm	Physiological Molecular	High
US	Sound	Several cm	10^{-8}	50 μm	Physiological Molecular	Low
PAT	Sound	<5 cm	10^{-12}	50 μm	Anatomical Physiological Molecular	Low

Adapted with permission from [36, 37]

compared to optical imaging [35]. Some fluorescent dyes such as Indocyanine-green (ICG), metal nanoparticles such as gold nanoparticles, and single-walled carbon nanotubes have been developed as PAT contrast agents [32]. Such wide variety of contrast agents expand PAT applications compared to ultrasound imaging which is limited to micrometer size contrast agents.

Magnetic particle imaging (MPI) is a new real-time imaging modality which has a potential to perform quantitative 3D imaging with high sensitivity and spatial resolution. MPI uses the signal directly generated from the magnetic nanoparticles, without any signal generated from surrounding tissues. When alternating magnetic field is applied to magnetic nanoparticles, the nanoparticles are excited and their time-varying magnetization response induces a signal in the pick-up coils [38]. Signal localization is achieved by an additional field gradient, which generates a single field-free point. Magnetic nanoparticles outside of field-free point are saturated and do not induce any signal [39]. Theoretical estimations predict that MPI can provide greater sensitivities compared to MRI and high spatial resolution comparable to X-ray CT and MRI due to its stronger signal-to-noise ratios compared to the proton signal used in MRI [39, 40]. Each individual imaging modality has its advantages and limitations (Table 4.1). By integrating multiple imaging modalities, multimodal imaging is expected to overcome the limitations of each imaging modality and improve reliability for clinical applications. Since the commercialization of PET/CT instrument more than a decade ago, multimodal imaging has received increasing attention. Also the development of multifunctional nanoparticles enabled diagnostics using various combinations of imaging contrast agents. For example, radiolabeled nanoparticles such as ^{89}Zr-iron oxide nanoparticles and ^{69}Ge-iron oxide nanoparticles have been developed for PET/MRI [28, 29]. Nanostructures containing iron oxide and gold nanoparticles and FePt nanoparticles have shown promising results for MRI/CT [41, 42]. Furthermore, various multifunctional nanoparticles which have capabilities to combine more than three imaging modalities have been reported such as Cy5.5 labelled iron oxide nanoparticles for MPI/MRI/near infra-red fluorescent (NIRF) [43] and nanocomposite of ^{64}Cu absorbed MoS_2 nanosheet and iron oxide nanoparticles for PET/PAT/MRI [44].

3 Recent Advances in Image-Guided Therapy

3.1 Image-Guided Surgery

By adapting medical imaging systems such as CT, MRI and optical imaging in the operation room, image-guided surgery allowed surgeons to know the patient's anatomy and the precise position of instruments. In addition to the anatomical information, nanoparticles have potential to contrast target site intraoperatively and distinguish the disease sites (such as tumor margins) from normal tissues [45]. Zheng et al. developed a liposomal formulation which co-encapsulated CT contrast agent and indocyanine green (ICG) for near infra-red fluorescent (NIRF) (Fig. 4.2) [46]. Nanoparticles were accumulated in tumor lesions of ovarian cancer and metastatic breast cancer models by enhanced permeability and retention (EPR) effect which was first proposed in 1986 [47]. It resulted in highly sensitive monitoring of tumor site with anatomical information provided by CT images combined with high sensitivity of NIRF. Significant reduction in the recurrence rate of breast cancer has been reported by applying image guided surgery using photoacoustic imaging and NIRF [48]. Iron oxide nanoparticles labelled with NIRF molecules showed selective tumor accumulation using amino-terminal fragments of urokinase plasminogen activator, and imaging using photoacoustic and NIRF enabled efficient resection of tumor site. Image-guided surgery has also been applied for detection of sentinel lymph nodes. Sentinel lymph nodes are the first lymph nodes which receive lymphatic drainage from the primary tumors [49]. Sentinel lymph nodes mapping and biopsy are widely accepted, but there is still no accurate system available to do this intraoperatively. Nanoparticles have been used for such intraoperative mapping applications, showing promising results using real-time imaging systems including PET and fluorescence imaging [50].

3.2 Image-Guided Drug and Gene Delivery

Drug delivery has been the most promising therapy in nanomedicine field. The delivery of drugs to the disease site and its capability for sustained or controlled release minimize the side effects and give efficient treatment with minimum amount of administered drugs. Drug delivery can be achieved by passive targeting based on enhanced permeability and retention (EPR) effect in tumor and inflammatory tissues or active targeting using ligands which bind to specific receptors overexpressed at the targeting tissues [51]. To optimize efficacy of the treatment, capability to

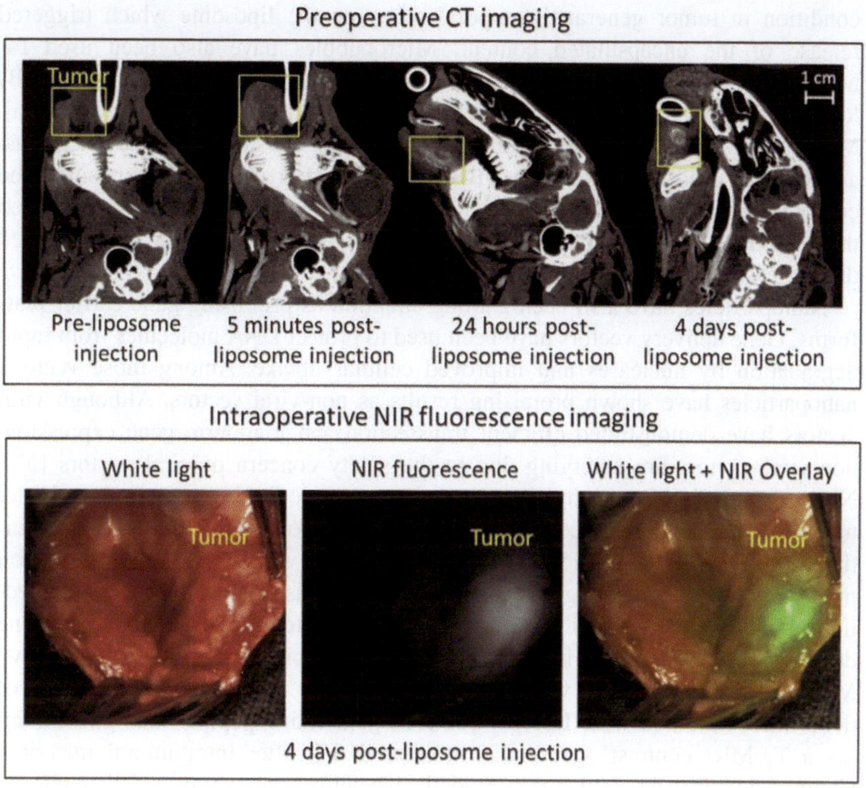

Fig. 4.2 Preoperative X-ray computed tomography and intraoperative NIR fluorescence imaging using a liposomal formulation which encapsulates CT contrast agent and indocyanine green (ICG) in rabbit cancer models. Reprinted under the Creative Commons Attribution-NonCommercial-NoDerivatives 4.0 International from [46].

visualize the drug distribution in real-time and quantify drug release is crucial. Image-guided drug delivery enables both visualization and drug release quantification by combining contrast agents or imaging tracers to nano-carriers. Various nano-formulations have been developed for image-guided drug delivery such as liposomal formulation which co-encapsulate contrast agents and drugs [52] and nanoparticles labeled by fluorescent dyes or radionuclides [53]. Furthermore, multifunctional nano-formulations with various capabilities such as controlled drug release and multimodal imaging have been developed. Chen et al. demonstrated controlled release of doxorubicin (DOX) from liposomal nano-formulations using mild hyperthermia [54]. The mild hyperthermia generated CO_2 bubbles within liposomal formulation which encapsulated DOX and ammonium bicarbonate, and allowed release of DOX from permeable defects induced by the bubbles. Liposomal nano-formulations modified with pH-responsive nanovalves have been developed for MR guided drug delivery with pH responsive release [55]. Accumulation of nano-formulations was monitored by encapsulated Gd-DTPA using MRI, and acidic pH condition in tumor generated temporal pores on the liposome which triggered release of the encapsulated content. Microbubbles have also been used for MR-guided focused ultrasound (FUS) to disrupt the blood–brain barrier (BBB) locally and deliver drugs to the brain by using biodegradable nanoparticles [56]. MRI pre-treatment scan was performed to identify the target, and MRI contrast agents were used to confirm the BBB disruption after ultrasound exposure. The combination of image-guided FUS with biodegradable nanoparticles demonstrated the potential of brain targeted drug delivery with deeper penetration and sustained drug release.

Nanoparticles have also been gaining attention as promising gene carrier platforms. Gene delivery vectors have been used to protect DNA molecules from rapid degradation by nucleases and improved cellular uptake. Among those vectors, nanoparticles have shown promising results as non-viral vectors. Although viral vectors have demonstrated efficient transfection and long-term gene expression, non-viral vectors are emerging due to the safety concern of viral vectors [57]. Nanoparticles based non-viral vectors such as cationic lipids and cationic polymer nanoparticles have improved the stability of DNA molecules, cellular internalization, and targeting ability [58, 59]. Magnetofection, which is a gene transfection method associated with magnetic nanoparticles, has also shown improved transfection efficiency in vitro and in vivo [60, 61]. The potential of image-guided gene delivery was demonstrated in vitro using up-conversion nanoparticles [62]. In vivo MRI tracking of non-viral vectors was conducted by Wang et al. [63]. Spread of dendrimers modified with TAT peptide (cell-penetrating peptide) and gadolinium (as a T_1 MRI contrast agent) was monitored 6 h after intratumoral injection. Compared to drug delivery, image-guided gene delivery has not been fully explored and there are less reports available in literature. However, monitoring the distribution and quantify the vectors in real-time are crucial due to the limited transfection efficiency of non-viral gene delivery.

3.3 *Image-Guided Hyperthermia and Photodynamic Therapy*

Hyperthermia helps to treat the tumors based on cell-killing effect at the temperatures above 41 °C and by increasing their temperature to 40–43 °C [64]. Significant decrease in cell viability and in vivo effect after hyperthermia treatment has been reported [65]. Although hyperthermia has great advantages such as minimum invasiveness and less side effects, it has not been accepted as primary cancer treatment due to the difficulty to deliver heat to deep tissues locally. Instead, hyperthermia has been used as adjunctive therapy with radiotherapy. It has been known that hyperthermia works as radiation sensitizer and improve outcome of radiation therapy significantly [66]. Some inorganic nanoparticles including magnetic nanoparticles and gold nanoparticles have been extensively used as heat sources for local hyperthermia (photothermal therapy). Magnetic nanoparticles induce heat under an alternating magnetic field through hysteresis and relaxation losses [67]. Their magnetic targeting ability allows targeting of nanoparticles to tumor site, and only the tumor site can be treated with heat by applying external AC magnetic field around the area. Gold nanoparticles have unique optical properties based on their strong light absorption at a specific wavelength and converting the light into heat [68]. Since gold nanoparticles have good surface functionalities, they can be delivered by active targeting using various ligands to target specific cells. Also, it has been reported that gold nanoparticles enhance the effect of radiation therapy as radiation sensitizers [69]. Image-guided hyperthermia can also improve spatial specificity of the treatments by avoiding side effects on surrounding healthy tissues and precise temperature control by quantifying heat source nanoparticles at targeted sites. Due to their intrinsic properties as heat sources and imaging contrast agents, inorganic nanoparticles are promising materials for image-guided hyperthermia. Radiolabeled poly(D,L-lactide-co-glycolide)-b-poly(ethylene glycol)-based nanocarriers containing iron oxide nanoparticles have been developed for hyperthermia with diagnostic functionalities as PET/MRI contrast agents [70]. Gold nanostars have been investigated as effective heat sources for photothermal therapy beside their multimodal imaging efficacy as contrast agents for surface-enhanced Raman scattering (SERS) detection, X-ray computed tomography (CT), and two-photon luminescence (TPL) imaging (Figs. 4.3) [71]. Yan et al. developed RGD peptide functionalized liposomes containing indocyanine green (ICG) which is an FDA approved near-infrared fluorescent dye [72]. The binding ability of RGD to some of the tumor cells and strong NIR absorbance of ICG allow efficient photothermal therapy and photodynamic therapy with guidance of NIR fluorescent imaging.

Photodynamic therapy (PDT) is a treatment which involves activation of photosensitizers by specific wavelength light sources. Upon the administration of photosensitizers and their localization at target sites, they are activated by visible or

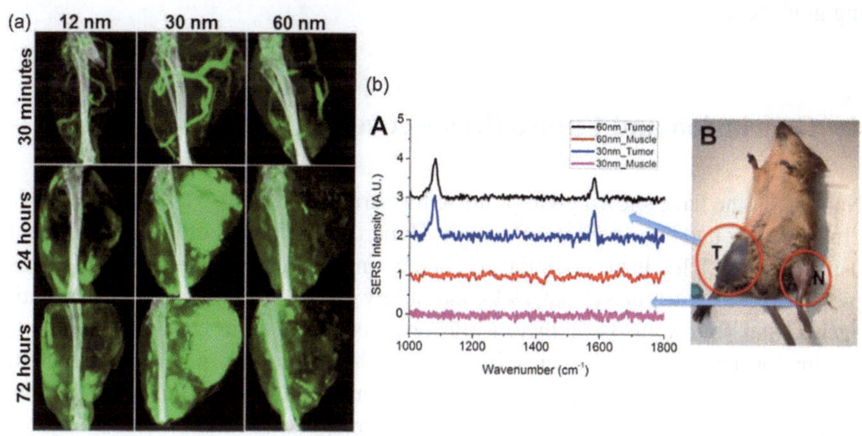

Fig. 4.3 CT images of hind leg primary sarcomas at 30 min, 24 h, and 72 h after injection of 12-nm gold nanoparticle, 30-nm and 60-nm gold nanostars (**a**). In vivo SERS spectra of 30-nm and 60-nm gold nanostars (**b**). Reprinted under the Creative Commons Attribution-NonCommercial-NoDerivatives 4.0 International from [71]

near-infrared light and induce local damage to tissues [73]. Photodynamic therapy is a less invasive treatment with minimal damage to healthy tissue and has ability to repeat treatment when needed. However, the limited penetration for currently available photosensitizers and long skin sensitivity to light after the administration of photosensitizers limit applications of this technique [74]. Water insolubility of FDA approved photosensitizers such as silicon phthalocyanine, limited targeting ability, and poor pharmacokinetics are other drawbacks of this approach [75]. To improve its solubility and localization to the target site, various nanoparticles have been introduced as carriers for photosensitizers. Due to the intrinsic fluorescent property of photosensitizers, image-guided photodynamic therapy has been emerging rapidly. Taratula et al. developed polymer nanoparticles containing silicon naphthalocyanine (SiNc), showing simultaneous florescent, photodynamic, and photothermal properties [76]. The efficient tumor accumulation of the nanoparticles was observed by fluorescence signals generated from SiNc under NIR, and the eradication of ovarian tumor tissues was achieved by combination of photodynamic therapy and hyperthermia. A photosensitizer, sinoporphyrin sodium (DVDMS), loaded on PEGylated graphene oxide was also delivered to tumor site with fluorescent imaging guidance, and demonstrated the elimination of tumor by photodynamic therapy [77]. Furthermore, multifunctional core-shell nanoparticles have been developed for simultaneous dual-modal imaging and multiple therapies (Fig. 4.4) [78]. A core-shell structure of up-conversion nanoparticles and mesoporous silica allowed image guiding using up-conversion luminescent and X-ray CT. A synergistic multiple-stages therapy was achieved by loading photosensitizers (thiolate-protected gold nanoclusters) and doxorubicin into silica shells. The release of doxorubicin was achieved by using a pH and temperature-responsive polymer which was triggered by low pH condition inside cancer cells and NIR irradiation. This multifunctional nanoparticle system demonstrated efficient cancer therapy with multi-modal imaging guidance.

4 Conclusion and Future Prospective

Nanomedicine have shown great promise for various therapies. Nanoparticle-based drug delivery had improved its efficiency. Hyperthermia and photodynamic therapy using nanoparticles have potential to achieve minimally-invasive localized therapies. However, current system lacks the ability to track and quantify therapeutic agents, and follow up their therapeutic response. By adding image guidance to existing or developing treatments, image-guided therapy has a potential to open its way to personalized medicine. The development of multifunctional nanoparticles enabled combination of real-time non-invasive imaging and local treatment by one nano-formulation system. These multifunctional nanoparticles have a potential to achieve efficient treatments without unnecessary side-effects to healthy tissues with an imaging guidance. Although various multifunctional nanoparticles have been reported and showed their potential, image-guided therapy is still a new technique

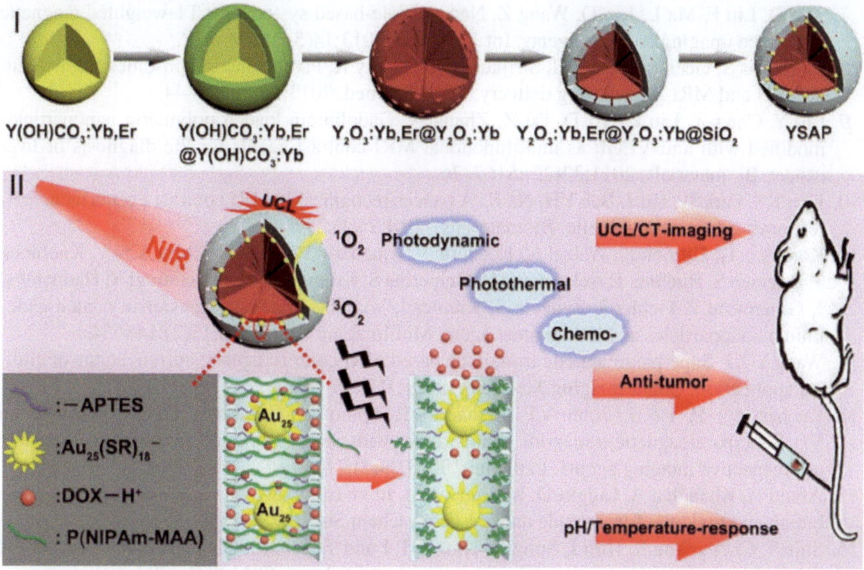

Fig. 4.4 Schematic illustration of multifunctional core-shell nanoparticles for imaging-guided synergistic multimodal anticancer therapy (Reprinted with permission from [78]).

and faces various challenges such as lack of nano-formulation optimization and in vivo data, its cost, and multidisciplinality. More research will be necessary to optimize the nano-formulations properties for each specific imaging modality. To visualize distribution of therapeutic agents, quantitative and qualitative multimodal imaging with the combination of anatomical imaging and high sensitivity imaging need to be considered. Therefore, selecting optimal combination of imaging systems for the application and developing suitable nano-formulations will be the key to achieve personalized image-guided therapies.

References

1. Lammers T, Kiessling F, Hennink WE, Storm G. Nanotheranostics and image-guided drug delivery: current concepts and future directions. Mol Pharm. 2010;7(6):1899–912.
2. Cormode DP, Skajaa T, Fayad ZA, Mulder WJM. Nanotechnology in medical imaging: probe design and applications. Arterioscler Thromb Vasc Biol. 2009;29(7):992–1000.
3. Barenholz (Chezy) Y. Doxil®—the first FDA-approved nano-drug: lessons learned. J Control Release. 2012;160(2):117–34.
4. Bin Na H, Song IC, Hyeon T. Inorganic nanoparticles for MRI contrast agents. Adv Mater. 2009;21(21):2133–48.
5. James ML, Gambhir SS. A molecular imaging primer: modalities, imaging agents, and applications. Physiol Rev. 2012;92(2):897–965.
6. Zhou Z, Lu Z-R. Gadolinium-based contrast agents for MR cancer imaging. Wiley Interdiscp Rev Nanomed Nanobiotechnol. 2013;5(1):1–18.

7. Zhu D, Liu F, Ma L, Liu D, Wang Z. Nanoparticle-based systems for T1-weighted magnetic resonance imaging contrast agents. Int J Mol Sci. 2013;14(5):10591–607.
8. Langereis S, Geelen T, Grüll H, Strijkers GJ, Nicolay K. Paramagnetic liposomes for molecular MRI and MRI-guided drug delivery. NMR Biomed. 2013;26(7):728–44.
9. Liu Y, Chen Z, Liu C, Yu D, Lu Z, Zhang N. Gadolinium-loaded polymeric nanoparticles modified with anti-VEGF as multifunctional MRI contrast agents for the diagnosis of liver cancer. Biomaterials. 2011;32(22):5167–76.
10. Kim KS, Park W, Hu J, Bae YH, Na K. A cancer-recognizable MRI contrast agents using pH-responsive polymeric micelle. Biomaterials. 2014;35(1):337–43.
11. Korkusuz H, Ulbrich K, Welzel K, Koeberle V, Watcharin W, Bahr U, Chernikov V, Knobloch T, Petersen S, Huebner F, Ackermann H, Gelperina S, Kromen W, Hammerstingl R, Haupenthal J, Gruenwald F, Fiehler J, Zeuzem S, Kreuter J, Vogl TJ, Piiper A. Transferrin-coated gadolinium nanoparticles as MRI contrast agent. Mol Imaging Biol. 2013;15(2):148–54.
12. Wang Y-XJ. Superparamagnetic iron oxide based MRI contrast agents: current status of clinical application. Quant Imaging Med Surg. 2011;1(1):35–40.
13. Kucheryavy P, He J, John VT, Maharjan P, Spinu L, Goloverda GZ, Kolesnichenko VL. Superparamagnetic iron oxide nanoparticles with variable size and an iron oxidation state as prospective imaging agents. Langmuir. 2013;29(2):710–6.
14. Arami H, Khandhar A, Liggitt D, Krishnan KM. In vivo delivery, pharmacokinetics, biodistribution and toxicity of iron oxide nanoparticles. Chem Soc Rev. 2015;44:8576–607.
15. Shin T, Choi J, Yun S, Kim I, Song H, Kim Y. T 1 and T 2 dual-mode MRI contrast agent for enhancing accuracy by engineered nanomaterials. ACS Nano. 2014;8(4):3393–401.
16. Szpak A, Fiejdasz S, Prendota W, Strączek T, Kapusta C, Szmyd J, Nowakowska M, Zapotoczny S. T1–T2 dual-modal MRI contrast agents based on superparamagnetic iron oxide nanoparticles with surface attached gadolinium complexes. J Nanopart Res. 2014;16(11):1–11.
17. Lee N, Choi SH, Hyeon T. Nano-sized CT contrast agents. Adv Mater. 2013;25(19):2641–60.
18. Liu Y, Ai K, Lu L. Nanoparticulate X-ray computed tomography contrast agents: from design validation to in vivo applications. Acc Chem Res. 2012;45(10):1817–27.
19. Lusic H, Grinstaff MW. X-ray-computed tomography contrast agents. Chem Rev. 2013;113(3):1641–66.
20. Badea CT, Athreya KK, Espinosa G, Clark D, Ghafoori AP, Li Y, Kirsch DG, Johnson GA, Annapragada A, Ghaghada KB. Computed tomography imaging of primary lung cancer in mice using a liposomal-iodinated contrast agent. PLoS One. 2012;7(4):1–7.
21. de Vries A, Custers E, Lub J, van den Bosch S, Nicolay K, Grüll H. Block-copolymer-stabilized iodinated emulsions for use as CT contrast agents. Biomaterials. 2010;31(25):6537–44.
22. Ai K, Liu Y, Liu J, Yuan Q, He Y, Lu L. Large-scale synthesis of bi 2S 3 nanodots as a contrast agent for in vivo X-ray computed tomography imaging. Adv Mater. 2011;23(42):4886–91.
23. Liu Y, Ai K, Liu J, Yuan Q, He Y, Lu L. A high-performance ytterbium-based nanoparticulate contrast agent for in vivo X-ray computed tomography imaging. Angew Chemie Int Ed Engl. 2012;51(6):1437–42.
24. Kim D, Jeong YY, Jon S. A drug-loaded aptamer-gold nanoparticle bioconjugate for combined CT imaging and therapy of prostate cancer. ACS Nano. 2010;4(7):3689–96.
25. Sun IC, Na JH, Jeong SY, Kim DE, Kwon IC, Choi K, Ahn CH, Kim K. Biocompatible glycol chitosan-coated gold nanoparticles for tumor-targeting CT imaging. Pharm Res. 2014;31(6):1418–25.
26. Townsend DW. Physical principles and technology of clinical PET imaging. Ann Acad Med Singap. 2004;33(2):133–45.
27. Phelps ME. Positron emission tomography provides molecular imaging of biological processes. Proc Natl Acad Sci USA. 2000;97(16):9226–33.
28. Thorek DLJ, Ulmert D, Diop N-FM, Lupu ME, Doran MG, Huang R, Abou DS, Larson SM, Grimm J. Non-invasive mapping of deep-tissue lymph nodes in live animals using a multimodal PET/MRI nanoparticle. Nat Commun. 2014;5:3097.

29. Chakravarty R, Valdovinos HF, Chen F, Lewis CM, Ellison PA, Luo H, Meyerand ME, Nickles RJ, Cai W. Intrinsically germanium-69-labeled iron oxide nanoparticles: synthesis and in-vivo dual-modality PET/MR zimaging. Adv Mater. 2014;26(30):5119–23.
30. Chan V, Perlas A. Basics of ultrasound imaging. In: Narouze SN, editor. Atlas of ultrasound-guided procedures in interventional pain management. Berlin: Springer; 2011. p. 13–20.
31. Khokhlova TD, Haider Y, Hwang JH. Therapeutic potential of ultrasound microbubbles in gastrointestinal oncology: recent advances and future prospects. Therap Adv Gastroenterol. 2015;8(6):384–94.
32. Wu D, Huang L, Jiang MS, Jiang H. Contrast agents for photoacoustic and thermoacoustic imaging: a review. Int J Mol Sci. 2014;15(12):23616–39.
33. Tomitaka A, Arami H, Gandhi S, Krishnan KM. Lactoferrin conjugated iron oxide nanoparticles for targeting brain glioma cells in magnetic. Nanoscale. 2015;7:16890–8.
34. Xu M, Wang LV. Photoacoustic imaging in biomedicine. Rev Sci Instrum. 2006;77(4):041101.
35. Xia J, Yao J, Wang LV. Photoacoustic tomography: principles and advances. Prog Eletromagnic Res. 2014;147(May):1–22.
36. Weissleder R, Pittet MJ. Imaging in the era of molecular oncology. Nature. 2008;452:580–9.
37. Hahn MA, Singh AK, Sharma P, Brown SC, Moudgil BM. Nanoparticles as contrast agents for in-vivo bioimaging: current status and future perspectives. Anal Bioanal Chem. 2011;399:3–27.
38. Panagiotopoulos N, Duschka RL, Ahlborg M, Bringout G, Debbeler C, Graeser M, Kaethner C, Lüdtke-Buzug K, Medimagh H, Stelzner J, Buzug TM, Barkhausen J, Vogt FM, Haegele J. Magnetic particle imaging: current developments and future directions. Int J Nanomedicine. 2015;10:3097–114.
39. Goodwill PW, Saritas EU, Croft LR, Kim TN, Krishnan KM, Schaffer DV, Conolly SM. X-space MPI: magnetic nanoparticles for safe medical imaging. Adv Mater. 2012;24(28):3870–7.
40. Knopp T, Buzug TM. Magnetic particle imaging: an introduction to imaging principles and scanner instrumentation. Berlin: Springer; 2012.
41. Zhu J, Lu Y, Li Y, Jiang J, Cheng L, Liu Z, Guo L, Pan Y, Gu H. Synthesis of au-Fe3O4 heterostructured nanoparticles for in vivo computed tomography and magnetic resonance dual model imaging. Nanoscale. 2014;6(1):199–202.
42. Liang SY, Zhou Q, Wang M, Zhu YH, Wu QZ, Yang XL. Water-soluble l-cysteine-coated FePt nanoparticles as dual MRI/CT imaging contrast agent for glioma. Int J Nanomedicine. 2015;10:2325–33.
43. Arami H, Khandhar AP, Tomitaka A, Yu E, Goodwill PW, Conolly SM, Krishnan KM. Vivo multimodal magnetic particle imaging (MPI) with tailored magneto/optical contrast agents. Biomaterials. 2015;52:251–61.
44. Liu T, Shi S, Liang C, Shen S, Cheng L, Wang C, Song X, Goel S. Iron oxide decorated MoS2 nanosheets with double PEGylation for chelator-free radiolabeling and multimodal imaging guided photothermal therapy. ACS nano. 2015;9:950–60.
45. Nguyen QT, Tsien RY. Fluorescence-guided surgery with live molecular navigation-a new cutting edge. Nat Rev Cancer. 2013;13(9):653–62.
46. Zheng J, Muhanna N, De Souza R, Wada H, Chan H, Akens MK, Anayama T, Yasufuku K, Serra S, Irish J, Allen C, Jaffray D. A multimodal nano agent for image-guided cancer surgery. Biomaterials. 2015;67:160–8.
47. Matsumura Y, Maeda H. A new concept for macromolecular therapeutics in cancer chemotherapy: mechanism of tumoritropic accumulation of proteins and the antitumor agent smancs. Cancer Res. 1986;46:6387–92.
48. Xi L, Zhou G, Gao N, Yang L, a Gonzalo D, Hughes SJ, Jiang H. Photoacoustic and fluorescence image-guided surgery using a multifunctional targeted nanoprobe. Ann Surg Oncol. 2014;21:1602–9.
49. Scoggins CR, Chagpar AB, Martin RCG, McMasters KM. Should sentinel lymph-node biopsy be used routinely for staging melanoma and breast cancers? Nat Clin Pract Oncol. 2005;2(9):448–55.

50. Bradbury MS, Phillips E, Montero PH, Cheal SM, Stambuk H, Durack JC, Sofocleous CT, Meester RJC, Wiesner U, Patel S. Clinically-translated silica nanoparticles as dual-modality cancer-targeted probes for image-guided surgery and interventions. Integr Biol (Camb). 2013;5(1):74–86.
51. Bertrand N, Wu J, Xu X, Kamaly N, Farokhzad OC. Cancer nanotechnology: the impact of passive and active targeting in the era of modern cancer biology. Adv Drug Deliv Rev. 2014;66:2–25.
52. de Smet M, Langereis S, van den Bosch S, Grüll H. Temperature-sensitive liposomes for doxorubicin delivery under MRI guidance. J Control Release. 2010;143(1):120–7.
53. Chen F, Hong H, Zhang Y, Valdovinos HF, Shi S, Kwon GS, Al CET. In Vivo tumor targeting and image-guided drug delivery with antibody-conjugated, radiolabeled mesoporous silica nanoparticles. ACS Nano. 2013;7(10):9027–39.
54. Chen K-J, Chaung E-Y, Wey S-P, Lin K-J, Cheng F, Lin C-C, Liu H-L, Tseng H-W, Liu C-P, Wei M-C, Liu C-M, Sung H-W. Hyperthermia-mediated local drug delivery by a bubble-generating liposomal system for tumor specific chemotherapy. ACS Nano. 2014;8(5):5105–15.
55. Pacheco-Torres J, Mukherjee N, Walko M, López-Larrubia P, Ballesteros P, Cerdan S, Kocer A. Image guided drug release from pH-sensitive ion channel-functionalized stealth liposomes into an in vivo glioblastoma model. Nanomedicine. 2015;11(6):1345–54.
56. Nance E, Timbie K, Miller GW, Song J, Louttit C, Klibanov AL, Shih TY, Swaminathan G, Tamargo RJ, Woodworth GF, Hanes J, Price RJ. Non-invasive delivery of stealth, brain-penetrating nanoparticles across the blood–brain barrier using MRI-guided focused ultrasound. J Control Release. 2014;189:123–32.
57. Lin G, Zhang H, Huang L. Smart polymeric nanoparticles for cancer gene delivery. Mol Pharm. 2015;12(2):314–21.
58. Kozielski KL, Tzeng SY, Hurtado De Mendoza BA, Green JJ. Bioreducible cationic polymer-based nanoparticles for efficient and environmentally triggered cytoplasmic siRNA delivery to primary human brain cancer cells. ACS Nano. 2014;8(4):3232–41.
59. Liang Y, Liu Z, Shuai X, Wang W, Liu J, Bi W, Wang C, Jing X, Liu Y, Tao E. Delivery of cationic polymer-siRNA nanoparticles for gene therapies in neural regeneration. Biochem Biophys Res Commun. 2012;421(4):690–5.
60. Namgung R, Singha K, Yu MK, Jon S, Kim YS, Ahn Y, Park IK, Kim WJ. Hybrid superparamagnetic iron oxide nanoparticle-branched polyethylenimine magnetoplexes for gene transfection of vascular endothelial cells. Biomaterials. 2010;31(14):4204–13.
61. Kami D, Kitani T, Kishida T, Mazda O, Toyoda M, Tomitaka A, Ota S, Ishii R, Takemura Y, Watanabe M, Umezawa A, Gojo S. Pleiotropic functions of magnetic nanoparticles for ex vivo gene transfer. Nanomedicine. 2014;10(6):1165–74.
62. He L, Feng L, Cheng L, Liu Y, Li Z, Peng R, Li Y, Guo L, Liu Z. Multilayer dual-polymer-coated upconversion nanoparticles for multimodal imaging and serum-enhanced gene delivery. ACS Appl Mater Interfaces. 2013;5(20):10381–8.
63. Wang Q, Li J, An S, Chen Y, Jiang C, Wang X. Magnetic resonance-guided regional gene delivery strategy using a tumor stroma-permeable nanocarrier for pancreatic cancer. Int J Nanomedicine. 2015;10:4479–90.
64. Hildebrandt B. The cellular and molecular basis of hyperthermia. Crit Rev Oncol Hematol. 2002;43(1):33–56.
65. Dewey WC. Arrhenius relationships from the molecule and cell to the clinic. Int J Hyperth. 2009;25(1):3–20.
66. Horsman MR, Overgaard J. Hyperthermia: a potent enhancer of radiotherapy. Clin Oncol. 2007;19(6):418–26.
67. Tomitaka A, Takemura Y. Measurement of specific loss power from intracellular magnetic nanoparticles for hyperthermia. J Pers Nanomedicine. 2015;1(1):33–7.
68. Huang X, El-Sayed M a. Gold nanoparticles: optical properties and implementations in cancer diagnosis and photothermal therapy. J Adv Res. 2010;1(1):13–28.

69. Zhang X-D, Wu D, Shen X, Chen J, Sun Y-M, Liu P-X, Liang X-J. Size-dependent radiosensitization of PEG-coated gold nanoparticles for cancer radiation therapy. Biomaterials. 2012;33:6408–19.
70. Baldi G, Ravagli C, Mazzantini F, Loudos G, Adan J, Masa M, Psimadas D, Locatelli E, Innocenti C, Sangregorio C. In vivo anticancer evaluation of the hyperthermic efficacy of anti-human epidermal growth factor containing magnetic nanoparticles. Int J Nanomedicine. 2014;9(1):3037–56.
71. Liu Y, Ashton JR, Moding EJ, Yuan H, Register JK, Fales AM, Choi J, Whitley MJ, Zhao X, Qi Y, Ma Y, Vaidyanathan G, Zalutsky MR, Kirsch DG, Badea CT, Vo-Dinh T. A plasmonic gold nanostar theranostic probe for in vivo tumor imaging and photothermal therapy. Theranostics. 2015;5(9):946–60.
72. Yan F, Wu H, Liu H, Deng Z, Liu H, Duan W, Liu X, Zheng H. Molecular imaging-guided photothermal / photodynamic therapy against tumor by iRGD-modified indocyanine green nanoparticles. J Control Release. 2016;224:217–28.
73. Dougherty TJ, Gomer CJ, Henderson BW, Jori G, Kessel D, Korbelik M, Moan J, Peng Q. Photodynamic therapy. J Natl Cancer Inst. 1998;90(12):889–905.
74. Bozzini G, Colin P, Betrouni N, Nevoux P, Ouzzane A, Puech P, Villers A, Mordon S. Photodynamic therapy in urology: what can we do now and where are we heading? Photodiagn Photodyn Ther. 2012;9(3):261–73.
75. Kim H, Mun S, Choi Y. Photosensitizer-conjugated polymeric nanoparticles for redox-responsive fluorescence imaging and photodynamic therapy. J Mater Chem B. 2013;1:429–31.
76. Taratula O, Doddapaneni BS, Schumann C, Li X, Bracha S, Milovancev M, Alani AWG, Taratula O. Naphthalocyanine-based biodegradable polymeric nanoparticles for image-guided combinatorial phototherapy. Chem Mater. 2015;27(17):6155–65.
77. Yan X, Niu G, Lin J, Jin AJ, Hu H, Tang Y, Zhang Y, Wu A, Lu J, Zhang S, Huang P, Shen B, Chen X. Enhanced fluorescence imaging guided photodynamic therapy of sinoporphyrin sodium loaded graphene oxide. Biomaterials. 2015;42:94–102.
78. Lv R, Yang P, He F, Gai S, Yang G, Dai Y, Hou Z, Lin J. Biomaterials an imaging-guided platform for synergistic photodynamic/photothermal/chemo-therapy with pH/temperature-responsive drug release. Biomaterials. 2015;63:115–27.

Chapter 5
Nanomaterials for Drug Delivery

Krati Sharma

Abstract Nanotechnology in medicine and drug delivery are currently investigating many nanomaterials for drug delivery. Several nanomaterials have been approved for medicinal use and treatment of life threatening diseases. Nanomaterials have applications from binding to the right target to the delivery of drug molecules at the target site. Nanomaterials are the potential candidates in site specific delivery due to their unique properties, size and shape. There versatile use brings an attention to know more about their size, shape and different chemical behavior. The detailed study can direct the ongoing research in a manner to know the potential hazards of nanoparticles (NPs). Several formulations include size and surface characteristics, chemical and physical interactions, increase in permeability, retention effect and other major influencers have been taken into account in development strategies of drug delivery methods. Nano materials may or may not be soluble in biological matrices and can lead to toxicity, thus cell toxicity majorly influence the potential exposure to biological systems. For NPs, the small size allow access to thin and very narrow cellular components such as a need of potential NPs to cross the blood brain barrier. A multitude of substances are recently under study for drug delivery applications such as liposomes, gene based conjugated materials, polymers micelles etc. Various research efforts have been safe and efficient in using NPs drug delivery methods. This chapter provides recent research protocols used along with different conjugation methods. Chemically modified nanomaterials and their derivatives used in drug delivery research are also discussed to evaluate the usefulness of these systems in delivering the bioactive molecules.

Keywords Nanostructures • Biomedical Applications • Drug delivery

1 Introduction

Nanoparticle (NPs)-based platforms for gene therapy and drug delivery are gaining popularity for treatment of various life threatening diseases. Cutting edge research and limitless efforts in development of drug delivery systems have gained

K. Sharma (✉)
Scientific Technician-2, Fox Chase Cancer Center, Philadelphia, PA, USA
e-mail: medstruct.katy@gmail.com

A. Kaushik et al. (eds.), *Advances in Personalized Nanotherapeutics*,
https://doi.org/10.1007/978-3-319-63633-7_5

success from conventional chemotherapy to the most recent research methods. There were times when only a fraction of the administered drug was reaching the tumor site or cancer cells and several other limitations such as nonspecific biodistribution and site specific targeting, lack of water solubility, poor oral bioavailability and low therapeutic indices. All these limitations have been addressed for rapid progressing and improvement due to novel drug delivery methods. For successful translation of developed formulations, it was imperative to evaluate the design and properties of these NPs. There have been many research studies done to improve circulation time, retention effect, reduce cell toxicity and enhance biocompatibility [1].

The "theranostic properties" associated with NPs have shown enhanced delivery of therapeutic drugs with superior imaging capabilities, minimal toxicities, enhanced drug delivery, efficient tumor targeting, treatment monitoring, and diagnostics. Such methods were categorized under passive mode of drug delivery systems. Other than passive methods, active targeting strategies involve other therapeutic chemical compounds and antibodies in conjugation with nano particles to increase drug loading capacity and efficiency in drug delivery without altering properties of therapeutic agents. Certain drugs on administration are always challenged and often encountered by the body defense mechanism. Certain limitations such as cell toxicity are always remain a challenge in not only conventional methods but also for the newer molecularly targeted therapeutics [2].

NPs with the help of combined active and passive strategies i.e. conjugation of drug with other chemical carriers resulted into multifunctional benefits, facilitated personalized and tailored many treatment. Chemical therapeutic agents have been able to distinguish between normal to diseased cell with the help of these conjugated methods. These conjugated NPs can increase retention time and also avoid increasing cell toxicity. Encapsulation is another effecting strategy of drug delivery in which NPs are encapsulated under the endosomes via receptor mediated endocytosis to avoid recognition and elimination by body defense mechanism. In this chapter, we will discuss recent novel NPs in drug delivery methods and there characteristics with some applications in improving therapeutic efficiency and function in life threatening diseases treatments [1, 2].

Nanomaterials are biomaterials used in tissue engineering and as a drug delivery channels. Developing surface engineered materials can enhance major properties which can favor the interaction of biological system with the biomaterial. Nanomaterials are categorized into nano-structured and nano-crystalline materials. Nano-structured materials are shaped raw nanomaterials with their unique functionalities for example, quantum dots, dendrimers, fullerenes and carbon nanotubes. In contrast, nano-crystalline materials are non-polymer materials which are highly demanded for their small size and wide surface area with biocompatible characteristics. Nanostructure materials are further distinguished into Polymer and Non-polymer forms [3].

Transferring genetic materials into the affected cells for treatment through inducing artificial expression of associated proteins is fast gaining acceptance in drug delivery systems. Nucleic acid based therapies use DNA and RNA as a drug. However, issues with cellular uptake, quick biodegradation and escape, poor stability associated with these biomolecules can limit the use in clinical trials. Though recent research efforts have combine the viral and non-viral vectors to overcome these limitations [4].

2 Inorganic-NPs

2.1 Carbon Nanotubes

Carbon based tubular structures with benzene ring are main constituent for biological application applied in sensors for detecting DNA, RNA and other biomolecules. They are hexagonally shaped branched network of carbon atoms varying from the size 1 to 100 nm. These branched networks are made of layer of graphite which are rolled into cylindrical form. Carbon natures are of two types-single walled nanotubes (SWNTs) and multi walled nanotubes (MWNTs). The size, geometry and surface features increase their demand for carrying different proteins from serum samples and delivery [5]. CNTs are completely insoluble in all solvent. They are also needed to treat for reducing cell toxicity. They are directed by endocytosis for cellular uptake or by insertion through the cell membrane. In the recent novel study, PEGylated multi wall carbon nanotubes (MWNTs) based drug delivery method was developed using both physical and chemical method. This method was applied for loading of Ibuprofen drug. Further, surface was modified with the use of several acylated groups. Cytotoxicity of PEGylated MWNTS were examined by MTT assay. The results showed improvement in drug loading by both physical and chemical based methods, though chemical loaded MWNTs showed more clear release of drug in comparison to the physically loaded one and could be used as a controlled release system for various drugs [6].

SWNTs have advantage over metallic NPs as they include high drug loading capacity, intrinsic stability, easy surface modification which can enhance circulation time and also bioavailability of the drug molecules [7]. They have been successful in delivery of nucleic acid [8–10], proteins [11, 12] and drug molecules [9, 13–15]. On conjugation with antibodies and low molecular weight targeting agents, they have high efficiency for nanotube internalization into cells. They also have higher loading capacity and allow incorporation of targeting agents in such a way that they can deceive the evading system of body [16, 17].

CNTs are usually needed in gene and other biomolecules delivery. A technique of molecular dynamics made it more convenient by inducing small molecule transportation with the help of flowing water and surface—functionalized carbon nanotubes. It has become an important tool in gene therapy. The antibody linked to the nanotubes are easily encapsulated into mammalian cells without producing any toxic effects in comparison to the naked antibody delivery methods [18]. The most recent research used polyethylenimine (PEI) with polyethylene glycol (PEG) linker to the carboxylated single-walled carbon nanotubes (SWNT) to facilitate delivery of shRNA. This whole molecule was attached with an aptamer AS1411 along with DOX to test targeted tumor cells. These tumor cells showed strong selectivity towards the whole conjugation and inhibited growth of nucleolin-abundant gastric cancer cells. Also, they found very less amount of DOX which is favorable condition for cells as the toxicity is reduced [19].

Drug loading can be external and internal or can be active or passive. Active targeting requires functionalization with tumor specific binding sites to selectivity bind to tumor cells. Passive targeting depends on the inertness and physical size of the macromolecule. Van der Waals forces hold internalized or encapsulated for insertion into the CNT. This is the best way to keep the drugs protected from the external environment [20]. CNTs have major advantages in reducing cell toxicity, increased circulation time, active targeting circulation and reduced side effects. There are many applications on different drugs which were administered in conjugation with other molecules such as PEG and Folate (Table 5.1)

Table 5.1 Drugs delivered using CNTs

Name of drug	Modification/Functionalization	CNTs	Advantage
Doxorubicin	PEG conjugation	SWCNT	Reduced toxicity
Mitoxantorne	PEG conjugation	SWCNT	Reduced toxicity
Paclitaxel	PEG conugation	SWCNT	Increased circulation time
Carboplastin		SWCNT	
Cisplatin	Non functialized	SWCNT	Decreased toxicity
Doxorubicin	Conjugated with folate	MWCNT	Active targeting
Paclitaxel	Folate conjugate	MWCNT	Increased circulation period
Methotrexate	PEGylated	MWCNT	Controlled toxicity
Quercitin	PEGylated	SWCNT	Reduced side effect
Folic acid	PEGylated	MWCNT	Active targeting circulation period
Gamicitabine	NA	SWCNTs	NA
HCPT(10-hydroxycamptothecin)	NA	MWCNT	NA

2.2 Gold NPs

Gold (Au) NPs provide novel platform for site specific drug delivery. They possess unique chemical and physical properties for releasing and loading drugs [21]. Such unique properties lead to different yet unique applications in medical field. AuNPs are found in different forms among which, bulk is a yellow solid and less reactive form otherwise they are wine red solution and are anti-oxidant in nature. These distinct characteristics of these AuNPs are due to their internal interactions [22]. The AuNPs can be easily prepared and are of size between 1 nm and 8 μm. The commonly used shapes are spherical, decahedral, icosahedral multiple twined, irregular shape, tetrahedral, nano-triangle, nano-prisms and hexagonal platelets and nano-rods (Table 5.2).

Various techniques have been employed to increase optical properties and purification of GNPS to enhance their medicinal properties (Fig. 5.1). The AuNPs are also widely used as a contrast agent in site specific delivery [25]. These particles have various applications such as biomolecules ultra-sensitive detention, cancer cells treatment, labeling of cells and proteins and delivering drugs. Surface Plasmon resonance (SPR) has been very effective method in identification of pathogens in clinical field [26]. Conjugated assemblies of GNPs with biomolecules improved loaded capabilities and make them most efficient method in gene delivery [27]. Gene used as a biomolecule in conjugation with AuNPs are used in epidermal delivery of DNA vaccines and this method is one of the most recognized in delivery of DNA based vaccines. The latest research in gene or drug laden near-infrared (NIR) absorbing Au-NPs in complex with NIR light showed special mechanism of DNA release. Researchers used plasmonic NPs complexes with continuous wave (CW) and femtosecond pulsed lasers which excited dehybridized single-stranded DNA whereas pulsed-laser excitation resulted in double stranded DNA release by cleavage of the Au–S bond. This study showed that different type of laser excitations can cause wide difference in characteristic of DNA which is bounded with the same DNA-NPs complex [23]. The vascular endothelial growth factor (VEGF) were also seen as a primary factor in preventing skin injury progression but delivery of it at injury site is challenging due to various skin barriers. By modifying surface of AuNPs and conjugate with negative surface VEGF resulted in site specific transdermal delivery [24].

Table 5.2 Shapes of Au-NPs and their applications

Shape	Size	Application
Nano rod	2–5 nm	Drug Delivery and photo thermal therapy [12].
Triangular particle	3.85–7.13 nm	Used effectively again *E.coli* strains and K.pneumonia [20]
Nano cage	50 nm	Highly effective contrast agent in endomicroscopy imaging [16] and in vivo medical applications [23].
Hollow particle	25 nm	Photo-electronics, catalysis and cancer therapy [9].
Branched particle	90 nm	Surface enhanced Raman Spectroscopy based imaging of Kidney cells [24].

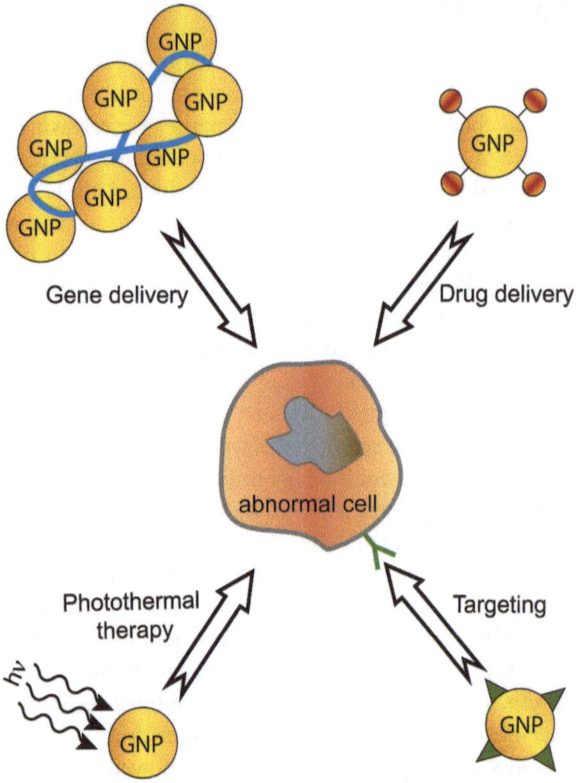

Fig. 5.1 Various biomedical applications of AuNPs

Coated Au nano-cages with temperature sensitive polymer in conjugation with near-infrared radiations further improved drug releasing capacity [28]. Colloidal AuNPs have been used tremendously due to their easy preparation and can be easily incorporated into tissues and cells due to their small size [29]. GNPs are also used in conjugation with organic molecules which drastically reduced cell toxicity. Also, on conjugation with organic molecules, the physical and chemical properties are very tunable and surface can be modified based on the requirements. GNPs are also used in analytical chemistry laboratories specifically in the tests where the native structure and enzymatic activity of attached proteins or enzymes need to be unchanged [30].

Drug delivery systems are mainly in demand due to favorable solubility, in-vivo stability and bio-distribution. The GNPs also favor unfavorable bio-distribution of some free drugs [31, 32]. Anti-tumor agents and antibiotics are the most commonly used in targeted drug delivery. The GNPs conjugated with anti-tumor drugs are paclitaxel [33], methotrexate [34], daunorubicine [35], hemcytabin [36], 6-mercaptopurine [37], dodecylcysteine [38], sulfonamide [39], 5-fluorouracil [40], platinum complexes [41], kahalalide [42], tamoxifen [43], herceptin [44], doxorubicin [45], and prospidin [46]. These conjugations were developed using simple adsorption of the drugs onto GNPs and in some cases, by using alkanthiol linkers with drug-GNPS. The validation of these conjugated agents was done in vitro models and in vivo, in mice suffering from different types of cancer cells. The focus is never limited to therapeutic agents but also to the targeting cells in order to avoid complications and challenges after administration of the drug. Many side effects are observed of these commonly used drugs in Cancer (Table 5.3). So this drawback brought necessary focus on improvement in developmental strategies for providing better support and binding with the target cells. The target modifications also enhanced reactivity with several therapeutic agents (hydrophilic, hydrophobic and auxiliary agents etc. [47].

2.3 Silica NPs

Silica NPs due to their small size and tunable surface modification are good candidates for carrying drugs and better dispersion. Mesoporous silica (Table 5.4), such as MCM-41 and SBA-15 are solid materials which are comprised of porous structure with hundreds of empty pores (mesopores). These pores are able to encapsulate large amount of biological molecules. Silica NPs have high surface area, large pore volume, adjustable pore size, good chemical, and thermal stability. These properties

Table 5.3 Commonly used drugs with their side effects

Drugs	Cancer type	Short term possible side effects
Carboplatin (Paraplatin)	Cancers of the ovary, head and neck, and lungs	Low blood counts, confusion, nausea and hair loss
Cisplatin	Bladder, ovary and testicales	Low blood counts, allergy reaction, ringing in ears and hearing loss, fluctuations in blood erythrocytes and kidney damage.
Flurururacil	Cancers of the colon, breast , stomach, head and neck	Low blood counts, diarrhea, mouth ulcers and photosensitivity
Doxorobucin	Breast cancer, lymphoma and multiplemyeloma.	Low blood counts, ulcers in mouth, nausea and heart damage.
Paclitaxel	Cancers of the breast, ovary and lung	Low blood count, allergic reactions, nausea, joint pain and numbness in fingers or toes.

Table 5.4 Types of mesoporous Silica NPs with their internal structure and pore diameter

Type	Internal Structure	Pore Diameter	References
MCM-41	2D hexagonal	1.5–3.5	[11]
MCM-41	Hexagonal structure with unidirectional	3.70′	[12]
SBA-15	2D hexagonal	6.0–10.0	[11]
SBA-15	3D cubic cage like	4.0–9.0	[13]
SBA-15	2D hexagonal	7.80	[12]
MCM-48	3D cubic	2.5–3.5	[14]

Table 5.5 List of various types of drugs delivered through mesoporous silica NPs

Classes	Drugs	Delivery agents	References
Anti-inflammatory	Ibuprofen	SBA-14	[10, 46]
	Naproxen	SBA-15	[46]
	Naproxen	MCM-4	[46]
	Naproxen	Amine-modified MCM-41	[49]
Chemotherapy	DOX	Folic-acid conjugate MSN	[47]
	DOX	DOX-hydrozone MSN-FA	[23]
	Camptothecin	Galactose-functionalized MSN	[29]
Estrogenic	Alendronate	SBA 15 with phosphorous	[48, 49]
	Alendronate	Amine modified MCM-41	[48, 49]
	Alendronate	MCM-41	[53, 54]
Antibiotics	Erythromycin	Octadecyl-functionalized SBA-15	[10]
	Erythromycin	SBA-15	[10]
	Vancomycin	CDS-capped MCM-41	[57]

make them ideal candidate for control release of drug at targeted sites as presented in Table 5.5 [48].

Mesoporous silica microspheres of micrometer was synthesized based on the Stober reaction [49]. The reaction involved the co-hydrolysis and condensation of tetraethoxysilane (TESO) and an alkyltrialkoxysilane in a mixture of ethanol, water and aqueous ammonia [50]. This resulted into micrometer-sized mesoporous silica spheres with a narrow size distribution [51]. Later, mesoporous silica nanosphere (MSN) were discovered which were more suitable for biotechnological and biomedical applications. MSN particle size is very compatible with biological agents such as bacteria which can make them efficient carriers for gene transfection or in intracellular drug delivery. It was synthesized using MCM-41 type of mesoporous silica, where TEOS was used as a silica source, a cationic surfactant cetyltrimethylammonium bromide (CTAB). MSN has an average particle dimeter of around 100 nm, surface areas around 900.2/g, pore sizes around 2 nm, pore volumes around 0.9 cm^3/g, abundant surface silanol groups (around 30 mol %), and a hexagonal porous channel structure [52].

The following unique properties of MSN have increased tremendous use in controlled release delivery applications

1. Tunable particle size—Size of the particle can be adjusted from 50 to 300 nm allowing an endocytosis without any considerable cytotoxicity [48].
2. Stable and rigid framework—MSN is more resistant to pH change, mechanical stress, heat and hydrolysis-induced degradations [45].
3. Uniform and tunable pore size—The pore size can be tuned between 2 and 6 nm [48].
4. High surface area and large pore volume—Due to high surface area (N 900 m^2/g) and large volume (N 0.9 cm^3/g), they have high loading capacities for drug molecules. The general total surface area and pore volume [48].
5. Two functional surfaces—MSN has an internal and external surface properties. These characteristics let them carry different moieties [48].

6. Unique porous structure-Many drug delivery materials have interconnecting porous structure such as dendrimers with branching porous structures and liposomes with a large void core and a porous shell [48].

MSN has been used in various shapes. Spheres, ellipsoids, rods and tubes have been seen as a good carriers for drugs delivery. Further, improvement of surface enhanced binding capacity of MSN with many drugs which also improved drug loading capacity. There are three methods: co-condensation, grafting and imprint coating method [53, 54].

The NPs prepared by microemulsion method provided inner hydrophobic core and the outer hydrophilic region. The hydrophobic core carries drug docetaxel, a chemotherapy drug while the outer hydrophilic region binds with water molecules to increase distribution of drug—NPS conjugates. Docetaxel is known to have radio sensitizing properties which makes it popular theranostic NPs in prostate cancer [28]. It is well defined that an efficient delivery system should have good loading capacity to transport the desired candidate drug without any loss of therapeutic characteristics. Any premature leakage of drug is a major challenge along with other considerable issues such as hydrolysis induced erosion of the carrier structure and require of organic solvents for drug loading that can induce undeniable modifications of the structure. Use go mesoporous silica materials offer several convenient features such as stable mesoporous structures, large surface areas, easily modified pore sizes with varying volumes and well defined surface properties for site specific drug delivery and for carrying molecules of various sizes, shapes, and functionalities [29, 30]. Recent research used hydrophobic interactions to attach lipid membrane to the surface of chlorodimethyloctadecylsilan -e-modified hollow mesoporous silica NPs. The main characteristics were uniform structure, comparable drug loading efficiency, desirable stability and strengthened controlled release. The prepared nano-vehicles Arg-Gly-Asp (RGD)-conjugated liposome showed good compatibility and low toxicity on HepG2, MCF-7 and LO2 cells. In comparison of RGD-LP-CHMSN ATO group to the free ATO, prolongation in half time and increase area under curve (AUC) were observed [55]. Another study used large-pore mesoporous silica NPs were utilized due to high loading capacity for [PtCl(en)(*N*-[acridin-9-ylaminoethyl]-*N*-methylpropionamidine)] dinitrate salt (P1A1). P1A1 is a dicationic and hydrophilic carrier for platinum-acridimide (PA) anticancer agents. This conjugation created artificial favorable pH inside the cells lysosomes and coating in phopholipd bilayers resulted in better colloidal stability. This method also caused S phase arrest and inhibited cell proliferation in pancreatic cancer. The most attractive result was that this conjugation withstand acidic environment of lysosomes vesicles into the cytoplasm [56].

2.4 Quantum Dots

Quantum dots (QDs) are semiconducting materials of size between 10 and 100 Å with other chemical characteristics such as fluorescence and narrow emission (Table 5.6). They show broad spectrum of excitation on exposure to UV rays and

Table 5.6 List of various QDs type and coating

Quantum Dot type and coating	Core/shell shape	Core/shell diameter (nm)	Expected surface charge
QD 565-PEG	Spherical	4.6	Neutral/uncharged
QD 565-amine	Spherical	4.6	Positive
QD 565-carboxylic acid	Spherical	4.6	Negative
QD 655-PEG	Ellipsoid	6 nm	Neutral/uncharged
QD-655-PEGamine	Ellipsoid	6 nm	Positive
QD 655-carboxylic acid	Ellipsoid	6 nm	Negative

also possess high photo stability. Quantum dots are nano crystals of semiconducting materials consisting of semiconductor inorganic core (CdSe) and organic coated shell (ZnS) to improve optical properties. They are heterogeneous NPs that consist of a colloidal core surrounded by one or more surface coatings. Surface coatings in single or multiple layers are applied to modify QDs as needed for experiments. Hydrophilic coatings increase solubility in a biological medium. This prevents leaking of metals from the core and bounded molecules such as receptor–ligands, antibodies, therapeutic and diagnostic macromolecules [58, 59]. The size of QDs are highly adjustable from 2–10 nm to 5–20 nm on polymerization. Particles small in size can be wiped out by renal filtration where as large size can be engulfed by the reticuloendothelial cellular system before reaching to the targeted site [60]. The recent improvement has lead drastic applications due to highly tunable sizes of QDs. Further, high surface-to-volume ratio of nanomaterials, it is possible to link multiple functional groups on single QDs while keeping the surface-volume constant. Hydrophilic materials fit well with QD such as siRNA, aptamers and can be immobilized on to hydrophilic side or amphiphilic polymer [61, 62].

They are successfully integrated in receptor mediated endocytosis, as a cancer marker, DNA hybridization, immunoassays and multiple color imaging. The outer surface of shell involves in interaction whereas core determine the color emission. These features can be modified based on the targeting biomarkers [63]. Current applications of QDs in drug delivery are focused on two major areas: QDs as carriers and labeling therapeutics or drug carriers QDs. The majority of drug carriers are Poly(lactic-co-glycolic acid) and polyethyleneimine (PEI). The QDs with LipofectamineTM 2000 was used for siRNA delivery [64]. Researchers found that amount of QD should 1:1 rations as excess of QDs can reduce the silencing efficiency whereas lower concentration can fail to detect fluorescence [57]. For antisense OD delivery, highly uniform QD-doped chitosan nano-beads for traceable siRNA delivery [65] and PEI-coated carbon nanotubes with QDs were used [66]. They improved quantification of proteins of interest by detecting their differential expression levels. They co-related their expressions with fluorescence and can be isolated using florescence-activated cell sorting and if multicolor QDs are used for screening of siRNA sequences. The potential application of QDs in drug delivery are shown in Fig. 5.2.

Most recently, for synergistic therapy, controlled drug release, magnetic hyperthermia and photo thermal therapy, combined approach using magnetic mesoporous

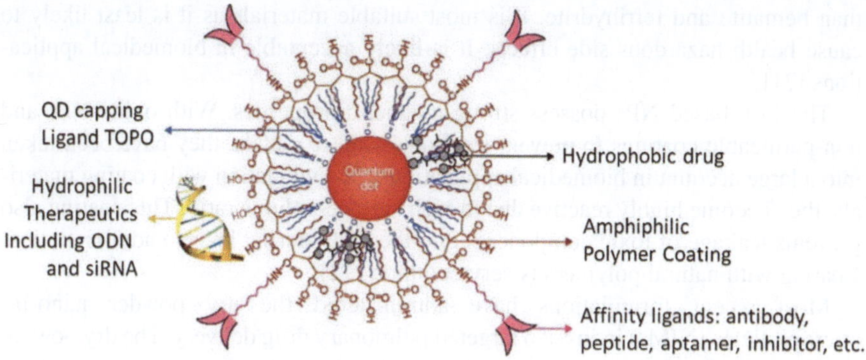

Fig. 5.2 QDs for drug delivery and therapy

silica NPs (MMSN) as drug carriers, magnetic thermos-seeds, local photo thermal generators and graphene quantum dots has been proposed. The results show that in low pH doxorubicin release can be triggered and loading of the drug can also be increased using 100 nm particle. Further, generated hyper-thermal temperature improved chemotherapy of 4T1 cells. Researchers also compared chemotherapy, magnetic hyperthermia or photo-thermal therapy when used alone, the combined chemo-magnetic hyperthermia therapy or chemo-photo-thermal therapy with the DOX-loaded MMSN/GQDs nano-system, resulted in a higher efficacy to kill cancer cells. Therefore, the MMSN/GQDs multifunctional platform has important application in cancer therapy for enhancing the therapeutic efficiency [67].

2.5 *Magnetic NPs*

Nano scale dimension particles, their magnetic properties and their capacity of carryon active biomolecules for specific tasks. Due to small size, such nano particles easily pass through the narrowest blood vessels and also penetrate through cell membranes if required [68]. If these particles are ferromagnetic or superparamagnetic, they can be manipulated by an external magnetic field which can further enhance efficiency in drug delivery [69, 70]. So the magnetic NPs are composed of number of components the magnetic core, the protective coating and the modified surface based on functionality. The challenge in using magnetic NPs is to bring all of these components into one system which can fit best for required therapeutic characteristics.

The magnetic core materials are Magnetite Fe_3O_4, Maghemite γ-Fe_2O_3, Iron-based metal oxides, Iron alloys and some rare earth metal alloys and transition metal clusters. The magnetite exhibits ferro (ferri) magnetic properties which has strong magnetization properties. Maghemite has the same lattice structure but all iron atoms in third oxidation states. It has 100 time stronger magnetization properties than hematite and ferrihydrite. It is most suitable materials as it is least likely to cause health hazardous side effects. It is likely acceptable in biomedical applications [71].

The iron based NPs possess strong magnetic properties. With oxide NPs and non-permeable coatings to prevent leaching of these metals, they have been taken into a large account in biomedical applications. In conjugation with coating materials, they become highly reactive due to high surface-volume ratio. This coating also prevents leakage of toxic components into the body during in-vivo administration. Coating with natural polymers is very common [72].

Most recent formulations have administered the dry powder nano-in-microparticles (NIMs) in in-vivo targeted pulmonary drug delivery. The dry powder NIMs containing fluorescent NPs and magnetically-responsive superparamagnetic iron oxide NPs were more efficient in targeting left lung of mice [73]. Superparamagnetic magnetite NPs with temperature-responsive poly(*N*-isopropylacrylamide) (PNIPAm) hydrogel core were prepared to enhance binding with targeting ligand. This system was further integrated with folic acid (Fig. 5.3) for targeting cervical cancer cell line (HeLa) which shown good capacity of intracellular uptake due to both magnetic properties and receptor mediate endocytosis [74]. The various linker used to modified surface of magnetic nanostructures and polymers to improve applications of magnetic NPs are shown in Tables 5.7 and 5.8, respectively.

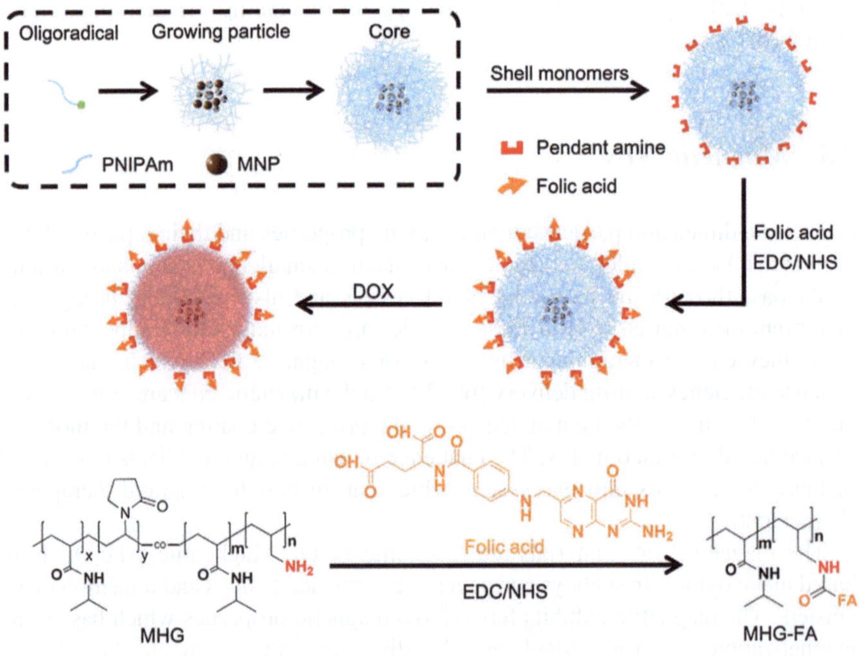

Fig. 5.3 Illustration of MHG-FA hydrogel synthesis

Table 5.7 List of various linkers used with Magnetic NPs

Linkers	Property	Target biomolecules	Examples
Amine	Positively charged	Drugs	Ibuprofen Aspirin DNA
Carboxylic acid	Negativcly charged Can form amide bond with $-NH_2$ Can form ester bond with -OH	Proteins	Lysozyme Antibodies
Aldehyde	Form imide bond with $-NH_2$	Proteins	Enzymes
Thiol	-SH	Proteins	Cytochrome C Trypsin Lipase

Table 5.8 Polymer of magnetic NPs.

Polymer	Surface behavior	Applications
	Natural	polymers
Carbohydrates Dextran	Hydrophilic	Drug delivery Radioimmunoassay MR imaging Hyperthermia Tumor targeting, Mr imaging, Xray imaging
Starch	Hydrophilic	Tumor targeting, MR imaging and X ray imaging.
Proteins Albumin RGD	Hydrophilic Hydrophilic	MR imaging Fluorescent imaging and MR imaging
Lipids	Hydrophobic	Immunoassay
	Synthetic	Polymers
PEG	Hydrophilic	MR imaging Drug delivery
Polyvinyl alcohol (PVA)	Hydrophilic	Drug delivery

3 Liposomes

Liposomes are self-assembling spherical shape colloidal structures composed of lipid bilayers. The outer layer of liposomes surround a central aqueous space. Depending upon their size and number of bilayers they are classified into three basic types

1. Multilamellar vesicles—These are made of several lipid bilayers separated from one another by aqueous spaces.
2. Small unilamellar vesicles—These vesicles are consist of a single bilayer surrounding the entrapped aqueous space. The diameter is less than 100 nm.
3. Large unilamellar vesicles—These vesicles consist of a single bilayer surrounding the entrapped aqueous space with diameters greater than 100 nm.

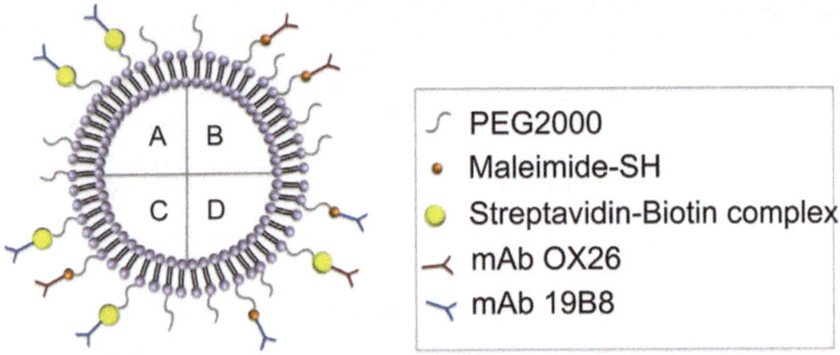

Fig. 5.4 Pegylated liposomes: (**a**) conjugated to 19B8MAb through biotin–streptavidin complex, (**b**) coupled to OX26MAb through maleimide, (**c**) coupled to both OX26MAb and 19B8MAb through, respectively, maleimide and biotin–streptavidin complex, and (**d**) conjugated to OX26MAb and 19B8MAb through, respectively, biotin–streptavidin complex and maleimide. *Colloids Surf B Biointerfaces*, *2017*

Several cancer based drugs such as anthracyclines doxorubicin and daunorubicin are used for the treatment of metastatic breast cancer and AIDS-related Kaposi's sarcoma [75–77] with the help of liposomes PEGylated conjugation method. Immunoliposomes are actively in use for treating brain affected areas. Pegylated liposomes functionalized with two antibodies, the anti-transferring receptor monoclonal antibody (OX26MAb) and the anti-amyloid beta peptide antibody (19B8MAb) as nano-carriers of drugs for Alzheimer's disease. Fluorescence spectroscopy showed cellular take of the immunobased liposomes is more efficient if OX26MAb is used in conjugation with streptavidin-biotin complex Fig. 5.4 [78].

4 Polymer Based Nano-materials

The method of preparation for polymers based nano-material utilize entrapment of drug either physically trapped in or covalently bound to the polymer matrix [79]. Polymers used as a drug conjugates are natural and synthetic in nature. This conjugation result into the polymeric NPs (capsules), polymeric micelles (amphiphilic core/shell) or hyper branched macromolecules (dendrimers) and drug-conjugates. Polymeric NPs are biodegradable, biocompatible and well recognized for their protective nature to drug molecules. They are considered as a good source for controlled release and sustained delivery of drugs. They can be easily integrated in active and passive delivery methods of drugs. However, such drugs have always been limited due to cytotoxicity and adverse side effects. To overcome this drawback, an approach called target-specific drug delivery system (DDS) was introduced which can transport a drug to the targeted cells and tissues. To make a general efficient drug delivery system, several perquisites which are required to be incorporated are

1. A biocompatibility of a nanomaterial,
2. High loading capacity of the desirable drug,
3. No leaking or zero premature release of the drug,
4. Cell or tissue specificity and capability for site directing release,
5. Accuracy in controlled release and rate of release to achieve an effective local concentration.

4.1 Micelles

Copolymers dependent micelles are nano-sized core/shell copolymer structure in aqueous media. The hydrophobic core region offers scaffold to hydrophobic drugs and hydrophilic shell stabilizes the hydrophobic core and make association with aqueous medium to enhance drug delivery [80]. The physical and chemical interactions are the most popular reactions seen in such conjugate studies [81, 82].

Most recent research applications have utilized diamond based NPs. The large surface area of biocompatible NDs with different properties is remarkable agent for carrying therapeutic agents. It has a good loading capacity and protects attached molecular entities from disruption. It needs less amount of the carrying agents even to load drug with high concentration. The highly potential agents with good releasing and carrying capacity are in continuous demand of drug delivery methods and research protocols. Diamond NPs have advantageous applications in controlled and sustained-release delivery. They have been explored for improving delivery of therapeutic drugs and several studies showed participating of surfaces in chemical bonding [83, 84] and physical adsorption procedures [84–86]. Such physical interactions avoid the use of complex chemical reactions which are not only cost driving but also alter therapeutic activity. It has been observed that structural alterations due to chemical interactions are often challenging to meet therapeutic requirements which is more strongly handled by physical interactions by NDs. In previous studies, doxorubicin hydrochloride was studied with NDs to find potential capabilities of conjugated molecule as a drug delivery agent. The study was developed based on the interaction between carboxylic and hydroxyl group of detonation NDs which can interact with the basic group of DOX via ionic forces in presence of aqueous medium [83]. Nano diamonds form loose clusters with DOX in such a manner that small amount of absorbed DOX on the NDs surface resides within the cavity of cluster [83]. Further, on adding 1% sodium chloride in aqueous medium increased loading capacity from 0.5 to 10 wt% and removal of salt favored release of DOX molecule [83]. A lower cytotoxicity was also observed in macrophages and human colorectal cancer cells compared with the systems utilized only DOX [83]. Thus, ND based delivery systems could overcome the limitation to the use of high concentrations of chemotherapeutic drugs in cancer treatments. Another study published NDs in conjugation with 10-hydroxycamptothecin (HCPT) which shown good amount of increase in drug loading capacity.

This study also found change in pH from salivation to alkaline [87]. Such multifunctional polymeric micelles are being actively developed for carrying heavy loads and therapeutic agents [88].

4.2 Dendrimers

They are highly branched polymers of size less than 10 nm generated under controlled polymerization process. They are consists of three main components, core, branch and surface. They are very good source for active and passive delivery of bioactive agents such as gene, protein, peptide and others. They are also known for their long sustaining feature in circulatory systems. Dendrimers are made of small monomers and fabricated into converged or diverged step growth polymerization. Such polymerization can be controlled process. These structures carry cavities which are used in drug transportation. The other end of dendrimer can be used for conjugation or attachment with molecules which can tune up with the requirement [89]. Such branches can be tailored into biocompatible compounds with low cytotoxicity and high permeability towards active biomolecules. Dendrimers are good vehicles for drug delivery, gene therapy dendrimer based drug delivery, and immunoassay imaging agent.

4.3 Drug Conjugates

Polymers in drug delivery method are natural and synthetic. Natural forms are albumin, heparin, and chitosan which are very popular form of agents for the delivery of dendrimers and other macromolecules. In treatment of metastatic breast cancer, serum albumin in conjugation with paclitaxel is used [91]. Another component, Abraxen has been explored in treatment of many cancer and advanced non hematologic malignancies [92, 93]. Among synthetic polymers such as *N*-(2-hydroxypropyl)-methacrylamide copolymer, polyethylene glycol (PEG) and poly-l-glutamic acid (PGA) have been extensively utilized as drug delivery channels. PGA is the first biodegradable polymer used in conjugated applications [94]. The outcomes of various studies (Fig. 5.5) confirmed that HPMA and PEG conjugated are most popular non-biodegradable synthetic polymers [95].

5 Conclusions

With continuous demand of novel protocols in drug delivery systems, diagnose and treatments of life threatening disease, the synthesis of nano-material using biological is fast, non-toxic and biocompatible in nature. Various nano-materials have been

Compound	Name	Company	Linker	Status of development
HPMA copolymer–doxorubicin	PK1; FCE28068	CRC/Pharmacia	Amide	Phase II
HPMA copolymer– doxorubicin-galactosamine	PK2; FCE28069	CRC/Pharmacia	Amide	Phase I/II
HPMA copolymer–paclitaxel	PNU166945	Pharmacia	Ester	Phase I
HPMA copolymer–camptothecin	MAG-CPT, PNU166148	Pharmacia	Ester	Phase I
HPMA copolymer–platinate	AP5280	Access Pharmaceuticals	Malonate	Phase I
Polyglutamate–paclitaxel	CT-2103, XYOTAX	Cell Therapeutics	Ester	Phase II/III
Polyglutamate–camptothecin	CT-2106	Cell Therapeutics	Ester	Phase I
PEG–camptothecin	PROTHECAN	Enzon	Ester	Phase II
PEG–aspartic acid-doxorubicin micelle	NK911	National Cancer Institute Japan	Amide/free drug	Phase I

CRC, UK Cancer Research Campaign; HPMA, *N*-(2-hydroxypropyl)methacrylamide; PEG, polyethylene glycol.

Fig. 5.5 Polymer-anticancer-drug-conjugates tested clinically

used to enhance efficiency with the help of favorable, yet different physicochemical parameters. It seems that the nano-enabled drug delivery systems hold great potential to overcome efficient targeting of the affected cells. It is exciting to have progress in facilitating movement of drugs across barriers such as those in the brain. The challenge, to understand the fate of the drugs once delivered in-vitro and the later biological impact is still a major concern.

Acknowledgment The authors appreciate Fox Chase Cancer Center, Philadelphia, USA for support and motivation.

References

1. Kaur PK, et al. Penetration enhancers and ocular bioadhesives: two new avenues for ophthalmic drug delivery. J Drug Develop Ind Pharm. 2002;28:353–69.
2. Bianco A, Kostarelos K, Partidos CD, Prato M. Biomedical applications of functionalized carbon nanotubes. Chem Commun. 2005:571–7.
3. Habibizadeh M, Rostamizadeh K, Dalali N, Ramazani A. Preparation and characterization of PEGylated multiwall carbon nanotubes as covalently conjugated and non-covalent drug carrier: a comparative study. Mater Sci Eng C Mater Biol Appl. 2017;74:1–9.
4. Liu Z, Davis C, Cal WB, He L, Chen XY, Dal HJ. Circulationand long-term fate of functionalized, biocompatible single-walled carbon nanotubes in mice probed by Raman spectroscopy. Proc Natl Acad Sci. 2008;105:1410–5.
5. Singh R, et al. Tissue biodistribution and blood clearance rates of intravenously administered carbon nanotube radiotracers. Proc Natl Acad Sci. 2006;103(9):3357–62.

6. Worle-Knirsh JM, Pulskamp K, Krug HF. Oops they did it again! Carbon nanotubes hoax scientists in viability assays. Nano Lett. 2006;6(6):1261–8.
7. Wang H, et al. Biodistribution of carbon single-wall carbon nanotubes in mice. J Nanosci Nanotechnol. 2004;4:1019–24.
8. Wong N, et al. Nanotube molecular transporters: internalization of carbon nanotube-protein conjugates into mammalian cells. J Am Chem Soc. 2004;126:6850–1.
9. Kam NW, Dai H. Carbon nanotubes as intracellular protein transporters: generality and biological functionality. J Am Chem Soc. 2005;127:6021–6.
10. Prato M, Kostarelos KAB. Functionalized carbon nanotubes in drug design and discovery. Acc Chem Res. 2008;41:60–8.
11. Chen JY, et al. Functionalized single-walled carbon nanotubes as rationally designed vehicles for tumor targeted drug delivery. J Am Chem Soc. 2008;130:16778–85.
12. Liu Z, Sun XM, Nakayama-Ratchford N, Dai HJ. Supramolecular chemistry on water soluble carbon nanotubes for drug loading and delivery. ACS Nano. 2007;1:50–6.
13. Mu QX, Broughton DL, Yan B. Endosomal leakage and nuclear translocation of multiwalled carbon nanotubes: developing a model for cell uptake. Nano Lett. 2009;9:4370–5.
14. Heister E, et al. Triple functionalization of single-walled carbon nanotubes with doxorubicin, a monoclonal antibody, and a fluorescent marker for targeted cancer therapy. Carbon. 2009;47:2152–60.
15. Kam NWS, Dai H. Carbon nanotubes as intracellular protein transporters: generality and biological functionality. J Am Chem Soc. 2005;127(16):6021–6.
16. Bianco A, Kostarelos K, Prato M. Applications of carbon nanotubes in drug delivery. Int J Pharm. 2017;456:143.
17. Kushwaha SKS, Ghoshal S, Rai AK, Singh S. Carbon nanotubes as a novel drug delivery system for anticancer therapy: a review. Braz J Pharm Sci. 2013;49(4):629–43.
18. Connor EE, Mwamuka J, Gole A, Murphy CJ, Wyatt MD. Gold nanoparticles are taken up by human cells but do not cause acute cytotoxicity. Small 1. 2005;1:325–7.
19. Deb S, et al. Multistability in platelets and their response to gold nanoparticles. Nanomedicine. 2011;7:376.
20. Ganeshkumar M, et al. Sun light mediated synthesis of gold nanoparticles as carrier for 6-mercaptopurine: preparation, characterization and toxicity studies in zebrafish. Mater Res Bull. 2012;47:2113–9.
21. Lee K, Lee H, Bae KH, Park TG. Heparin immobilized gold nanoparticles for targeted detection and apoptotic death of metastatic cancer cells. Biomaterials. 2010;31:6530.
22. Bianco A, Kostarelos K, Prato M. Applications of carbon nanotubes in drug delivery. Curr Opin Chem Biol. 2005;9(6):674–9.
23. Radin S, Ducheyne P, Kamplain T, Tan BH, Silica J. Sol-gel for the controlled release of antibiotics. Synthesis, characterization, and in vitro release. Biomed Mater Res. 2001;57:313.
24. Slowing II, Trewyn BG, Giri S, Lin VS-Y. Mesoporous silica nanoparticles for drug delivery and biosensing applications. Adv Funct Mater. 2007;17:1225.
25. Goodman AM, et al. Understanding resonant light-triggered DNA release from Plasmonic nanoparticles. ACS Nano. 2017;11:171.
26. Chen Y, et al. Transdermal vascular endothelial growth factor delivery with surface engineered gold nanoparticles. ACS Appl Mater Interfaces. 2017;9:5173–80.
27. Belz J, Castilla-Ojo N, Sridhar S, Kumar R. Radiosensitizing silica nanoparticles encapsulating docetaxel for treatment of prostate cancer. Methods Mol Biol. 2017;1530:403–9.
28. Hostetler MJ, et al. Alkanethiolate gold cluster molecules with core diameters from 1.5 to 5.2 nm: core and monolayer properties as a function of core size. Langmuir. 1998;14:17.
29. Gibson JD, Khanal BP, Zubarev ER. Paclitaxel-functionalized gold nanoparticles. J Am Chem Soc. 2007;129:11653.
30. Paciotti GF, Kingston DGI, Tamarkin L. Colloidal gold nanoparticles: a novel nanoparticle platform for developing multifunctional tumor-targeted drug delivery vectors. Drug Dev Res. 2006;67:47.

31. Chen Y-H, et al. Methotrexate conjugated to gold nanoparticles inhibits tumor growth in a syngeneic lung tumor model. Mol Pharm. 2007;4(5):713–22.
32. Li J, Wang X, Wang C, Chen B, Dai Y, Zhang R, Song M, Lv G, Fu D. The enhancement effect of gold nanoparticles in drug delivery and as biomarkers of drug-resistant cancer cells. ChemMedChem. 2007;2:374.
33. Patra CR, Bhattacharya R, Wang E, Katarya A, Lau JS, Dutta S, Muders M, Wang S, Buhrow SA, Safgren SL, Yaszemski MJ, Reid JM, Ames MM, Mukherjee P, Mukhopadhyay D. Targeted delivery of gemcitabine to pancreatic adenocarcinoma using cetuximab as a targeting agent. Cancer Res. 2008;68:1970.
34. Podsiadlo P, Sinani VA, Bahng JH, Kam NW, Lee J, Kotov NA. Gold nanoparticles enhance the anti-leukemia action of a 6-mercaptopurine chemotherapeutic agent. Langmuir. 2008;24(2):568–74.
35. Azzam EMS, Morsy SMI. Enhancement of the antitumour activity for the synthesised Dodecylcysteine surfactant using gold nanoparticles. J Surf Deterg. 2008;11:195–9.
36. Agasti SS, et al. Photoregulated release of caged anticancer drugs from gold nanoparticles. J Am Chem Soc. 2009;131(16):5728–9.
37. Dhar S, et al. Polyvalent oligonucleotide gold nanoparticle conjugates as delivery vehicles for platinum (IV) warheads. J Am Chem Soc. 2009;131(41):14652–3.
38. Huang-Chiao H, et al. Simultaneous enhancement of photothermal stability and gene delivery efficacy of gold nanorods using polyelectrolytes. Acs Nano. 2009;3(10):2941–52.
39. Podsiadlo P, Sinani VA, Bahng JH, Kam NW, Lee J, Kotov NA. Gold nanoparticles enhance the antileukemia action of a 6-mercaptopurine chemotherapeutic agent. Langmuir. 2008;24(2):568–74.
40. Dreaden EC, et al. Tamoxifen–poly (ethylene glycol)–thiol gold nanoparticle conjugates: enhanced potency and selective delivery for breast cancer treatment. Bioconjug Chem. 2009;20(12):2247–53.
41. Eghtedari M, et al. Engineering of hetero-functional gold nanorods for the in vivo molecular targeting of breast cancer cells. Nano Lett. 2008;9(1):287–91.
42. Asadishad B, Vossoughi M, Alemzadeh I. In vitro release behavior and cytotoxicity of doxorubicin-loaded gold nanoparticles in cancerous cells. Ind Eng Chem Res. 131(16): 5728–9.
43. Staroverov SA, et al. Gold nanoparticles in biology and medicine: recent advances and prospects. Rossiyski bioterapevticheski zhurnal. 2010;31(41):14652–3.
44. Kim CK, Ghosh P, Rotello VM. Multimodal drug delivery using gold nanoparticles. Nanoscale. 2009;1(1):61–7.
45. Igor I, et al. Mesoporous silica nanoparticles as controlled release drug delivery and gene transfection carriers. Adv Drug Deliv Rev. 2008;60:1278.
46. Gruen M, Lauer I, Unger KK. The synthesis of micrometer- and submicrometersize spheres of ordered mesoporous oxide MCM-41. Adv Mater. 1997;9:254–7.
47. Unger KK, et al. Synthesis of spherical porous silicas in the micron and submicron size range: challenges and opportunities for miniaturized high-resolution chromato- graphic and electrokinetic separations. J Chromatogr A. 2000;892(1):47–55.
48. Huo Q, Feng J, Schüth F, Stucky GD. Preparation of hard mesoporous silica spheres. Chem Mater. 1997;9(1):14–7.
49. Huh S, et al. Organic functionalization and morphology control of mesoporous silicas via a co-condensation synthesis method. Chem Mater. 2003;15(22):4247–56.
50. Burleigh MC, et al. Stepwise assembly of surface imprint sites on MCM-41 for selective metal ion separations. ACS Symposium Series; 2001. p. 146–158.
51. Chen HT, Huh S, Lin VS. In: Regalbuto J, editor. Fine tuning the functionalization of mesoporous silica. New York: CRC/Taylor & Francis; 2007.
52. Fei W, et al. RGD conjugated liposome-hollow silica hybrid nano-vehicles for targeted and controlled delivery of arsenic trioxide against hepatic carcinoma. Int J Pharm. 2017;519(1):250–62.

53. Zheng Y, et al. Large-pore functionalized mesoporous silica nanoparticles as drug delivery vector for a highly cytotoxic hybrid platinum-acridine anticancer agent. Chemistry. 2017;23(14):3386–97.
54. Huh S, Wiench JW, Yoo J-C, Pruski M, Lin VSY. Organic functionalization and morphology control of mesoporous silicas via a co-condensation synthesis method. Chem Mater. 2003;15(22):4247–56.
55. Michalet X, et al. Quantum dots for live cells, in vivo imaging, and diagnostics. Science. 2005;307(5709):538–44.
56. Gao X, et al. In vivo cancer targeting and imaging with semiconductor quantum dots. Nature Biotechnol. 2004;22(8):969–76.
57. Willard MA, et al. Chemically prepared magnetic nanoparticles. Int Mater Rev. 2004;49(3–4):125–70.
58. Choi HS, et al. Renal clearance of quantum dots. Nat Biotechnol. 2007;25(10):1165–70.
59. McNamara JO, et al. Cell type–specific delivery of siRNAs with aptamer-siRNA chimeras. Nature Biotechnol. 2006;24(8):1005–15.
60. Medarova Z, et al. In vivo imaging of siRNA delivery and silencing in tumors. Nat Med. 2007;13(3):372–7.
61. Tan WB, Jiang S, Zhang Y. Quantum-dot based nanoparticles for targeted silencing of HER2/neu gene via RNA interference. Biomaterials. 2007;28(8):1565–71.
62. Jia N, et al. Intracellular delivery of quantum dots tagged antisense oligodeoxynucleotides by functionalized multiwalled carbon nanotubes. Nano Lett. 2007;7(10):2976–80.
63. Remaut K, et al. Pegylation of liposomes favours the endosomal degradation of the delivered phosphodiester oligonucleotides. J Control Release. 2007;117(2):256–66.
64. Yao X, et al. Graphene quantum dots-capped magnetic mesoporous silica nanoparticles as a multifunctional platform for controlled drug delivery, magnetic hyperthermia, and Photothermal therapy. Small. 2017;13(2):1602225.
65. Wilson MW, et al. Hepatocellular carcinoma: regional therapy with a magnetic targeted carrier bound to doxorubicin in a dual MR imaging/conventional angiography suite—initial experience with four patients. Radiology. 2004;230(1):287–93.
66. Plank C, et al. The magnetofection method: using magnetic force to enhance gene delivery. Biol Chem. 2003;384(5):737–47.
67. Dobson J. Magnetic properties of biological materials. In: Barnes S, Greenebaum B, editors. Handbook of biological effects of electromagnetic fields: bioengineering and biophysical aspects of electromagnetic fields. Boca Raton: Taylor and Francis/CRC Press; 2007.
68. Shafi KVPM, et al. Sonochemical preparation and size-dependent properties of nanostructured CoFe2O4 particles. Chem Mater. 1998;10(11):3445–50.
69. Wilson MW, Kerlan RK, Fidleman NA. Hepatocellular carcinoma: regional therapy with a magnetic targeted carrier bound to doxorubicin in a dual MR imaging/conventional angiography suite—initial experience with 4 patients. Radiology. 2004;230:287–93.
70. Kim H, et al. Synergistically enhanced selective intracellular uptake of anticancer drug carrier comprising folic acid-conjugated hydrogels containing magnetite nanoparticles. Sci Rep. 2017;7:41090.
71. Markman M. Pegylated liposomal doxorubicin in the treatment of cancers of the breast and ovary. Expert Opin Pharmacother. 2006;7(11):1469–74.
72. Rivera E. Current status of liposomal anthracycline therapy in metastatic breast cancer. Clin Breast Cancer. 2003;4:S76–83.
73. Rosenthal E, et al. Phase IV study of liposomal daunorubicin (DaunoXome) in AIDS-related Kaposi sarcoma. Am J Clin Oncol. 2002;25(1):57–9.
74. Loureiro JA, et al. Dual ligand immune liposomes for drug delivery to the brain. Colloids Surf B Biointerfaces. 2015;134:213.
75. Rawat M, et al. Nanocarriers: promising vehicle for bioactive drugs. Biol Pharm Bull. 2006;29(9):1790–8.
76. Adams ML, Lavasanifar A, Kwon GS. Amphiphilic block copolymers for drug delivery. J Pharm Sci. 2003;92(7):1343–55.

77. Batrakova EV, et al. Anthracycline antibiotics non-covalently incorporated into the block copolymer micelles: in vivo evaluation of anti-cancer activity. Br J Cancer. 1996;74(10):1545.
78. Nakanishi T, et al. Development of the polymer micelle carrier system for doxorubicin. J Control Release. 2001;74(1):295–302.
79. Li X, et al. TAT-conjugated nanodiamond for the enhanced delivery of doxorubicin. J Mater Chem. 2011;21(22):7966–73.
80. Liu K-K, et al. Covalent linkage of nanodiamond-paclitaxel for drug delivery and cancer therapy. Nanotechnology. 2010;21(31):315106.
81. Huang H, et al. Active nanodiamond hydrogels for chemotherapeutic delivery. Nano Lett. 2007;7(11):3305–14.
82. Li J, Zhu Y, Li W, Zhang X, Peng Y, Huang Q. Nanodiamonds as intracellular transporters of chemotherapeutic drug. Biomaterials. 2010;31(32):8410–8.
83. Chow EK, Zhang XQ, Chen M, et al. Nanodiamond therapeutic delivery agents mediate enhanced chemoresistant tumor treatment. Sci Transl Med. 2011;3(73):73ra21.
84. Chen M, Pierstorff ED, Lam R, et al. Nanodiamond-mediated delivery of water-insoluble therapeutics. ACS Nano. 2009;3(7):2016–22.
85. Chang YR, Lee HY, Chen K, et al. Mass production and dynamic imaging fluorescent nanodiamonds. Nat Nanotechnol. 2008;3(5):284–8.
86. Gradishar WJ, et al. Phase III trial of nanoparticle albumin-bound paclitaxel compared with polyethylated castor oil–based paclitaxel in women with breast cancer. J Clin Oncol. 2005;23(31):7794–803.
87. Green MR, et al. Abraxane®, a novel Cremophor®-free, albumin-bound particle form of paclitaxel for the treatment of advanced non-small-cell lung cancer. Ann Oncol. 2006;17(8):1263–8.
88. Nyman DW, et al. Phase I and pharmacokinetics trial of ABI-007, a novel nanoparticle formulation of paclitaxel in patients with advanced nonhematologic malignancies. J Clin Oncol. 2005;23(31):7785–93.
89. Li C. Poly(l-glutamic acid)-anticancer drug conjugates. Adv Drug Deliv Rev. 2002;54:695–713.
90. Duncan R. The dawning era of polymer therapeutics. Nat Rev Drug Discov. 2003;16:347–60.
91. Cai D, et al. Carbon nanotube-mediated delivery of nucleic acids does not result in non-specific activation of B lymphocytes. Nanotechnology. 2007;18:101–10.
92. Liu Z, Winters M, Holodniy M, Dai H. siRNA delivery into human T cells and primary cells with carbon-nanotube transporters. Angewandte Chem. 2007;46:2023–7.
93. Pantarotto D, et al. Functionalized carbon nanotubes for plasmid DNA gene delivery. Angew Chem Int Ed Engl. 2004;43:5242–6.

Chapter 6
Nanoformulations for Therapeutics

PSS Rao and Christopher RT Stang

Abstract In the recent decade and a half, research in the area of nanoformulation has vastly expanded, including research dedicated to discovery of novel drugs as well as the reformulation of tried and tested drugs. The utilization of nanoformulations has yielded positive results in areas that have typically been regarded as therapeutic challenges. Nanoformulations, by in large, aim to alter the pharmacokinetic properties of a medication, improving absorption, distribution, selectivity, and altering metabolism. The blood-brain barrier has largely been an obstacle in the treatment of neurodegenerative disorders and tumors located within the central nervous system. Several drugs have been discovered that are effective in treating these pathologies. However, reaching appropriate therapeutic concentration in the central nervous system, without excessive systemic drug concentrations and side effects, has remained a limitation to these medications' efficacy. Nanoformulating such medications has yielded improved drug delivery and efficacy to the central nervous system despite the influences of multi-drug resistance proteins. Various methods of nanoformulation have been employed to facilitate the selective transport of drugs across the blood-brain barrier including the use of focused ultrasound, magnetic targeting, macrophage loaded drug formulations, carbon nanotubules, polymer coatings, and receptor-mediated transcytosis. These formulations have resulted in the efficient delivery of drug molecules, proteins, and DNA to the relevant tissue in the central nervous system. The continued research into nanoformulations allows for the development of more selective and effective medications while possibly salvaging medications that were deemed not useful due to their sub-therapeutic properties.

Keywords Nanoformulations • Personalized nanotherapeutics • Nanobiotechnology

P. Rao (✉) • C.R. Stang
Department of Pharmaceutical Sciences, College of Pharmacy, The University of Findlay, Findlay, OH 45840, USA
e-mail: rao@findlay.edu

A. Kaushik et al. (eds.), *Advances in Personalized Nanotherapeutics*,
https://doi.org/10.1007/978-3-319-63633-7_6

1 Introduction

The establishment of National Nanotechnology Initiative (NNI) by the U.S. government in 2000 has provided tremendous impetus to the research and development efforts in the field of nanotechnology. With the infusion of critical financial support, a rapid expansion of nanotechnology directed towards the improvement of human health has been observed. The exceeding interest and successful implementation of nanotechnology in the pharmaceutical industry has resulted in a plethora of nanopharmaceutical products which are available on the market [1], and a significant number of other promising nanopharmaceuticals currently in the pipeline [2]. These novel nanoformulations utilize the salient features of nanotechnology for effective drug delivery and such applications are expected to increase in the future.

The implementation of nanotechnology in pharmaceutical product development offers distinct advantages in terms of incorporating desirable pharmacokinetic properties. The use of appropriate nanostructures, for instance, can mitigate common concerns associated with a drug's intrinsic properties like low solubility, gastrointestinal instability, and low lipophilicity. Such advantages have led to the recognition of nanoformulations as attractive drug delivery tools for new product development [3].

Given the widespread application of nanotechnology in pharmaceutical preparations, this chapter aims to describe and highlight the success of nanoformulations in the effective treatment of various neurological ailments.

2 Targeting Neurodegenerative Inflammation Utilizing Nanoformulation

Neurodegenerative disorders and diseases are often the result of progressive destruction of neuronal cells. Frequently, the destruction of neuronal cells is the result of inflammation and secretion of inflammatory cytokines secondary to the induction of oxidative stress. This effect is often mitigated by the glial cells, which, while in the process of trying to protect neural cells, release signaling molecules leading to overall destruction of neurons they are trying to protect [4]. Given the inability of neuronal cells to regenerate, protecting these cells from death, or at least slowing cell death, still remains a major treatment goal of modern medicine. Diseases that are considered neurodegenerative include amyotrophic lateral sclerosis (ALS), Parkinson's (PD), Alzheimer's (AD) and Multiple Sclerosis (MS). Currently, the majority of treatment options are limited to managing the symptoms of each disease after the destruction of neural cells. For example, developing dopamine mimetics or dopamine receptor agonists to replace the lack of secreted dopamine in PD. Designing formulations that are able to cross the blood-brain barrier (BBB) has been a major obstacle in the development of neuroprotective drugs. Many molecules are very effective at reducing inflammation, and mitigating some of the causes of inflammation and cell death in vitro. However, poor pharmacokinetics and bioavailability frequently hinder the in vivo suitability of several neuroprotective compounds. With the rise of nanoparticles and other nanoformulations, new methods to package these neuroprotective molecules and deliver them to the central nervous system are actively being discovered and explored.

An effective nanoformulation is able to take biologically active molecule with poor pharmacokinetic activity and effectively distribute, maintain concentration, and deliver it to the desired tissues [1]. Essentially, the nanoformulation should give the molecule new or improved properties than if the molecule was administered alone [1]. The following sub-sections outline disease specific nanoformulation advances being researched and used in treatment of neurodegenerative ailments.

2.1 Parkinson's Disease

The leading proposition into the pathogenic mechanism of PD involves the presence and generation of free radicals, leading to the destruction of dopaminergic neurons [5]. Targeting oxidative stress prior to the destruction of neuronal cells can therefore confer neuroprotective effect. However, efficient delivery of protective drugs to such targets require passage through the BBB. Utilizing nanoformulations, researchers have been able to encapsulate biologically active molecule in nanoparticles, thus altering its bioavailability and pharmacokinetic profile, while largely retaining the molecules biological activity. One such approach is using polymers as an outer shell for the molecule. The subsequent nanoformulated molecule is then administered and able to cross the BBB. Brynskikh et al. were able to attenuate oxidative stress and decrease neural cell death seen in a mouse model of PD by conjugating the enzyme catalase with a polymer of with poly(ethylene imine) (PEI) and polyethylene glycol (PEG) [6]. Catalase, being a protein, exhibits poor crossing of the BBB given its polarity and relatively large size. It has been noted that in a Parkinsonian brain, there is a decrease in catalase activity [7], and such a nanoformulation would provide a type of replacement therapy. The same PEI polymer coating was used in conjunction with dextran sulfate to create a nanoformulation for the neuroprotective molecule retinoic acid (RA) as evident by an increase in substantia nigra dopaminergic neuron viability in the same animal model of PD [8]. RA is a molecule with many essential roles in brain, many of which are outlined by Maden [9]. While Esteves et al. demonstrated the benefits of nanoformulated RA in PD, further research should be completed on the benefits of RA nanoformulation given its role in other neurological disorders [8].

In addition to the possibilities of nanoformulating newer drugs for treating PD, effective delivery of current drugs of choice for PD, across the gastrointestinal tract or BBB, can also been achieved through the design of targeted nanoformulations. For instance, a recent study reported the potential impact of prolonging the residence time of levodopa using a gastroretentive nanosystem on its bioavailability [10]. The drug release profile, obtained from this in vitro/ex vivo study, was much more consistent and prolonged compared to conventional dosing, supporting further evaluation of gastroretentive nanosyetem as a therapeutic tool for the treatment of PD. In another study, to augment the central bioavailability of L-DOPA, the principal drug used for treating PD, Gambaryan et al. have demonstrated the effectiveness of poly (lactic-co-glycolic acid) [PLGA] nanoparticles in a rodent model of PD [11]. To minimize the side effects from conventional dosing, ensure sustained-release of encapsulated L-DOPA, and bypass the hepatic metabolism, this nanoformulation approach was efficiently combined with intranasal delivery of nanoformulated L-DOPA. The intranasal delivery of nanoparticles resulted in a significantly higher and long-lasting effect of L-DOPA in the tested rodent model, rationalizing the potential application of similar approaches for clinical use in the future.

2.2 *Alzheimer's Disease*

The formation of amyloid plaques in the pathophysiology of AD involves enhanced oxidative stress and glial secretion of pro-inflammatory cytokines leading to downstream cognitive dysfunction among other imbalances in the brain [12]. Such cascade of events makes for many potential targets in the treatment of AD. In one of the early studies, Cui et al. were able to utilize components of microemulsions to formulate nanoparticles capable of carrying the copper chelator D-penicillamine [13]. D-penicillamine alone is too hydrophilic to cross the BBB, limiting its chelator properties to areas outside the CNS. Preliminary in vitro testing of the nanoformulation revealed the ability to resolubilize Cu-Aβ aggregates, which could potentially lead to a decrease in the downstream effects of aggregate formation. The same nanoparticles have been also used and researched in other applications and demonstrated long circulation in the blood as well as the ability to cross the BBB in vivo [14]. A less targeted therapy being researched utilizes a polymer combination of poly(D,L-lactide-co-glycolide) to deliver neuroprotective effects of survivin mutant SurR9-C84A in vitro [15]. Importantly, it was demonstrated that the naïve form of SurR9-C84A was inactive in the model of AD. Further work performed by Songlin et al. was able to build upon the PEI polymer coating discussed above by conjugating the PEI with biscarbamate to form a novel poly(ethylene imine) biscarbamate conjugate (PEIC) [16]. Such a modification was able to further reduce the size of the nanoparticle formed while increasing the activity of gene transfection with no cytotoxic effect observed.

Interesting pre-clinical results have been obtained upon nanoformulating investigational drugs as well, supporting a promising future for the treatment and management of Alzheimer's. Piperine (PIP), an active alkaloid found in *Piper nigrum*, is one such promising molecule which has demonstrated effectiveness in in vitro models of AD [17]. While this compound has beneficial antioxidant and acetylcholinesterase inhibiting properties, PIP suffers from low-systemic bioavailability owing to high hepatic metabolism. To overcome low bioavailability, while maintaining a low dose and masking its pungency, a novel nanoparticle formulation intended for intranasal delivery of PIP was reported recently [18]. The PIP-loaded chitosan nanoparticles were able to deliver higher amounts of drug to the brain compared to free drug formulation. In addition, PIP nanoparticles were as effective as donepezil, the standard drug of choice, in reducing memory deficits in the tested animal model of AD. Similarly, tarenflurbil, a selective Aβ(42)-lowering agent, was sought as a promising drug candidate for the treatment of AD. However, owing to its low bioavailability in the brain, this promising drug candidate failed to clear the phase III clinical trial [19]. To maximize its beneficial pharmacological effects, tarenflurbil was nanoformulated as part of a nanoparticle and the intranasal delivery of this formulation was evaluated [20]. Compared to oral and intravenously administered free drug, the brain concentration of tarenflurbil administered as nanoparticles was significantly higher and remained elevated over a longer duration of time, supporting the potential use of this promising drug candidate in the future.

Improved pathophysiological understanding of AD has revealed the involvement of low-density lipoprotein (LDL) receptors in the uptake and clearance of amyloid beta in the brain [21]. A direct correlation between the expression levels of LDL receptors and the amyloid beta levels in the brain has been established wherein, over expression of LDL receptors causes increased clearance of amyloid beta and lack of LDL receptors is associated with amyloid deposition [22, 23]. A polyphenolic xanthone derivative from mangosteen, α-mangostin (α-M), was found to inhibit and dissociate the amyloid beta aggregation and attenuate neurotoxicity [24]. However, owing to this drug's low aqueous solubility, α-M suffers from low bioavailability upon oral administration [25]. To enhance the drug's bioavailability across BBB, and to maximize its efficacy in AD, α-M was encapsulated as nanoparticles in a recent study [26]. Administering α-M loaded nanoparticles to a rodent model of AD resulted in enhanced clearance of amyloid beta in a LDL receptor-dependent manner (summarized in Fig. 6.1). Moreover, increased delivery of α-M to the brain was associated with reduced neuroinflammatory responses and reversal of behavioral deficits in the rodent model for AD. In a similar approach, Apolipoprotein E (ApoE)-based biomimetic nanocarrier was employed to facilitate the delivery of α-M across the BBB [27]. This novel approach exploits the previously reported physiological preference of ApoE towards amyloid beta [28, 29]. The amyloid beta binding affinity of ApoE carrying biomimetic nanocarrier was able to deliver α-M and decrease amyloid deposition in the tested animal model. These results were also accompanied by memory improvement in the AD mouse model. In combination, these results have further highlighted the advantages of nanoencapsulating potent drugs for the potential treatment of AD.

2.3 *Amyotrophic Lateral Sclerosis*

Amyotrophic Lateral Sclerosis (ALS) occurs at a rate of 1.5–2.5 in 100,000 [30] and the pathophysiology of the disease has not been elucidated entirely [31]. Furthermore, there is currently only one approved treatment option, riluzole [32],

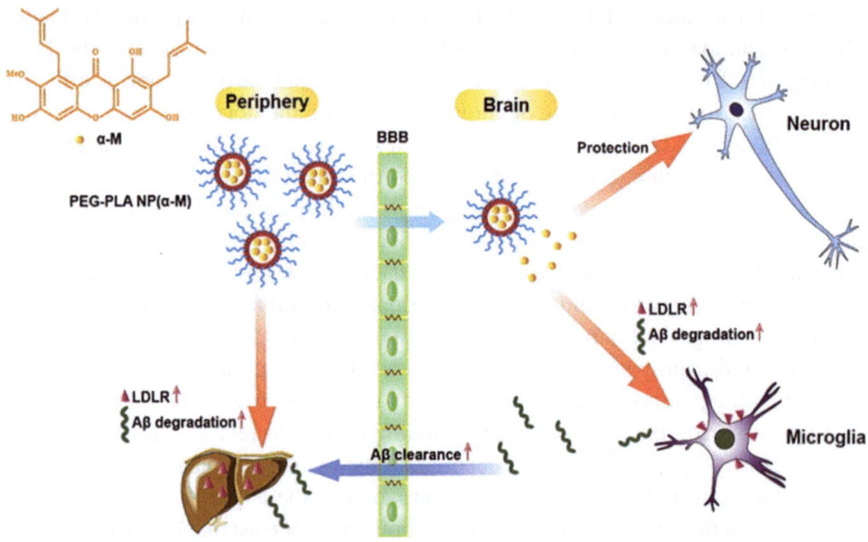

Fig. 6.1 A schematic representation of the therapeutic benefits of nanoformulating alpha-mangostin. Encapsulating the active molecule in a nanoformulation masked the low aqueous solubility of the compound and enhanced its CNS bioavailability. Overall, upon nanoformulation, alpha-mangostin was found to effectively clear the amyloid beta in brain through a LDL receptor-dependent mechanism. This image is reprinted from the Journal of Controlled Release, Volume No. 226 authored by Lei Yao, Xiao Gu, Qingxiang Song, Xiaolin Wang, Meng Huang, Meng Hu, Lina Hou, Ting Kang, Jun Chen, Hongzhuan Chen, and Xiaoling Ga, Nanoformulated alpha-mangostin ameliorates Alzheimer's disease neuropathology by elevating LDLR expression and accelerating amyloid-beta clearance, Pages 1–14, Copyright (2016), with permission from Elsevier (http://www.sciencedirect.com/science/article/pii/S0168365916300529)

which targets the excess glutamate release in CNS, which is believed to be neurotoxic and involved in the development of ALS [33]. Currently, the therapeutic effects of riluzole are effective, but can be improved upon. Several attempts to increase efficacy have been performed via nanoformulation of riluzole. For instance, riluzole-loaded solid lipid nanoparticles were able to increase the delivery of the drug to rat brains [34]. In addition, compared to traditional dosing of the drug, the nanoparticles imparted a brain-selective biodistribution of riluzole along with a low plasma drug concentration. A different nanoformulation of riluzole was prepared via covalent linkage of the drug to carbon nanotubules and this resulted in the rapid internalization of the drug [35]. Furthermore, this was accomplished in a non-toxic manner, in vitro, which can be a common challenge with the usage of carbon nanotubules [35]. Studies utilizing nanoformulation of agents other than riluzole have also been conducted. Wiley et al. demonstrated that targeting the toll-like receptor 4 (TLR4) of mice expressing mutant type SOD1 using a modified liposome with LPS [36] could carry minocycline and deliver it to the microglial cells. This was sequentially shown to delay disease progression in the animal model, while also decreasing TNF-alpha secretion. Interestingly, similar amounts of efficacy were noted in liposomes containing the LPS and without LPS, possibly indicating the liposome itself as being a cornerstone of the nanoformulation in this study.

2.4 Multiple Sclerosis

As is the common theme amongst neurodegenerative disorders, recent studies in the pathophysiology of multiple sclerosis (MS) have demonstrated the role of oxidative stress in the demyelination and neuronal damage [37]. In order to combat such oxidative stress, Binyamin et al. were able to enhance the antioxidant effects of naturally occurring pomegranate seed oil (PSO). Alone, PSO would require large daily doses to be effective, therefore it was demonstrated that a properly sized nano-droplet of PSO could reduce the burden of disease in an experimental autoimmune encephalomyelitis model of MS [38]. In the same animal model of MS, it was shown that a polymerized form of nano-curcumin was better able to elicit neuroprotective effects as both therapeutic and prophylaxis treatment [39]. In this study, a dendrosome nanoparticle was used to package the curcumin, with the key features of the dendrosome being that it was neutral, amphipathic, and biodegradable in nature [39]. Similar to other approaches, this nanoformulation was aimed to alter the pharmacokinetic properties and bioavailability of the treatment and optimize its efficacy.

3 Cancer: Delivering Chemotherapeutic Drugs Across the BBB

Glioma is widely recognized as one of the most lethal malignant brain tumor with a limited survival rate [40, 41]. While the rapid growth and metastasis of tumor presents the primary obstacle towards effective management of this cancer, the limited accessibility across the BBB hinders the drugs from attaining therapeutic concentrations in the brain. The limited accessibility of currently administered chemotherapeutic agents to brain tissue is often accompanied by non-specific toxicity directed towards peripheral tissues. To improve the clinical management of gliomas, nanotechnology-based novel drug delivery systems have been proposed as a possible solution for targeted delivery of chemotherapeutic drugs across the BBB [42, 43].

Several methods have been implemented in various in vitro and in vivo studies to facilitate the targeted delivery of nanoformulation, encapsulating the chemotherapeutic agents, across the BBB. Increased entry of nanoformulations under external stimuli is one such technique wherein the physical properties of nanomaterial is utilized to force the entry of drug-loaded nanoformulation across the physical BBB. Both high-intensity focused ultrasound [44] and magnetic targeting [45, 46] of nanoparticles have been proposed and tested for enhancing the delivery of drugs targeting gliomas across the BBB. A representative figure depicting the utilization of high-intensity focused ultrasound for targeted delivery of liposomal doxorubicin to tumor cells has been included below (Fig. 6.2). The other prominent approach to facilitate nanoformulation transport across BBB involves the

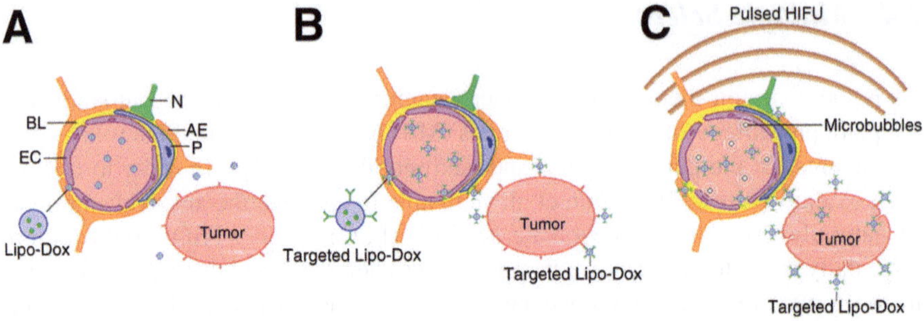

Fig. 6.2 Targeted delivery of liposomal doxorubicin to brain tumor in the presence of high-intensity focused ultrasound (HIFU). Compared to unconjugated liposome (**a**), the human atherosclerotic plaque-specific peptide-1 (AP-1)-conjugated liposomes (**b**) carries a marker for the targeted delivery of doxorubicin. The brain penetration of these conjugated liposomes was significantly enhanced in the presence of HIFU (**c**) providing a strategy for facilitating transport of nanoformulation across the blood-brain barrier. This image is reprinted from the Journal of Controlled Release, Volume No. 160, Issue 3, authored by Feng-Yi Yang, Tai-Tong Wong, Ming-Che Teng, Ren-Shyan Liu, Maggie Lu, Hsiang-Fa Liang, and Ming-Cheng Wei, Focused ultrasound and interleukin-4 receptor-targeted liposomal doxorubicin for enhanced targeted drug delivery and antitumor effect in glioblastoma multiforme, Pages 652–658, Copyright (2012), with permission from Elsevier (http://www.sciencedirect.com/science/article/pii/S0168365912001484)

receptor-mediated transcytosis. This approach takes advantage of the various receptors that are expressed on BBB including receptors for insulin, transferrin and folic acid. Nanoparticles are coated with ligands which can interact with these receptors and such interactions result in transcytosis of the nanoparticles across BBB [47, 48]. For instance, while free zoledronic acid is known to have no effect against glioblastoma, the cytotoxic effect of nanoformulated zoledronic acid, loaded in transferrin-labeled nanoparticles, in mice xenografted with glioblastoma was remarkable [49]. Free zoledronic acid tends to accumulate in bones, and limits the drug use to osteoporosis. Coating the nanoparticle surface with transferrin enhanced both the endocytosis of drug across BBB and preferential accumulation of zoledronic acid in transferrin-overexpressing glioblastoma cells.

Upon crossing the BBB, to further improve the delivery of drugs to gliomas, recent advances in nanotechnology have put forth the capability to include additional targeting tool within the nanoformulations. To encounter non-specific distribution of nanoparticles in the brain, Agrawal et al. targeted the specificity of integrins (αvβ3), which are overexpressed on glioma blood vessels, towards the arginyl–glycyl–aspartic acid (RGD) sequence in their dual targeted paclitaxel nanoformulation [50]. Similarly, targeting the fibronectins abundantly expressed in the glioma extracellular matrix by the cyclic decapeptide, CGLIIQKNEC (CLT1) [51], resulted in enhanced retention and uptake of paclitaxel loaded nanoparticles by the glioma [52]. To further improve tumor selectivity, recently, non-traditional tumor markers have also been targeted to increase the nanoparticle uptake in glioblastoma [53]. This study comprised of synthesizing nanoparticles that were coated with a novel peptide with dual targeting capabilities. A novel peptide, the Ft peptide, was synthesized by combining two peptides that individually targeted Tenascin C, a glycoprotein overexpressed in the extracellular matrix of glioblastomas [54], and Neuropilin-1, a non-tyrosine kinase receptor found in tumors [55]. Synthesizing Ft peptide provided a novel tool capable of interacting with two non-traditional targets which are overexpressed in glioblastomas. In animal studies, the dual-functionalized nanoparticles loaded with paclitaxel demonstrated remarkable accumulation of nanoparticles in gliomas and this observation translated into significant improvement in the mean survival time of treated animals.

In addition to serving as the physical barrier restricting the entry to xenobiotics into brain, the BBB has been shown to express various efflux transporters which can actively pump the drugs out of the brain [56]. Amongst these efflux transporters, the P-glycoprotein (P-gp) and multidrug-resistance-associated proteins (MRPs) are known to be expressed on the BBB and are held responsible for attributing resistance to chemotherapeutic approaches employed to treat gliomas [57–59]. In fact, P-gp mediated efflux has been determined as a prominent variable which differentiates CNS drugs from non-CNS drugs [60]. In this regard, nanoformulations have shown promising results in overcoming P-gp mediated efflux and providing substantial delivery of drug across BBB. Liu et al. have shown the enhanced ability of P123 micelles to deliver loaded drug across the BBB and overcoming P-gp mediated efflux [61]. An alternate approach, as reported by Tang et al. [62], involved loading of the active drug along with a known P-gp inhibitor in the nanoformulation to maximize the accumulation of nanoparticles in the glioma cells.

Temozolomide is an orally available DNA alkylating agent which is FDA approved for the treatment of newly diagnosed glioblastoma multiforme since 2005. Upon administration, temozolomide is activated non-enzymatically and methylates guanine residues causing DNA double strand breaks and subsequent apoptosis [63, 64]. However, resistance to temozolomide therapy is common and the therapeutic shortcomings are attributed to a variety of molecular mechanisms within glioblastomas [65, 66]. Nanoformulating temozolomide have provided encouraging results towards the treatment of glioblastoma multiforme in recent studies. These results further bolster the application of nanotechnology to improve the delivery of temozolomide across the BBB. In one recent study, for example, temozolomide was packaged in a liposome coated with single chain anti-transferrin antibody fragment and its efficacy in a rodent model of glioblastoma was evaluated [67]. Compared to the free drug, systemically administered liposome formulation of temozolomide showed enhanced bioavailability across the BBB, greater inhibition of tumor growth, and significant improvement in the survival of treated animals. The anti-transferrin receptor antibody on the surface of these temozolomide-loaded liposomes is rationalized for the greater accumulation of the drug across BBB followed by the improved delivery of drug inside the tumors (owing to the overexpression of transferrin receptors in glioblastoma tumors).

Another important factor that governs the clinical efficacy of current chemotherapeutic agent of choice for glioblastoma, temozolomide, is the degree of epigenetic silencing of the O[6]-methylguanine-DNA methyltransferase (MGMT) [68]. Resistance to temozolomide is attributed to the enhanced DNA repair protein MGMT. Studies have revealed that a higher degree of promoter methylation and subsequent silencing of the MGMT gene is associated with a greater drug efficacy and better survival rate in patients with glioblastoma [69, 70]. To achieve effective silencing of the MGMT gene, enhanced delivery of MGMT gene inhibitors across the BBB has been explored through the use of nanoformulations. One such inhibitor is the O[6]-benzylguanine (BG), which is known to directly inactivate MGMT, and enhance the cytotoxic effects of co-administered chemotherapeutic agents [71, 72]. As rationalized, nanoformulating BG, along with the choice of chemotherapeutic agent, was associated with an increased efficacy of anti-glioma treatment due to targeted delivery of BG across BBB and increased silencing of MGMT [73–75]. Similarly, nanoparticles carrying either the small interference RNA (siRNA) for silencing the MGMT gene or wild-type p53 plasmid, which can inhibit MGMT, were found to sensitize the glioblastoma to the co-delivered chemotherapeutic agents and enhance survival [76, 77]. A pictorial depiction of the above mentioned strategy for enhanced delivery of p53 gene has been included as Fig. 6.3. Overall, the combined approach of silencing/inactivating the effects of MGMT along with the efficient delivery of the chemotherapeutic agent using nanoformulations is a promising approach for treating glioblastomas.

In a different approach, validated in an in vitro model of glioblastoma cells, tumor cells were sensitized for temozolomide treatment through the inhibition of

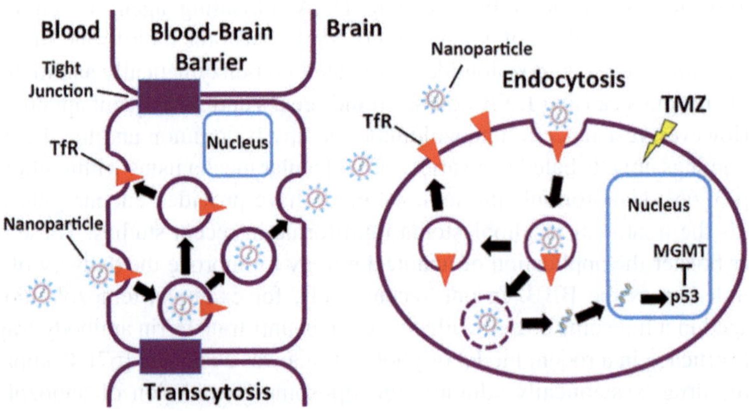

Fig. 6.3 Novel strategy to sensitize the temozolomide-resistant glioblastoma cells via enhanced delivery of nanoformulated tumor suppressor gene p53. Wild-type p53 gene was packaged within nanoparticles labeled with an antibody targeting the transferrin receptor. This ensured successfully transport of nanoparticles across the BBB into the glioblastoma cells. Enhanced delivery of p53 to cancer cells resulted in sensitization of these cells towards temozolomide and enhanced cytotoxicity. This image is reprinted from the published article by Sang-Soo Kim, Antonina Rait, Eric Kim, Kathleen F. Pirollo, Maki Nishida, Natalia Farkas, John A. Dagata, and Esther H. Chang, A Nanoparticle Carrying the p53 Gene Targets Tumors Including Cancer Stem Cells, Sensitizes Glioblastoma to Chemotherapy and Improves Survival, ACS Nano 2014, 8, 5494–5514, DOI: 10.1021/nn5014484 Copyright (2014), with permission from American Chemical Society (http://pubs.acs.org/doi/full/10.1021/nn5014484)

oncogenic microRNAs [78]. Given the fact that different microRNAs have been shown to attribute anti-apoptotic and pro-growth properties to glioblastomas [79–82], antisense oligonucleotides designed for two different microRNAs, antimiR-21 and antimiR-10b, were co-delivered to glioblastoma cells using nanoparticles [78]. As hypothesized, the co-delivery of antisense microRNA, targeted against two different anti-apoptotic factors, to glioblastoma cells conferred significantly higher sensitivity towards temozolomide treatment. A summarized depiction of this novel approach is included as Fig. 6.4. Interestingly, co-delivery of a chemotherapeutic agent along with a microRNA inhibitor, delivered in a nanoformulation, was also found to be effective in an animal model injected with glioblastoma cells [83]. In addition to improving the delivery of the anti-cancer drug, the microRNA inhibitors entrapped in the polymer micelles were found to escape lysosomal degradation thereby, achieving substantial inhibition of the anti-apoptotic microRNA. On the other hand, curcumin nanoformulation was reported to be effective in inducing apoptosis in glioblastoma cells via a microRNA, miR-145, dependent pathway [84]. Typically, the expression of miR-145 is maintained at a low level in tumor cells

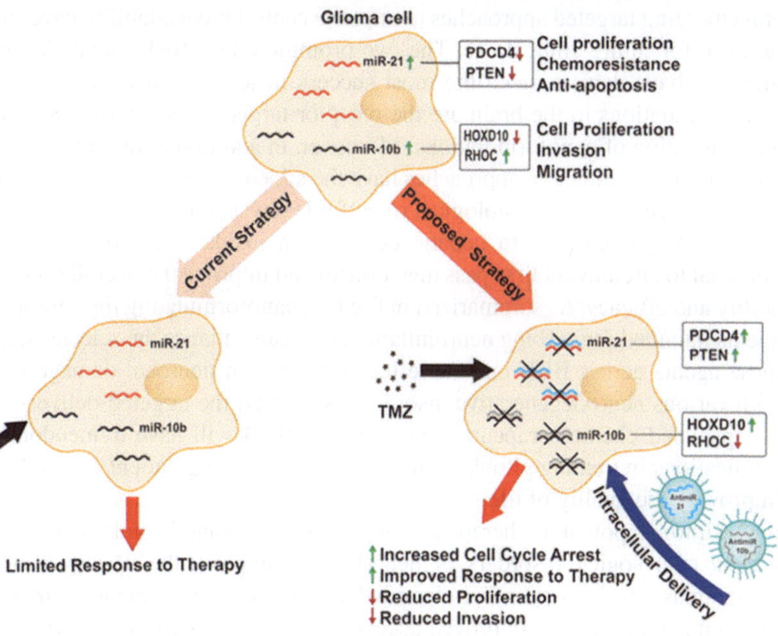

Fig. 6.4 The proposed strategy of blocking microRNAs that promote cancer growth and survival to increase the effectiveness of anti-tumor therapy. Nanoparticles were designed to deliver antisense microRNAs and made glioblastomas more susceptible to temozolomide. This image is reprinted with permission from Jeyarama S. Ananta, Ramasamy Paulmurugan, and Tarik F. Massoud, Tailored Nanoparticle Codelivery of antimiR-21 and antimiR-10b Augments Glioblastoma Cell Kill by Temozolomide: Toward a "Personalized" Anti-microRNA Therapy, Mol. Pharmaceutics, 2016, 13 (9), pp. 3164–3175. Copyright (2016) American Chemical Society (http://pubs.acs.org/doi/full/10.1021/acs.molpharmaceut.6b00388)

since this microRNA is known to inhibit cell growth and metastasis [85, 86]. By inducing the expression of miR-145, the curcumin-loaded nanoformulation was found to induce intracellular changes that led to growth inhibition and apoptosis in glioblastoma cells. Based on the promising observations from these studies, further evaluation of microRNA inhibitor/activator based strategies may tremendously improve the therapeutic efficacy of temozolomide and other chemotherapeutic agents in patients with glioblastoma.

4 The Future Therapeutic Implications of Nanoformulations

While the biologically active agent being delivered can range from chemical compounds to peptides, DNA, and microRNAs, there are several ways in which the pharmacokinetic properties of drugs can be improved by the use of nanotechnology. Some of the tried and tested physicochemical alterations to improve nanoparticle bioavailability across the BBB include the use of polymer coatings, carbon nanotubules, and droplet size optimization. Beyond the physicochemical properties, several emerging targeted approaches to improve central bioavailability have shown the promise for future clinical use. The two prominent non-BBB manipulation strategies that have demonstrated the most success in terms of delivering drug-loaded nanoformulations to the brain are the receptor-targeted endocytosis and intranasal administration of nanoformulations. Moreover, in addition to improving the central bioavailability, these two approaches have the added advantage of not disrupting the BBB and keeping the physiological role of this physical barrier intact. Also, encapsulating drugs sensitive to hepatic degradation inside nanoformulations has the potential to circumvent first-pass metabolism and improve the overall drug bioavailability and efficacy. As summarized in Fig.6.5, nanoformulating the neuroprotective agents intended for curbing neuroinflammation can enhance the selected delivery of these agents across BBB and can effectively abolish neuronal damage associated with various neurodegenerative disorders. Similarly, the targeted-delivery of nano-encapsulated chemotherapeutic agents across BBB will have tremendous clinical implications in treating glioblastomas thereby, increasing patient survival rates and improving the quality of life.

Additional potential therapeutic application of nanoformulations, in treating various neurological disorders, is their ability to improve the efficacy of promising compounds which were earlier classified as sub-therapeutic due to their poor pharmacokinetic properties. Improved access of drugs across BBB, as part of nanoformulations, has the potential to revive the drug development process of compounds with otherwise poor clinical acceptability. In addition, increased bioavailability of drugs in brain, compared to other organs, will have a substantial impact on reducing the adverse effects of existing and new drugs designed for the treatment of various neurological disorders. By achieving organ selectivity and higher bioavailability, CNS-targeting nanoformulations are expected to eliminate some of the major issues associated with drug development.

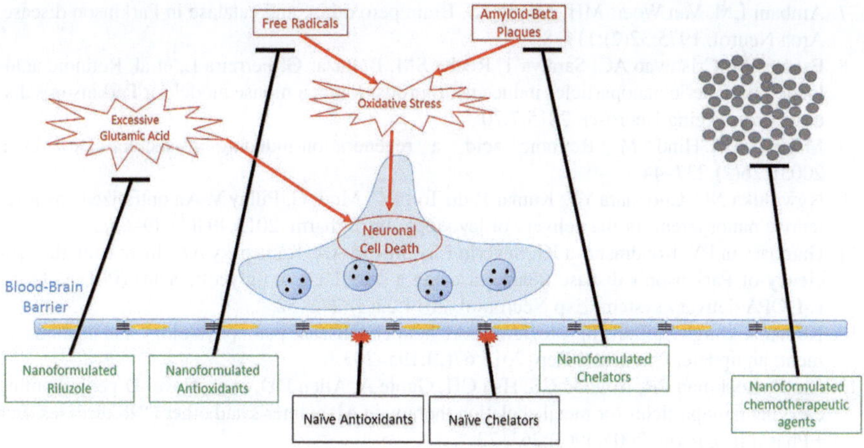

Fig. 6.5 Targeted-nanoformulations for treating neurological disorders. Encapsulating neuroprotective agents or chemotherapeutic drugs as nanoformulation provides a viable solution for enhancing drug delivery across the blood-brain barrier (BBB) . Reducing oxidative stress can directly attenuate neuronal damage associated with neurodegenerative diseases. Increased bioavailability of chemotherapeutic drugs in the brain, on the other hand, can improve survival rates in patients with glioblastoma

Overall, the recently developed and upcoming brain-targeting nanoformulations have tremendous scope for future clinical applications. Optimizing the drug loading and release parameters and further technological improvement in the field will help translate these promising nanoformulations from in vivo to clinical use.

Acknowledgment The authors would like to thank the original authors and Elsevier for providing the copyrights permission to use Figs. 6.1 and 6.2 included in this book chapter. The authors would also like to thank the original authors and the American Chemical Society for providing the copyrights permission to use Fig. 6.3 and Fig. 6.4.

References

1. Weissig V, Pettinger TK, Murdock N. Nanopharmaceuticals (part 1): products on the market. Int J Nanomedicine. 2014;9:4357–73.
2. Weissig V, Guzman-Villanueva D. Nanopharmaceuticals (part 2): products in the pipeline. Int J Nanomedicine. 2015;10:1245–57.
3. Devalapally H, Chakilam A, Amiji MM. Role of nanotechnology in pharmaceutical product development. J Pharm Sci. 2007;96(10):2547–65.
4. Amor S, Puentes F, Baker D, van der Valk P. Inflammation in neurodegenerative diseases. Immunology. 2010;129(2):154–69.
5. Fahn S, Cohen G. The oxidant stress hypothesis in Parkinson's disease: evidence supporting it. Ann Neurol. 1992;32(6):804–12.
6. Brynskikh AM, Zhao Y, Mosley RL, Li S, Boska MD, Klyachko NL, et al. Macrophage delivery of therapeutic nanozymes in a murine model of Parkinson's disease. Nanomedicine (Lond). 2010;5(3):379–96.

7. Ambani LM, Van Woert MH, Murphy S. Brain peroxidase and catalase in Parkinson disease. Arch Neurol. 1975;32(2):114–8.
8. Esteves M, Cristovao AC, Saraiva T, Rocha SM, Baltazar G, Ferreira L, et al. Retinoic acid-loaded polymeric nanoparticles induce neuroprotection in a mouse model for Parkinson's disease. Front Aging Neurosci. 2015;7:20.
9. Maden M, Hind M. Retinoic acid, a regeneration-inducing molecule. Dev Dyn. 2003;226(2):237–44.
10. Ngwuluka NC, Choonara YE, Kumar P, du Toit LC, Modi G, Pillay V. An optimized gastroretentive nanosystem for the delivery of levodopa. Int J Pharm. 2015;494(1):49–65.
11. Gambaryan PY, Kondrasheva IG, Severin ES, Guseva AA, Kamensky AA. Increasing the efficiency of Parkinson's disease treatment using a poly(lactic-co-glycolic acid) (PLGA) based L-DOPA delivery system. Exp Neurobiol. 2014;23(3):246–52.
12. Kumar A, Singh A, Ekavali. A review on Alzheimer's disease pathophysiology and its management: an update. Pharmacol Rep. 2015;67(2):195–203.
13. Cui Z, Lockman PR, Atwood CS, Hsu CH, Gupte A, Allen DD, et al. Novel D-penicillamine carrying nanoparticles for metal chelation therapy in Alzheimer's and other CNS diseases. Eur J Pharm Biopharm. 2005;59(2):263–72.
14. Koziara JM, Lockman PR, Allen DD, Mumper RJ. In situ blood-brain barrier transport of nanoparticles. Pharm Res. 2003;20(11):1772–8.
15. Sriramoju B, Kanwar RK, Kanwar JR. Nanoformulated mutant SurR9-C84A: a possible key for Alzheimer's and its associated inflammation. Pharm Res. 2015;32(8):2787–97.
16. Xu S, Chen M, Yao Y, Zhang Z, Jin T, Huang Y, et al. Novel poly(ethylene imine) biscarbamate conjugate as an efficient and nontoxic gene delivery system. J Control Release. 2008;130(1):64–8.
17. Chonpathompikunlert P, Wattanathorn J, Muchimapura S. Piperine, the main alkaloid of Thai black pepper, protects against neurodegeneration and cognitive impairment in animal model of cognitive deficit like condition of Alzheimer's disease. Food Chem Toxicol. 2010;48(3):798–802.
18. Elnaggar YS, Etman SM, Abdelmonsif DA, Abdallah OY. Intranasal Piperine-loaded chitosan nanoparticles as brain-targeted therapy in Alzheimer's disease: optimization, biological efficacy, and potential toxicity. J Pharm Sci. 2015;104(10):3544–56.
19. Green RC, Schneider LS, Amato DA, Beelen AP, Wilcock G, Swabb EA, et al. Effect of tarenflurbil on cognitive decline and activities of daily living in patients with mild Alzheimer disease: a randomized controlled trial. JAMA. 2009;302(23):2557–64.
20. Muntimadugu E, Dhommati R, Jain A, Challa VG, Shaheen M, Khan W. Intranasal delivery of nanoparticle encapsulated tarenflurbil: a potential brain targeting strategy for Alzheimer's disease. Eur J Pharm Sci. 2016;92:224–34.
21. Basak JM, Verghese PB, Yoon H, Kim J, Holtzman DM. Low-density lipoprotein receptor represents an apolipoprotein E-independent pathway of Abeta uptake and degradation by astrocytes. J Biol Chem. 2012;287(17):13959–71.
22. Castellano JM, Deane R, Gottesdiener AJ, Verghese PB, Stewart FR, West T, et al. Low-density lipoprotein receptor overexpression enhances the rate of brain-to-blood Abeta clearance in a mouse model of beta-amyloidosis. Proc Natl Acad Sci USA. 2012;109(38):15502–7.
23. Katsouri L, Georgopoulos S. Lack of LDL receptor enhances amyloid deposition and decreases glial response in an Alzheimer's disease mouse model. PLoS One. 2011;6(7):e21880.
24. Wang Y, Xia Z, Xu JR, Wang YX, Hou LN, Qiu Y, et al. Alpha-mangostin, a polyphenolic xanthone derivative from mangosteen, attenuates beta-amyloid oligomers-induced neurotoxicity by inhibiting amyloid aggregation. Neuropharmacology. 2012;62(2):871–81.
25. Li L, Brunner I, Han AR, Hamburger M, Kinghorn AD, Frye R, et al. Pharmacokinetics of alpha-mangostin in rats after intravenous and oral application. Mol Nutr Food Res. 2011;55(Suppl 1):S67–74.
26. Yao L, Gu X, Song Q, Wang X, Huang M, Hu M, et al. Nanoformulated alpha-mangostin ameliorates Alzheimer's disease neuropathology by elevating LDLR expression and accelerating amyloid-beta clearance. J Control Release. 2016;226:1–14.

27. Song Q, Song H, Xu J, Huang J, Hu M, Gu X, et al. Biomimetic ApoE-reconstituted high density lipoprotein Nanocarrier for blood-brain barrier penetration and amyloid Beta-targeting drug delivery. Mol Pharm. 2016;13:3976.
28. Patil SP, Ballard R, Sanchez S, Osborn J, Santangelo D Jr. ApoE: the link between Alzheimer's-related glucose hypometabolism and Abeta deposition? Med Hypotheses. 2012;78(4):494–6.
29. Thal DR, Capetillo-Zarate E, Schultz C, Rub U, Saido TC, Yamaguchi H, et al. Apolipoprotein E co-localizes with newly formed amyloid beta-protein (Abeta) deposits lacking immunoreactivity against N-terminal epitopes of Abeta in a genotype-dependent manner. Acta Neuropathol. 2005;110(5):459–71.
30. Zaldivar T, Gutierrez J, Lara G, Carbonara M, Logroscino G, Hardiman O. Reduced frequency of ALS in an ethnically mixed population: a population-based mortality study. Neurology. 2009;72(19):1640–5.
31. Gordon PH. Amyotrophic lateral sclerosis: pathophysiology, diagnosis and management. CNS Drugs. 2011;25(1):1–15.
32. Miller RG, Jackson CE, Kasarskis EJ, England JD, Forshew D, Johnston W, et al. Practice parameter update: the care of the patient with amyotrophic lateral sclerosis: drug, nutritional, and respiratory therapies (an evidence-based review): report of the quality standards Subcommittee of the American Academy of neurology. Neurology. 2009;73(15):1218–26.
33. Doble A. The pharmacology and mechanism of action of riluzole. Neurology. 1996;47(Issue 6, Supplement 4):233S–41S.
34. Bondi ML, Craparo EF, Giammona G, Drago F. Brain-targeted solid lipid nanoparticles containing riluzole: preparation, characterization and biodistribution. Nanomedicine (Lond). 2010;5(1):25–32.
35. Chigumbu N, Iyuke S, Pillay V, Ndlovu S. In vitro evaluation of the physicochemical effects of drug loaded carbon nanotubes on toxicity. J Nanomed Nanotechnol. 2012;03(04):135.
36. Wiley N, Madhankumar A, Mitchell R, Neely E, Rizk E, Douds G, et al. Lipopolysaccharide modified liposomes for Amyotropic lateral sclerosis therapy: efficacy in SOD1 mouse model. Adv Nanopart. 2012;01:44–53.
37. Haider L, Fischer MT, Frischer JM, Bauer J, Hoftberger R, Botond G, et al. Oxidative damage in multiple sclerosis lesions. Brain. 2011;134(7):1914–24.
38. Binyamin O, Larush L, Frid K, Keller G, Friedman-Levi Y, Ovadia H, et al. Treatment of a multiple sclerosis animal model by a novel nanodrop formulation of a natural antioxidant. Int J Nanomedicine. 2015;10:7165–74.
39. Mohajeri M, Sadeghizadeh M, Najafi F, Javan M. Polymerized nano-curcumin attenuates neurological symptoms in EAE model of multiple sclerosis through down regulation of inflammatory and oxidative processes and enhancing neuroprotection and myelin repair. Neuropharmacology. 2015;99:156–67.
40. Laws ER Jr, Thapar K. Brain tumors. CA Cancer J Clin. 1993;43(5):263–71.
41. Mesti T, Ocvirk J. Malignant gliomas: old and new systemic treatment approaches. Radiol Oncol. 2016;50(2):129–38.
42. Chen Y, Liu L. Modern methods for delivery of drugs across the blood-brain barrier. Adv Drug Deliv Rev. 2012;64(7):640–65.
43. Dinda SC, Pattnaik G. Nanobiotechnology-based drug delivery in brain targeting. Curr Pharm Biotechnol. 2013;14(15):1264–74.
44. Yang FY, Wong TT, Teng MC, Liu RS, Lu M, Liang HF, et al. Focused ultrasound and interleukin-4 receptor-targeted liposomal doxorubicin for enhanced targeted drug delivery and antitumor effect in glioblastoma multiforme. J Control Release. 2012;160(3):652–8.
45. Shevtsov MA, Multhoff G. Recent developments of magnetic nanoparticles for theranostics of brain tumor. Curr Drug Metab. 2016;17:737–44.
46. Yi GQ, Gu B, Chen LK. The safety and efficacy of magnetic nano-iron hyperthermia therapy on rat brain glioma. Tumour Biol. 2014;35(3):2445–9.
47. Ulbrich K, Knobloch T, Kreuter J. Targeting the insulin receptor: nanoparticles for drug delivery across the blood-brain barrier (BBB). J Drug Target. 2011;19(2):125–32.

48. Ulbrich K, Hekmatara T, Herbert E, Kreuter J. Transferrin- and transferrin-receptor-antibody-modified nanoparticles enable drug delivery across the blood-brain barrier (BBB). Eur J Pharm Biopharm. 2009;71(2):251–6.
49. Salzano G, Zappavigna S, Luce A, D'Onofrio N, Balestrieri ML, Grimaldi A, et al. Transferrin-targeted nanoparticles containing Zoledronic acid as a potential tool to inhibit glioblastoma growth. J Biomed Nanotechnol. 2016;12(4):811–30.
50. Agrawal U, Chashoo G, Sharma PR, Kumar A, Saxena AK, Vyas SP. Tailored polymer-lipid hybrid nanoparticles for the delivery of drug conjugate: dual strategy for brain targeting. Colloids Surf B Biointerfaces. 2015;126:414–25.
51. Pilch J, Brown DM, Komatsu M, Jarvinen TA, Yang M, Peters D, et al. Peptides selected for binding to clotted plasma accumulate in tumor stroma and wounds. Proc Natl Acad Sci USA. 2006;103(8):2800–4.
52. Zhang B, Shen S, Liao Z, Shi W, Wang Y, Zhao J, et al. Targeting fibronectins of glioma extracellular matrix by CLT1 peptide-conjugated nanoparticles. Biomaterials. 2014;35(13):4088–98.
53. Kang T, Zhu Q, Jiang D, Feng X, Feng J, Jiang T, et al. Synergistic targeting tenascin C and neuropilin-1 for specific penetration of nanoparticles for anti-glioblastoma treatment. Biomaterials. 2016;101:60–75.
54. Leins A, Riva P, Lindstedt R, Davidoff MS, Mehraein P, Weis S. Expression of tenascin-C in various human brain tumors and its relevance for survival in patients with astrocytoma. Cancer. 2003;98(11):2430–9.
55. Ellis LM. The role of neuropilins in cancer. Mol Cancer Ther. 2006;5(5):1099–107.
56. Golden PL, Pollack GM. Blood-brain barrier efflux transport. J Pharm Sci. 2003;92(9):1739–53.
57. Agarwal S, Sane R, Oberoi R, Ohlfest JR, Elmquist WF. Delivery of molecularly targeted therapy to malignant glioma, a disease of the whole brain. Expert Rev Mol Med. 2011;13:e17.
58. Haimeur A, Conseil G, Deeley RG, Cole SP. The MRP-related and BCRP/ABCG2 multidrug resistance proteins: biology, substrate specificity and regulation. Curr Drug Metab. 2004;5(1):21–53.
59. Thiebaut F, Tsuruo T, Hamada H, Gottesman MM, Pastan I, Willingham MC. Immunohistochemical localization in normal tissues of different epitopes in the multidrug transport protein P170: evidence for localization in brain capillaries and crossreactivity of one antibody with a muscle protein. J Histochem Cytochem. 1989;37(2):159–64.
60. Mahar Doan KM, Humphreys JE, Webster LO, Wring SA, Shampine LJ, Serabjit-Singh CJ, et al. Passive permeability and P-glycoprotein-mediated efflux differentiate central nervous system (CNS) and non-CNS marketed drugs. J Pharmacol Exp Ther. 2002;303(3):1029–37.
61. Liu JS, Wang JH, Zhou J, Tang XH, Xu L, Shen T, et al. Enhanced brain delivery of lamotrigine with Pluronic((R)) P123-based nanocarrier. Int J Nanomedicine. 2014;9:3923–35.
62. Tang B, Fang G, Gao Y, Liu Y, Liu J, Zou M, et al. Lipid-albumin nanoassemblies co-loaded with borneol and paclitaxel for intracellular drug delivery to C6 glioma cells with P-gp inhibition and its tumor targeting. Asian J Pharm Sci. 2015;10(5):363–71.
63. Danson SJ, Middleton MR. Temozolomide: a novel oral alkylating agent. Expert Rev Anticancer Ther. 2001;1(1):13–9.
64. D'Atri S, Tentori L, Lacal PM, Graziani G, Pagani E, Benincasa E, et al. Involvement of the mismatch repair system in temozolomide-induced apoptosis. Mol Pharmacol. 1998;54(2):334–41.
65. Chamberlain MC. Temozolomide: therapeutic limitations in the treatment of adult high-grade gliomas. Expert Rev Neurother. 2010;10(10):1537–44.
66. Ramirez YP, Weatherbee JL, Wheelhouse RT, Ross AH. Glioblastoma multiforme therapy and mechanisms of resistance. Pharmaceuticals (Basel). 2013;6(12):1475–506.
67. Kim SS, Rait A, Kim E, DeMarco J, Pirollo KF, Chang EH. Encapsulation of temozolomide in a tumor-targeting nanocomplex enhances anti-cancer efficacy and reduces toxicity in a mouse model of glioblastoma. Cancer Lett. 2015;369(1):250–8.
68. Hegi ME, Diserens AC, Gorlia T, Hamou MF, de Tribolet N, Weller M, et al. MGMT gene silencing and benefit from temozolomide in glioblastoma. N Engl J Med. 2005;352(10):997–1003.

69. Donson AM, Addo-Yobo SO, Handler MH, Gore L, Foreman NK. MGMT promoter methylation correlates with survival benefit and sensitivity to temozolomide in pediatric glioblastoma. Pediatr Blood Cancer. 2007;48(4):403–7.
70. Idbaih A, Omuro A, Ducray F, Hoang-Xuan K. Molecular genetic markers as predictors of response to chemotherapy in gliomas. Curr Opin Oncol. 2007;19(6):606–11.
71. Kokkinakis DM, Ahmed MM, Chendil D, Moschel RC, Pegg AE. Sensitization of pancreatic tumor xenografts to carmustine and temozolomide by inactivation of their O6-Methylguanine-DNA methyltransferase with O6-benzylguanine or O6-benzyl-2′-deoxyguanosine. Clin Cancer Res. 2003;9(10 Pt 1):3801–7.
72. Rhines LD, Sampath P, Dolan ME, Tyler BM, Brem H, Weingart J. O6-benzylguanine potentiates the antitumor effect of locally delivered carmustine against an intracranial rat glioma. Cancer Res. 2000;60(22):6307–10.
73. Liu C, Yao S, Li X, Wang F, Jiang Y. iRGD-mediated core-shell nanoparticles loading carmustine and O6-benzylguanine for glioma therapy. J Drug Target. 2016;25:235–46.
74. Qian L, Zheng J, Wang K, Tang Y, Zhang X, Zhang H, et al. Cationic core-shell nanoparticles with carmustine contained within O(6)-benzylguanine shell for glioma therapy. Biomaterials. 2013;34(35):8968–78.
75. Stephen ZR, Kievit FM, Veiseh O, Chiarelli PA, Fang C, Wang K, et al. Redox-responsive magnetic nanoparticle for targeted convection-enhanced delivery of O6-benzylguanine to brain tumors. ACS Nano. 2014;8(10):10383–95.
76. Kim SS, Rait A, Kim E, Pirollo KF, Nishida M, Farkas N, et al. A nanoparticle carrying the p53 gene targets tumors including cancer stem cells, sensitizes glioblastoma to chemotherapy and improves survival. ACS Nano. 2014;8(6):5494–514.
77. Yoo B, Ifediba MA, Ghosh S, Medarova Z, Moore A. Combination treatment with theranostic nanoparticles for glioblastoma sensitization to TMZ. Mol Imaging Biol. 2014;16(5):680–9.
78. Ananta JS, Paulmurugan R, Massoud TF. Tailored nanoparticle codelivery of antimiR-21 and antimiR-10b augments glioblastoma cell kill by Temozolomide: toward a "personalized" anti-microRNA therapy. Mol Pharm. 2016;13(9):3164–75.
79. Chan JA, Krichevsky AM, Kosik KS. MicroRNA-21 is an antiapoptotic factor in human glioblastoma cells. Cancer Res. 2005;65(14):6029–33.
80. Ciafre SA, Galardi S, Mangiola A, Ferracin M, Liu CG, Sabatino G, et al. Extensive modulation of a set of microRNAs in primary glioblastoma. Biochem Biophys Res Commun. 2005;334(4):1351–8.
81. Gabriely G, Yi M, Narayan RS, Niers JM, Wurdinger T, Imitola J, et al. Human glioma growth is controlled by microRNA-10b. Cancer Res. 2011;71(10):3563–72.
82. Karsy M, Arslan E, Moy F. Current progress on understanding MicroRNAs in glioblastoma multiforme. Genes Cancer. 2012;3(1):3–15.
83. Qian X, Long L, Shi Z, Liu C, Qiu M, Sheng J, et al. Star-branched amphiphilic PLA-b-PDMAEMA copolymers for co-delivery of miR-21 inhibitor and doxorubicin to treat glioma. Biomaterials. 2014;35(7):2322–35.
84. Tahmasebi Mirgani M, Isacchi B, Sadeghizadeh M, Marra F, Bilia AR, Mowla SJ, et al. Dendrosomal curcumin nanoformulation downregulates pluripotency genes via miR-145 activation in U87MG glioblastoma cells. Int J Nanomedicine. 2014;9:403–17.
85. Sachdeva M, Mo YY. miR-145-mediated suppression of cell growth, invasion and metastasis. Am J Transl Res. 2010;2(2):170–80.
86. Sachdeva M, Mo YY. MicroRNA-145 suppresses cell invasion and metastasis by directly targeting mucin 1. Cancer Res. 2010;70(1):378–87.

Chapter 7
Targeted Drug Delivery for Personalized Cure

Rashmi Chaudhari and Abhijeet Joshi

Abstract Human population bears a similarity to an extent of 99.9%, however mere 0.1% of variability also creates large differences in personal make-up owing to several hereditary and environmental factors. Conventionally, development of healthcare modalities like diagnostics and therapeutics has been by mass production assuming uniformity in characteristics of recipient population. Due to the inherent variability in each individual clinical response to diagnostic and therapeutic products vary greatly ranging from inefficient therapeutic compliance, adverse reactions and hypersensitivity reactions. Personalization of medications is the need of hour to avoid perils of both modalities so that efficient healthcare is provided to human population. Advances in delivery like targeted drug formulations, companion diagnostics and triggered drug formulations provide necessary dimensions to personalized medicines. Formulations with these capabilities can be described as theranostics and have been instrumental in this decade to establish the field of personalized medicine. The current chapter reviews different types, mechanisms of targeted and triggered drug delivery systems that have helped to improve the diagnostic and therapeutic compliance in disease management.

Keywords Targeted drug delivery • Passive targeting • Active targeting • Triggered delivery

---

R. Chaudhari
Department of Biosciences and Bioengineering, Indian Institute of Technology, Mumbai 400076, India

A. Joshi (✉)
Centre for Biosciences and Biomedical Engineering, Indian Institute of Technology Indore, Indore 453552, India
e-mail: abhijeet.joshi@iiti.ac.in

A. Kaushik et al. (eds.), *Advances in Personalized Nanotherapeutics*,
https://doi.org/10.1007/978-3-319-63633-7_7

1 Introduction to Personalized Medicine

Personalized medicine as a new development in health care is the result of the inherent 0.1% variations in human genomic structure. The small variation leads to huge differences some of which are harmless (e.g. phenotypic changes) and others can be harmful (e.g. causing various diseases like diabetes, cancer, heart diseases etc.). Disease management which is the primary goal of any healthcare professional have been facing several challenges due to these variations between individuals. Management of several diseases have shown significant advances in recent times due to the developments in molecular, genetic, genomic, proteomic and clinical understandings. Use of such information aids in development and administration of new patient oriented therapeutic regimens depending on the stage of disease, genomics of the patients and response of patients to the therapy. Personalized medicine can be understood as congregate of methodologies which aim at improving the functionality of diagnostic and therapeutic systems in order to efficiently treat diseases and disorders. It comprises of new therapeutic regimens which include novel approaches of delivery of conventional drugs, or delivery of new drugs in a spatially or temporally controlled manner, using different diagnostic techniques as companion to drug delivery and formulations for stimuli responsive delivery. Scientists believe that personalized medicine will bring a paradigm shift in economics of healthcare industry in coming years and will replace the conventional drug delivery systems which are now perceived unsuitable due to their generic usage. Further, usage of mass produced drug delivery systems on entire population with a large diversity in genetic makeup, disease status and patient responsiveness has been proved to be unsuitable. Instead the individualized drug therapy seems to be a more suitable option to move ahead to reduce the overall fate of adverse effects, cost and time of therapy, therapeutic compliance etc. The current chapter deals with approaches and mechanisms of developing personalized medicines using targeted/triggered delivery systems that help improve diagnostic and therapeutic compliance.

2 Personalized Medicine: History, Definition and Economics

A few decades of research in the field of personalized medicine still has not provided a concrete definition. Personalized medicine offers the ability to offer right kind of drug, to the right patient for the right disease with right dosage. The study of genetic and metabolic data influences the way drugs behaves inside the body with respect to the pharmacokinetics or pharmacodynamics. Several terms which are included in the purview of personalized medicine as described by Jain K. are customized drug therapy, genomic medicine, genotype-based therapy, individualized medicine or individual based therapy, information based medicine, integrated healthcare, pharmacogenomics/pharmacogenetics/pharmaco-proteomics/pharmaco-metabolomics, precision medicine, rational drug selection, stratified medicine, systems medicine, tailored therapy and translational medicine [1].

Traditionally the delivery of drug was believed to be based on the concept that *"one drug-fits all"*. Evaluation of drugs for their physico-chemical properties and clinical manifestations was determined as an average of multiple data and did not consider the individualized variation that patients inherit. Although individualization of certain treatments has been performed to some extent by modifying the dose of drug based on race, ethnicity, body weight, age and condition of diseases or localization of drug to a site of requirement by modifying physico-chemical properties but a comprehensive understanding and experimentations are still required in order to move towards the ultimate goal of personalized medicine. It is estimated that about half of all the patients fail to respond to drugs they are prescribed by physicians. The statistics of drugs not working in patients accounts for Antidepressants 38%, Anti-asthmatics 40% Anti-arthritic drugs 50% and anti-cancer drugs 75%. These percentages suggest that huge expenditures get wasted ($350 billion in 2011) which brings great limitations for efficiency of treatment in resource poor settings. Increasing focus in personalized medicine can be validated by the fact that 35% of new oncology drugs are categorized in this class have been approved by FDA [2].

Among several advantages of personalized medicine, patient and clinician oriented advantages include high chances of desired therapeutic effect and low chances of side effects, possibility of employing preventive strategies, focussed therapy, reduced cost and better therapeutic compliance in patients. Industry will also face following advantages by developing personalized medicines like increased efficiency and reduce cost of target and lead discovery in reduced time periods. Cost of clinical trial will also decrease and industries can get advantage of product differentiation in the market. The basic challenge in understanding the economics of personalized medicine is to weigh patient's benefit, financial cost and clinical value for biomarker-based diagnostics. Such biomarker-based diagnostics will improve the research productivity by reducing clinical trial sizes, enhancing the speed to reach the market and controlling the higher drug prices. According to a recent news, pharmaceutical companies save about 130 million USD per approved drug molecule using companion diagnostics. Cancer and Infectious disease may well be the primary beneficiaries of personalized medicine formulations.

3 Personalized Medicine and Drug Discovery

Traditionally the drug discovery process was based on serendipitous identification of active ingredients from traditional remedies. However with increased understanding of patho-physiological processes and information at the atomic/molecular level concept of rational drug design has emerged. Along with the serendipitous discovery, currently used drug discovery methods include random screening, molecular manipulation, rational drug designing and use of drug metabolites.

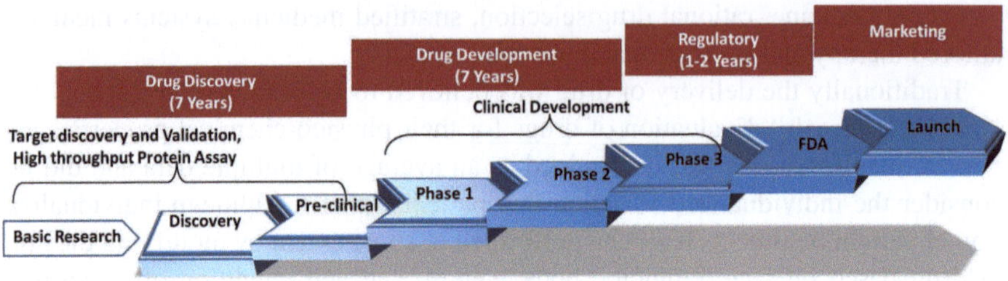

Fig. 7.1 Drug discovery process and time line

Rational drug design and discovery along with the knowledge of specific biomarkers has now become the mainstay for deciding upon the diagnosis and therapy of patients. The drug discovery timeline shows that for a drug molecule to be approved and reach market takes about 15 years (Fig. 7.1). Further to that the process of drug discovery has been a costly affair, as per one of the statistics in 2003 about USD 500–800 million were the costs for drug discovery companies to develop new pharmaceutical products. The great part of expenses go in vain as 75% of the products fail in the drug development stage.

Drug discovery was a conservative process based on screening of chemically synthesized molecules and testing them for therapeutic activity without any inputs from the clinical developments. Such a process was time consuming and labour intensive. The approach was useful when researchers and industry did not know the drug targets. According to a rough estimate only 483 targets (which include cell receptors, enzymes, hormones, ion channels, nuclear receptors and DNA) account for all the drugs, which are commercially available. Knowledge about the targets has remained the main holdup of drug discovery process [3]. Several approaches of target discovery, biomarker discovery, and new drug development strategies will bring several loops in the drug discovery process which will provide feedback to each of the stages of development. Further, biomarker discovery and validation should be included well in advance in the drug discovery cycle so that these biomarkers can be used in during clinical trial for monitoring of progression of treatment. Genomic technologies have been successful in identifying genes regulated/co-regulated with drugs and both targets and co-regulated genes could be possible replacement biomarkers for pre-clinical and clinical studies [4]. For example gene encoding leptin which acts as a controller of bodily lipids was discovered using genomic methodologies which has been proven to be target and a biomarker for drug mediated weight gain [4]. Bioinformatics, proteomics and animal models can be used as validating tools before using chemical compounds for studying the effectiveness. Toxico-genomic markers can screen chemical compounds reaching to human patients to reduce the chances of adverse drug reactions detected after marketing of drugs. Pharmaco-dynamic and pharmaco-genomic markers can be used for study of drug toxicity to various organs like liver, kidney, heart or bone marrows in different phases of clinical trials to assess the safety and efficacy. Precautionary therapies have been fully incorporated by the medical fraternity in clinical practice e.g. use of selective estrogen-receptor modulators (SERMs) for osteoporosis and use of statins to prevent coronary artery disease [5]. High-risk people should be intermittently screened for preclinical ailment detection. Sub-optimal treatment or adverse drug reactions can be continuously monitored using molecular markers. In addition, targeted therapeutic molecule based therapy along with companion biomarker detection system, as companion diagnostics will open new options to continuously monitor the disease conditions.

4 Targeted Drug Delivery Systems

Targeted drug delivery is a method of administering medications to patients in a way that the medication reaches to a particular part of the body to a greater extent than to others. The main aim of such delivery mode is to concentrate the medication to a particular site with a view to improve the therapeutic effect or reduce the adverse effects. Several formulations have been investigated for achieving the targeted drug delivery. Figure 7.2 describes different kinds of drug delivery systems employed for delivery of drugs. The development to these formulations can be tuned in by selecting proper materials or attaching ligands specific to certain receptors present at a particular organ for bringing about targeting. Carbon nanotubes are recent additions to formulations for drug delivery due to its properties of high payload and good cell penetration capability. Single walled carbon nanotubes can be beneficial with respect to its near infra-red (NIR) stimulation which has been used to rupture endosomes for cancer therapy. Hydrogels are porous polymer networks having a high affinity for water which can be used for entrapment/encapsulation of actives like drugs and peptides. Several polymers poly lactide-co-glycolide (PLGA), chitosan, alginate, poly *N*-isopropyl acrylamide (PNIPAM), polyethylene oxide (PEO), polyvinyl-pyrorlidone (PVP) etc. have been used as hydrogels for drug delivery applications. They have been shown to be useful for sustained release of drugs, peptides and proteins, stimuli responsive delivery based on light, pH and temperature and as scaffolds for tissue engineering. Liposomes and micelles are vesicular nano-carriers structures formed from lipids. They are highly useful for loading lipophilic drugs in the bilayer of lipids, however hydrophilic drugs has also been encapsulated in the core of liposomes. Liposomes due to the flexibility of membranes have been useful to deliver drugs (like doxorubicin, amphotericin, vincristine, cytarabine etc.) to sites difficult to access (like tumours and brain) using other formulations. Silica nanoparticles or such other inorganic nanoparticles are mesoporous in nature and can encapsulate different drugs inside their pores. Drug nanocrystals are formed by size reduction, by techniques like high speed or high pressure homogenization of drug powders and thus are 100% drugs without any excipients. Nanocrystals are mainly utilized for improving the solubility of poorly soluble drugs.

Metallic nanoparticles like iron oxide, gold, silver, manganese etc. have an inclination of usage towards imaging either due to NIR properties or magnetic properties. Due to hyperthermia effect created due to irradiation or application of magnetic fields they find their usage in photo-thermal or magnetic hyperthermia. Drugs can be

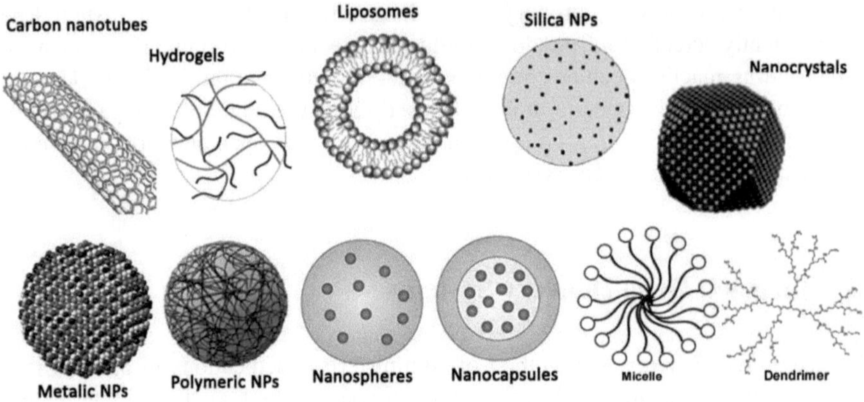

Fig. 7.2 Novel drug delivery systems used for targeting to different organs in the body

coated on metallic nanoparticles via layer by layer self-assembly using different polyelectrolytes. Polymeric nanoparticles or nanocapsules are the most common forms of drug delivery vehicles investigated which can be differentiated based on filled or unfilled core. Different kinds of polymers form nanaoparticles ranging from natural, semi synthetic and synthetic polymers which can encapsulate hydrophilic, hydrophobic drugs, proteins and peptides. Nanocapsules are formed by removing the cores by dissolving certain templates like calcium carbonate using acids or chelating agents. They have a great potential to encapsulate large molecules like proteins, peptides and enzymes. Dendrimers are highly symmetric, spherical compounds formed from branched polymers like Poly (amido-amine) (PAMAM). Similar to liposomes several hydrophilic or hydrophobic drugs can be linked to polymer branches based on functional groups present in the inner core and outer shell of the dendrimer. The linkage of drug and dendrimer can be covalent or electrostatic. Apart from these formulations several approaches of development are employed depending on the site of action and challenges pertaining to access that particular site and maintaining stability of encapsulated actives while passage after administration. For example delivery of drugs to the gastro-intestinal tract (GIT) will involve exposure to varying pH conditions ranging from pH 1 to 8 and hence protection is needed in unfavourable conditions. The following section describes many such approaches where in drug has been targeted to a particular site in the GIT.

4.1 *Oral Targeted Drug Delivery*

Oral drug delivery using solid and liquid dosage forms is by far the most important and frequently used route of administration. Personalized medicine in oral drug delivery approaches can revolutionize the oral therapy to improve the therapeutic compliance and reduce toxic effects. Some of the personalized medicine approaches employed in oral delivery includes partitioning for divided doses in both solids and liquids for paediatric and geriatric populations. Recently dosing cups, spoons, dropping pippetes and oral syringes have been introduced. Dose dispensing devices (e.g. powder dispensers, pellet dispensers, multi-particulate systems, solid dosage pen, buccal dosage forms etc.) have also become an important part of oral personalized drug delivery [6]. These systems have significantly enhanced the capabilities of oral delivery.

(a) Targeting to Stomach Drug delivery to stomach is required in cases where drug is preferentially absorbed through stomach or in cases of pathological conditions pertaining to stomach (GERD, peptic ulcers caused by acids/bacterial infection by *H. Pylori*).

Gastro-retentive drug delivery systems (GRDDS) have been fairly explored and applied in few decades for prolonging the gastric residence time with an aim to provide drug release in stomach and upper GIT for local and/or systemic effects. GRDDS can find applications in one or the other disease pertaining gastro-intestinal tract. Different approaches including floating, swelling, expanding, modified shape, high density, mucoadhesive, magnetic and other delayed gastric emptying devices have been employed for retaining drug delivery systems in the stomach (Fig. 7.3). Chitosan polymer and its several derivatives have been very commonly used as mucoadhesive systems. Thiolated chitosans are gaining popularity due to formation of disulphide bonds with cysteine rich parts of mucous polysaccharides.

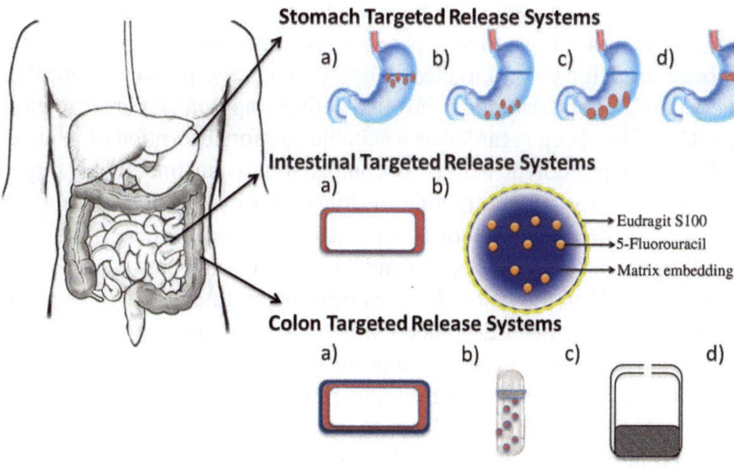

Fig. 7.3 Strategies of oral targeting for local or systemic drug delivery which include Stomach targeted release systems: (a) Low density systems, (b) High density systems, (c) Swellable systems, (d)Raft forming systems and (e) Muco-adhesive systems, Intestinal targeted release systems: (a) Enteric coated tablets/capsules, (b) enteric coated microspheres/nanoparticles and Colon targeted release systems: (a) Enteric-microbially degradable tablets or capsules (CODES technology), (b) Pulsincap®, (c) OROS-CT® and (d) Microspheres and nanoparticles

(b) Targeting to Small-Intestine Small intestine is the preferred site of absorption of most of the nutrients and drugs due to large surface area offered by villi and proteolytic enzymes present in the small intestine. The most common approach of enteric coating has been used for intestinal targeting of drugs providing protection to drugs susceptible to acidic pH. Different grades of Eudragits have been employed for development of dosage forms because of their pH sensitive properties which enable the drug release preferentially in the small intestine. Micro/nanocarriers can also be coated with Eudragit polymers to obtain intestinal release of drugs. Mucoadhesive properties of certain polymers like chitosan have been explored for intestinal release. Chitosan loaded with drugs in the form of nano/microcarriers, gels, films etc. can be used due to its mucoadhesive and permeation enhancing properties. Macrophages present in the intestine are key to maintain mucosal sensitivity and to maintain homeostasis. In cases IBD, Crohn's disease there is an uncontrolled inflammation and manipulating macrophage activity can act as an important strategy for treatment of such diseases. Intestinal targeting to macrophages has been described by Wang et al. using nano-vesicles released from grapefruit. Grapefruit derived nano-vesicles (GDN) were shown to be selectively taken up by intestinal macrophages in dextran induced mouse colitis model. Methotrexate and an anti-inflammatory drug has been encapsulated GDNs and showed to bear lower toxicity in comparison to free drug administered [7].

(c) Targeting to Colon Colon delivery of drugs can provide both local (e.g. inflammatory bowel disease, Crohn's disease, ulcerative colitis, amoebiasis, colon cancer etc.) as well as systemic delivery of therapeutic small and macromolecules. The main desirable features in a colon targeted drug delivery system are the protection of drug till it reaches the colon, selective release and absorption in colon. Primarily pH dependent systems, prodrugs, time dependent systems, and microbial triggered system have been used for colon targeted delivery. Recently, pressure controlled colonic delivery capsules and osmotic controlled delivery approaches are also gaining popularity. Colon has been regarded as a suitable site for absorption of peptides and proteins due to low proteolytic activity in comparison to intestine. Rectal targeting is the shortest route for colon targeting of drugs. However major limitation is that proximal part of the colon cannot be targeted using rectal route. After oral administration drug traverses between pH 1 and 2 in stomach and between pH 6 and 7.5 in small intestine. pH in the colon declines from ileum to colon and hence pH based release systems have been developed which can play an important role in colon targeting. Delayed release systems are promising which are developed on the basis of GI transit time. The use of GI transit time provides with a rough estimate of dosage forms in colon, however it is subject to lot of variabilities due to type and amount of food, GI movement, disease conditions can lead to accelerated or decelerated movement of food in certain sections of intestine. Site specific delivery of drugs in the colon can be improved by integrating pH and time dependent mechanisms in a single formulation. Enteric coated systems have been primarily applied for intestinal targeting. These systems consist of an acid resistant coating over a swellable time release polymer containing an immediate releasing tablet. Acid resistant layer does not allow the drug to be liberated inside the stomach. After gastric emptying enteric coating layer rapidly dissolves making available the swellable polymer which can provide timed release once it reaches small intestine. Polymers like HPC provide no drug release or lag phase of drug release in the small intestine. After certain period of time drug release occurs as the formulation reaches the colon. The anaerobic bacterial flora of colon ferment different kinds of substrates which are undigested in the small intestine. Several enzymes like glucuronidase, azoreductase, deaminase etc. can be used as trigger for degradation of drugs attached to specific functional groups. The enzyme induced degradation will release the drug and make it available for absorption. Amino-salicylic acid has been described to be conjugated to azo groups, amino groups or tyrosine/methionine amino acids for ulcerative colitis. Drugs like Vassopressin, insulin, 5FU have been investigated for their microbial triggered mechanism of release [8]. Fried and Chang have evaluated glycoside conjugated drugs like dexamethasone-21-glucoside and prednisolone-21-glucoside for delivery to the colon. Glycosidase activity cleaves the prodrug liberating the free form of drug specifically in the colon [9]. The pressure generated by muscular contractions throughout the GIT has led to development of pressure based systems. In these systems drug release occurs after disintegration of water insoluble polymer capsule due to the external luminal pressure in the colon. The systems can be tuned in to withstand or rupture at a particular pressure by modifications in size of capsule and thickness of shell wall. Researchers have developed pressure controlled colon-delivery capsules prepared using an ethyl cellulose, which is insoluble in water. In such systems drug release occurs following disintegration of a water-insoluble polymer capsule as a result of pressure in the lumen of the colon. The thickness of the ethyl cellulose membrane is the most important factor for disintegration of the formulation.

OROS-CT® developed by Alza Corporation was used to release the drug locally in the colon. The system constitutes the osmotic unit/s encapsulated in a hard gelatin capsules. Each osmotic layer can act as a push pull pump surrounded by a semi-permeable membrane. Drug release begins when the unit reaches the colon. OROS-CT units can maintain a constant release rate for up to 24 h in the colon. For treating ulcerative colitis, each push-pull unit is designed with a 3–4 h post gastric delay to prevent drug delivery in the small intestine. Another unique technology consisting of combined pH and microbial triggering was developed which involved lactulose for specific drug release coated over acid soluble Eudragit and enteric Eudragit materials. The enteric coating protects the tablet while it is in stomach and then dissolves in intestine. The acid soluble Eudragit protects the drug while passage in the intestine. The enzymes in the colon causes lowering of pH sufficient to dissolve the acid soluble coating material which will degrade Lactulose present in side [8, 10]. A combination of above methods can be used for colon based targeting using pH based targeting, enteric targeting and timed release as described by Yamamoto et al. who used chitosan capsules of five Amino-salicylic acid coated with enteric coating material. The results of this research indicated that capsules were able to reach the colon in 3.5 h after oral administration in 2,4,6-trinitrobenzenesulfonic acid-induced ulcerative rats [11]. Nano/Microcarrier mediated delivery of drugs were studied to investigate the effect of particle size of carriers on deposition in the inflamed colon in the trinitro benzene sulfonic acid induced (TNBS) rat model of colitis. Fluorescent polystyrene particles (0.1–10 μm) were developed and administered orally for 3 days. The results indicated that smaller particles were deposited to a greater extent in the colon. A high proportion of the small particles of 0.1 μm were found to be adhering to mucus layer in the colon. Colon targeting has yet to be explored to its full potential for delivery of drugs which are poorly bioavailable or cannot be given orally directly.

4.2 Brain Targeting

Delivery of drugs into the CNS has remained a challenge due to presence of blood-brain barrier (BBB), which is the main reason of higher therapeutic inefficacies of brain related disorders. BBB constitutes of endothelial cells lacking fenestrations and joined by tight junctions in blood vessels in the brain. Astrocytes foot processes encapsulate the blood vessels present in the brain, which also form an important component of BBB. The tight junctions lead to a very high trans-endothelial electrical resistance than other tissue membranes. Such structure reduces aqueous transport of polar small molecules and large polar molecules in the brain. Further, presence of efflux pumps (P glycoprotein) and metabolizing enzymes on the lining of the cerebral capillaries makes the barrier highly effective to destroy unwanted peptides and other small molecules in the blood as it flows through the brain. Essential nutrients and gasses are spared from such a hindrance of transport across the BBB. Researchers have explored different strategies to alter or bypass the BBB. Deploying such strategies will help improve the bioavailability of drugs in the brain. Some of the most commonly investigated techniques (like osmotic, chemical induced opening or using transport or carrier based systems) to improve the drug permeation across BBB have been described in following section.

(a) Prodrugs and Lipophilic Analogues Lipophilic drugs have been proven to have better bioavailability in the brain owing to a greater penetration across BBB. In order to improve the bioavailability to brain prodrugs have been designed which have improved pharmacokinetic profile and can be either active or inactive pharmacologically. The active drug can be produced after metabolism by hydrolysis or enzymatic cleavage at the site of action and thus will be useful in producing targeted delivery. An example of morphine can describe the role of lipophilicity and prodrug approach. Dopamine and Morphine do not permeate across the BBB however Levodopa and Heroin, respectively which act as prodrugs for dopamine and morphine can readily traverse across the BBB. Addition of hydrophobic moieties like acetyl group in case of Heroin makes it more lipophilic making it suitable for transport across the BBB. Development of lipophilic analogs increases the permeability across BBB however decreases the aqueous solubility and increase protein binding. These changes lead to a low free drug concentration in blood, which can permeate across BBB. Pro-drugs can also generate metabolites contributing to toxicity of compounds.

(b) BBB Alteration and Exploring Physiological Differences Physical disruption of BBB is one of the invasive methodologies of increasing the transport of drugs. This technique employs use of osmotic agents, natural chemical transporters chemotherapy, ultrasound and implants for transient opening of BBB. Administration of hypertonic mannitol or arabinose in cerebral arteries results in an opening for about half an hour to few hours during which the drugs can permeate freely to CNS. The mechanism of such an opening is thought to be the shrinking of endothelial cells and widening of tight junctions. Intra-carotid infusion of leukotriene C4 can be used for selective opening of BBB especially in cases of brain tumour. The mechanism for such a selective opening can be described by g-glutamyl transpeptidase (g-GTP) present in normal capillaries which is down regulated in tumours that result in reduction in enzymatic barriers in tumour tissues. Similar vasoactive amines like natural and synthetic bradykinin analogues like RMP-7, histamine etc. can selectively open the blood tumour barriers. Ultrasound when combined with microbubbles has shown to temporarily disrupt the BBB. The mechanism of this can be described as disassembly of tight junctional complexes transiently allowing agents to be transported into the brain parenchyma. The method has shown reproducible success for brain tumours and Alzheimer's disease. Higher concentration of microbubbles causes inertial cavitation which can result in damage to blood vessels. Recently nasal delivery has been investigated to a lesser extent for drug delivery to brain using olfactory pathway. The route can be used due to less efficient blood brain barrier at the olfactory pathway and drugs are transported across the sensory neurons to yield high concentrations in the olfactory bulb. Nasal delivery can also provide with non-invasive and harmless mode to delivery to brain reducing systemic exposure. This route can prove to be useful for protein drugs like brain delivered neurotropic factor for Alzheimer's disease. Some challenges in executing this route of brain delivery include enzymatic activity, low pH nasal epithelium, mucosal irritation and pathological conditions of patients in disease like cold.

(c) Carrier/Receptor Mediated Transport Carrier-mediated drug delivery to the brain relies on the interaction of a ligand with moieties expressed at the luminal surface of cerebral endothelial cells. Active targeting strategies researched for transport across BBB including brain-tumour treatment use absorption-mediated transcytosis, transporter mediated transcytosis and receptor mediated endocytosis. Absorptive methods of transcytosis provide drug delivery by cationic proteins or cell penetrating peptides. The absorption is mediated by electrostatic interaction of positively charged carriers and negatively charged endothelial membranes. BSA conjugated PEGylated nanoparticles developed by Lu et al. demonstrated 7.7 times higher penetration than BSA nanoparticles. Aclarubicin loaded BSA nanoparticles were able to prolong the survival of intracranial glioblastoma bearing mice [12, 13]. In addition, cell-penetrating peptides (e.g. TAT-derived peptides and cationic albumin proteins) have been commonly used to enhance brain drug delivery. Qin et al. covalently conjugated cell-penetrating peptide TAT (AYGRKKRRQRRR) to cholesterol for preparing doxorubicin-loaded liposomes for glioma therapy. The bio-distribution in the brain and heart demonstrated higher efficiency of brain delivery and lower cardiotoxicity. The survival time of the glioma-bearing rats treated with TAT-modified liposomes was significantly prolonged [14]. Cerebral endothelium provides brain with necessary nutrient transporter systems for glucose, neutral amino acids (phenyl alanine, leucine, etc.), acidic amino acid (glutamic acid and aspartic acid), monocarboxylic acid (lactate), short chain fatty acid (acetate, propionate), choline transporter system, amine, and nucleoside and peptide transport systems. Brain targeting using these transporter systems is described as transporter-mediated transcytosis. Glucose transporters (GLUT) are involved in transport of glucose from blood to brain physiologically. Thus glucose derivative bound nano-carriers can aid in targeting via these transporters. Liposomes incorporated with mannose derivatives could be transported across the BBB using GLUT1 in mouse brain. Same group of researchers developed glycosyl derivatives of cholesterol as components for liposome preparation. Pharmacokinetic study indicated that the formulation showed good targeting ability to brain. Apart from glucose transporters, choline transporter also can be targeted using quaternary ammonium ligands have been tested in vitro models of BBB (Fig. 7.4) [15]. Several studies in the literature suggest that absorption mediated and transporter mediated brain targeting strategies are inferior to receptor-mediated transcytosis owing to electrostatic charge dependence, potential toxicity of cell penetrating peptides and substrate selectivity with respect to the transporters. Among some of the receptors, which are expressed on capillary endothelium of the brain transferrin receptor (TFR), low-density lipoprotein receptor (LDR), insulin receptor and nicotinic acetylcholine receptor, transferrin receptor is highly expressed on endothelial cells of BBB. Transferrin modified paclitaxel loaded micelles were evaluated using in vitro and in vivo experiments and showed that survival time was increased bearing intra cranial U87-MG glioma. The competitive binding of natural transferrin and transferrin bound to drug delivery system may limit the permeation of drug delivery system in

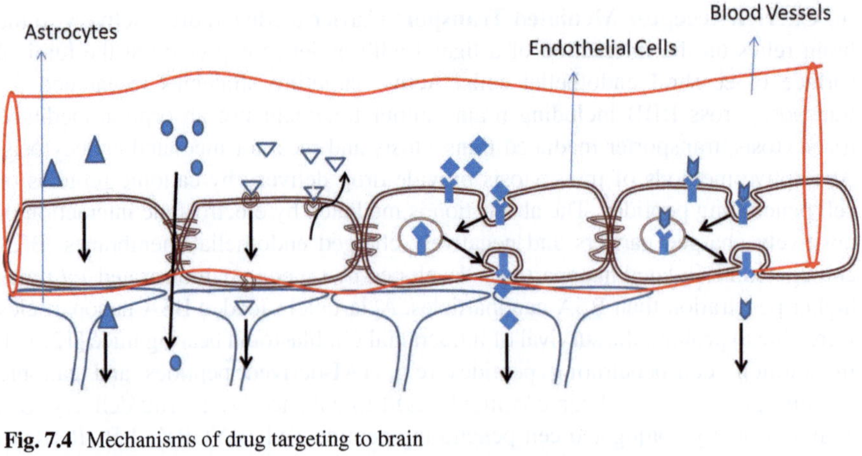

Fig. 7.4 Mechanisms of drug targeting to brain

the brain. Alternatively, a monoclonal antibody against transferrin (OX26) has been coupled to liposomes containing daunomycin and tested for brain targeting capability [16, 17]. Researchers have used a chimeric transferrin receptor monoclonal antibody (TfRMAb) with high activity for the mouse transferrin receptor (TfR) as a therapeutic fusion protein for BBB drug delivery by the receptor mediated transport. Insulin is also transported to brain tissues by insulin receptor via transcytotic mechanism present at vascular endothelial surface. Human anti-insulin receptor monoclonal antibody can be used for transport of drug delivery for effective targeting to brain. Aprotinin is a low-density lipo-protein which when evaluated showed more than ten fold greater penetration inside the brain in comparison to holo-transferrin. A derivative of aprotinin called Angiopep-2 showed further higher parenchymal accumulation of cationic liposomes containing a gene and paclitaxel to the brain [18, 19]. Leptin is an endogenous molecule liberated by GIT, which is transported to brain endothelium via a specific receptor called ObR on the luminal side of the brain capillary. Leptin conjugated liposomes with a size of 180 nm showed significantly enhanced uptake in brain endothelium [20]. Thiamine a water-soluble vitamin has many physiological functions in BBB. Expression of thiamine transporter can facilitate thiamine functionalized nanoparticles permeation across the BBB. This fact was investigated by Lockman et al. Nanoparticles prepared using emulsification method with distearoyl phosphatidylethanolamine (DSPE) anchor showed brain permeation after brain perfusion associated with thiamine carrier. Similar to thiamine receptors, PEGylated Liposomes coated with glutathione have also showed enhanced drug delivery to brain due to abundant glutathione receptors in the BBB [21, 22].

4.3 *Tumour Targeting*

Conventional chemotherapy is limited due to high toxicity, non-specific mechanism of action of drugs causing adverse effects. In addition to these factors anti-cancer chemotherapeutics are poorly soluble due to which use of organic excipients for development of formulations deems necessary. Use of organic solvent may cause venous irritability and distress. A successful cancer therapy is dependent on the success of targeting capability of the formulations so that normal cells are spared from the adverse effects of cancer chemotherapeutic agents.

Cancer chemotherapeutic targeting can be classified into passive targeting and active targeting. Passive targeting is mainly constituted by the enhanced permeability and retention effect due to the leaky vasculature and disorganized endothelium of tumour tissues, which is characterized by selective localization of high molecular weight compounds and nanoparticles (10–100 nm size) in tumour tissues. In addition to these the lymphatic drainage in the tumours is also poor. Researchers have studied this fact and developed macromolecular anti-cancer drug regimens like polymer drug conjugates, nano-particulates loaded with anticancer drugs, micelles and liposomes to localize the drugs to tumour tissues and reduce the toxic effects produced. Glycol-chitosan-DOX conjugates have been investigated with a spacer like *N-cis*-aconityl moiety formed using carbo-diimide coupling reaction. The polymer drug conjugate can be used to form nanoparticles like DOX-GC nanoparticles. According to a research systemic administration of DOx-GC nanoparticles preferentially accumulated in the tumour tissues due to EPR effect [23]. Using a similar approach paclitaxel was conjugated to low molecular weight chitosan through a succinate linker, which cleaves at physiological conditions. Low molecular weight chitosan (<10 kDa) can open the tight junctions between human epithelial colorectal adreno-carcinoma cells (Caco-2). LMWC-PTX was found to be absorbed through small intestines after oral administration (Fig. 7.5) [24].

Apart from the physical concentration of drug therapy, active targeting relies on interaction of drug with target site like cell membranes, cytoplasm or nucleus. Cancer cells express some specific antigens and receptors on their surfaces, which

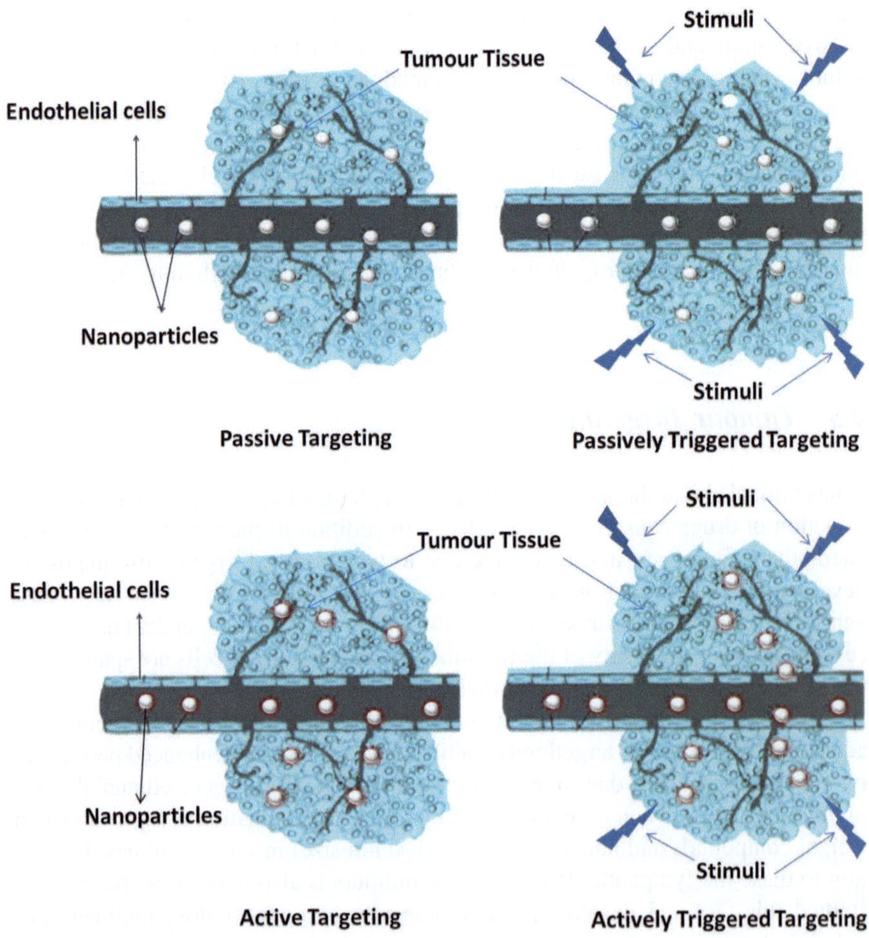

Fig. 7.5 Mechanisms of drug targeting to tumour tissues

can be targeted. Ligand drug conjugates were the first attempts to target tissues; however they suffered limitations of reduced therapeutic activity of drugs due to the conjugation process. Folate targeting is a method for localization of anticancer drugs to tumour tissues by attachment of folic acid. Folic acid interacts with folate receptors which are over expressed in many cancer tissues. Some of the common methods using folate include folate drug conjugates and folate conjugates of nanoparticles loaded with drugs. The attachment of folate functionality can induce endocytotic uptake process thereby delivering the molecules inside the cells. EPR method of targeting although being commonly used suffers with severe pathophysiological limitations of different tumour types, vascularity, fibrosis and necrosis of tissues. Use of magnetic field can prove to be an interesting approach for targeting of drugs to tumours. Such delivery systems require that the magnetic nanoparticles remain non-magnetic when external magnetic field is removed. Nano-sized magnetic nanoparticles function as desired owing to a single domain state in contrast to multiple domains present in larger particles. In order to attach drugs on to magnetic particles polymers, polyelectrolytes, metal/non-metal coatings can be established [25]. Chen et al. described the development of DOX conjugated Fe_3O_4 nanoparticles embedded in PEG modified porous silica shell. The silica shell provides protection to the entrapped drug and also produces a barrier for drug release. PEGylation of these carriers allow for stealth properties against the reticulo-endothelial system to prolong the drug release over a period of time [26]. Arias et al. investigated the competences of polycyanoacrylate nanoparticles coated over a core of magnetic nanoparticles. Anti-tumour drug 5-fluoro uracil has been loaded in these carriers and proved that targeting using magnetic nanoparticles can reduce undesired toxic effects [27]. Development of magnetic nanoparticles based anti-tumour therapies provide with capability of imaging of tissues. MNPs of iron oxide encapsulated with poly[aniline-co-*N*-(1-one-butyric acid)] aniline containing epirubicin drug was developed by Liu et al. These carriers were successfully utilized as a theranostic for focused ultrasound imaging and magnetic targeting. Apart from the MRI capabilities of MNPs, they prove to be useful in magnetic hyperthermia. Magnetic hyperthermia is a complementary anticancer therapy which involves heating up of tumour tissue by application of alternating magnetic field which finally leads to apoptosis. Currently available hyperthermia techniques involve ultrasound, radiofrequency, microwaves, infra-red radiations and magnetic particles. Among several advantages of MNPs, they can be targeted by cancer specific binding agents, altering frequencies of magnetic fields so that heat is generated in the tissues containing MNP's, smaller size and ability of permeate through BBB makes them suitable for brain tumours. Metallic nanoparticles having super-paramagnetism are more preferred over paramagnetic systems due to the absence of any remnant magnetism after removal of magnetism. Due to the possibility of administration of MNPs through different routes of administration a wide range of cancer tissues can be treated. Different coatings can be imparted over MNP's using layer by layer self-assembly to attach different kinds of drugs by entrapment or adsorption for simultaneous and controlled release of two or more drugs. Metallic nanoparticles with Mn, Fe, Co, Ni, Zn, Gd, Mg, and their oxides have the potential to be used as magnetic nanoparticles. Most commonly used among the above are iron oxide magnetite particles (Fe_3O_4) which are coated with carbohydrates, liposomes, hydrogels, lauric/citric acid etc. to obtain desired specificity and stability. The coating of nanoparticles provides stability and functional groups for attachment of ligands for targeting. Different types of ligands that can be attached to MNPs include small molecules like folate and large molecules like RGD peptide, bombesin, A54, Leutinizing hormone releasing hormone, protiens like EGFR, VEGF, antibodies, cell penetrating peptides, myristoylated poly arginine peptide, TAT peptide etc.

4.4 Liver, Kidney and Lung Targeting

The liver is an important organ involved in metabolism of xenobiotics. It is also associated with conditions like hepatitis, hepatoma, and deficit of metabolizing enzymes. Liver targeting systems use microparticles trapped by reticuloendothelial systems or based on hepatic receptor based ligand bearing particles [28]. Lactosaminated *N*-succinyl chitosan was used for liver specific drug delivery which was synthesized using reductive amination of *N*-succinyl chitosan and lactose along with sodium cyanoborohydride by Machida et al. This polymer showed specific binding to asialoglycoprotein receptors present in liver parenchymal cells [29]. Liu et al. developed polyion complex micelles (PIC micelles) based on methoxy poly (ethylene glycol) (PEG)-grafted on chitosan and lactose conjugated PEG-graft-chitosan for liver-targeted delivery. Diammonium glycyrrhizinate (DG) has been targeted to liver using these carriers for treatment of chronic hepatitis and immunodeficiency virus infection. Pharmacokinetic studies indicate that the micelles were superior in comparison to intravenous injection diammonium glycyrrhizinate [30].

Kidney targeting is important to reduce extra renal toxicity of drugs for treatment of kidney related diseases. Drug-polymer conjugates have been investigated for renal targeting. Zhang et al. reported 50% *N*-acetylated low molecular weight chitosan selectively localized in renal tubules after intravenous injection into mice [31]. Prednisolone conjugated to LMWC through a succinic acid spacer showed a 13 fold higher deposition in kidneys in comparison to prednisolone injection. The studies show that polymer molecular weight plays important role in targeting characteristics [31]. Targeting to lungs is critical for diseases like tuberculosis, pneumonia, etc. Microcarriers having diameter >5 μm have a potential to block blood capillaries and induce chronic obstructive pulmonary emphysema. Shim et al. prepared chitosan modified PLGA nanoparticles containing paclitaxel with a diameter 200–300 nm. The study showed that nanoparticles are taken up by lung cancer cell line (A549) was significantly increased by chitosan modification [32]. Bacterial respiratory infections are treated with orally administered or injectable drugs. The limitation of such a mode of therapy is that high dose of drugs need to be administered for long durations to maintain the therapeutic concentrations. The therapeutic efficiency is hampered for diseases like tuberculosis and asthma due to patient non-compliance. Nanoparticulate drug delivery systems show potential in addressing the spatial localization of therapy to a particular site. Inhalable nano/microparticles provide advantage of reducing the side effects, minimise drug resistance and lower the therapeutic dosage requirement. Use of polymeric carriers for drug delivery provides added advantage of temporal control of release in the form of controlled or burst or stimuli responsive nature. Stealth nature of nanoparticles can be imparted by functionalizing the nanoparticles by hydrophilic polymers. Spatial and temporal control of release from inhaled aerosolized nanoformulations has remained an unmet need for respiratory disease management. Ahmad et al. showed development of alginate nanoparticles loaded with anti-tubercular drugs and an oral administration of this formulation could control the drug release till 11 days. The alginate nanoparticle was found to improve the bioavailability significantly. The study involved use of pulmonary surfactant mimetic material (1, 2-dipalmitoyl-sn-glycero-3-phosphocholine) for developing paclitaxel-encapsulated nanovesicles to treat metastatic lung cancer. The nano-formulations were tested using B16F10 murine melanoma cells and lung fibroblasts MRC5 [33, 34].

Acknowledgement Abhijeet Joshi acknowledges the INSPIRE Faculty award and fellowship provided by Department of Science and Technology, Government of India. Authors would also like acknowledge financial support from DBT and SERB.

References

1. Jain KK. Textbook of personalized medicine. 2nd ed. Berlin: Springer; 2015.
2. Personalized Medicine Coalition. 2015. http://www.personalizedmedicinecoalition.org/News/Press_Releases/More_Than_1_in_4_Novel_New_Drugs_Approved_by_FDA_in_2015_are_Personalized_Medicines. Accessed Sept 2017.
3. Reiss T. Drug discovery of the future: the implications of the human genome project. Trends Biotechnol. 2001;19(12):496–9. doi:10.1016/S0167-7799(01)01811-X.
4. Ginsburg GS, McCarthy JJ. Personalized medicine: revolutionizing drug discovery and patient care. Trends Biotechnol. 2001;19(12):491–6. doi:10.1016/S0167-7799(01)01814-5.
5. Doren M, Samsioe G. Prevention of postmenopausal osteoporosis with oestrogen replacement therapy and associated compounds: update on clinical trials since 1995. Hum Reprod Update. 2000;6(5):419–26.
6. Wening K, Breitkreutz J. Oral drug delivery in personalized medicine: unmet needs and novel approaches. Int J Pharm. 2011;404:1–9. doi:10.1016/j.ijpharm.2010.11.001.
7. Wang B, Zhuang X, Deng Z-B, Jiang H, Mu J, Wang Q, Xiang X, Guo H, Zhang L, Dryden G, Yan J, Miller D, Zhang H-G. Targeted drug delivery to intestinal macrophages by bioactive nanovesicles released from grapefruit. Mol Therapeutics. 2014;22(3):522–34. doi:10.1038/mt.2013.190.
8. Tiwari G, Tiwari R, Wal P, Wal A, Rai AK. Primary and novel approaches for colon targeted drug delivery—a review. Int J Drug Deliv. 2010;2:1–11.
9. Friend DR, Chang GW. A colon-specific drug-delivery system based on drug glycosides and the glycosidases of colonic bacteria. J Med Chem. 1984;27(3):261–6. doi:10.1021/jm00369a005.
10. Hata T, Shimazaki Y, Kagayama A, Tamura S, Ueda S. Development of a novel drug delivery system, time-controlled explosion system (TES): V. Animal pharmacodynamic study and human bioavailability study. Int J Pharm. 1994;110(1):1–7. doi:10.1016/0378-5173(94)90369-7.

11. Tozaki H, Odoriba T, Okada N, Fujita T, Terabe A, Suzuki T, Okabe S, Muranishi S, Yamamoto A. Chitosan capsules for colon-specific drug delivery: enhanced localization of 5-aminosalicylic acid in the large intestine accelerates healing of TNBS-induced colitis in rats. J Control Release. 2002;82(1):51–61. doi:10.1016/S0168-3659(02)00084-6.
12. Lu W, Tan Y-Z, K-L H, Jiang X-G. Cationic albumin conjugated pegylated nanoparticle with its transcytosis ability and little toxicity against blood–brain barrier. Int J Pharm. 2005; 295(1-2):247–60. doi:10.1016/j.ijpharm.2005.01.043.
13. Lu W, Wan J, Zhang Q, She Z, Jiang X. Aclarubicin-loaded cationic albumin-conjugated pegylated nanoparticle for glioma chemotherapy in rats. Int J Cancer. 2007;120(2):420–31. doi:10.1002/ijc.22296.
14. Qin Y, Chen H, Zhang Q, Wang X, Yuan W, Kuai R, Tang J, Zhang L, Zhang Z, Zhang Q, Liu J, He Q. Liposome formulated with TAT-modified cholesterol for improving brain delivery and therapeutic efficacy on brain glioma in animals. Int J Pharm. 2011;420(2):304–12. doi:10.1016/j.ijpharm.2011.09.008.
15. Fenart L, Casanova A, Dehouck B, Duhem C, Slupek S, Cecchelli R, Betbeder D. Evaluation of effect of charge and lipid coating on ability of 60-nm nanoparticles to cross an in vitro model of the blood-brain barrier. J Pharmacol Exp Ther. 1999;291(3):1017–22.
16. Zhang P, Hu L, Yin Q, Zhang Z, Feng L, Li Y. Transferrin-conjugated polyphosphoester hybrid micelle loading paclitaxel for brain-targeting delivery: synthesis, preparation and in vivo evaluation. J Control Release. 2012;159(3):429–34. doi:10.1016/j.jconrel.2012.01.031.
17. Huwyler JWD, Pardridge WM. Brain drug delivery of small molecules using immunoliposomes. Proc Natl Acad Sci. 1996;93:14164–9.
18. Demeule M, Regina A, Che C, Poirier J, Nguyen T, Gabathuler R, et al. Identification and design of peptides as a new drug delivery system for the brain. J Pharmacol Exp Ther. 2008;324:1064–72.
19. Sun X, Pang Z, Ye H, Qiu B, Guo L, Li J, Ren J, Qian Y, Zhang Q, Chen J, Jiang X. Co-delivery of pEGFP-hTRAIL and paclitaxel to brain glioma mediated by an angiopep-conjugated liposome. Biomaterials. 2012;33(3):916–24. doi:10.1016/j.biomaterials.2011.10.035.
20. Tamaru M, Akita H, Fujiwara T, Kajimoto K, Harashima H. Leptin-derived peptide, a targeting ligand for mouse brain-derived endothelial cells via macropinocytosis. Biochem Biophys Res Commun. 2010;394(3):587–92. doi:10.1016/j.bbrc.2010.03.024.
21. Lockman PR, Oyewumi MO, Koziara JM, Roder KE, Mumper RJ, Allen DD. Brain uptake of thiamine-coated nanoparticles. J Control Release. 2003;93(3):271–82. doi:10.1016/j.jconrel.2003.08.006.
22. Gaillard PJ, de Boer AG. 2B-trans™ technology: targeted drug delivery across the blood-brain barrier. In: Jain KK, editor. Drug delivery systems. Totowa, NJ: Humana Press; 2008. p. 161–75. doi:10.1007/978-1-59745-210-6_8.
23. Son YJ, Jang J-S, Cho YW, Chung H, Park R-W, Kwon IC, Kim I-S, Park JY, Seo SB, Park CR, Jeong SY. Biodistribution and anti-tumor efficacy of doxorubicin loaded glycol-chitosan nanoaggregates by EPR effect. J Control Release. 2003;91(1-2):135–45. doi:10.1016/S0168-3659(03)00231-1.
24. Lee E, Lee J, Lee I-H, Yu M, Kim H, Chae SY, Jon S. Conjugated chitosan as a novel platform for oral delivery of paclitaxel. J Med Chem. 2008;51(20):6442–9. doi:10.1021/jm800767c.
25. Mody VV, Cox A, Shah S, Singh A, Bevins W, Parihar H. Magnetic nanoparticle drug delivery systems for targeting tumor. Appl Nanosci. 2013;4(4):385–92. doi:10.1007/s13204-013-0216-y.
26. Chen F-H, Zhang L-M, Chen Q-T, Zhang Y, Zhang Z-J. Synthesis of a novel magnetic drug delivery system composed of doxorubicin-conjugated Fe3O4 nanoparticle cores and a PEG-functionalized porous silica shell. Chem Commun. 2010;46(45):8633–5. doi:10.1039/C0CC02577A.
27. Arias JL, Gallardo V, Ruiz MA, Delgado ÁV. Magnetite/poly(alkylcyanoacrylate) (core/shell) nanoparticles as 5-fluorouracil delivery systems for active targeting. Eur J Pharm Biopharm. 2008;69(1):54–63. doi:10.1016/j.ejpb.2007.11.002.

28. Ogawara K-i, Yoshida M, Higaki K, Toshikiro K, Shiraishi K, Nishikawa M, Takakura Y, Hashida M. Hepatic uptake of polystyrene microspheres in rats: effect of particle size on intrahepatic distribution. J Control Release. 1999;59(1):15–22. doi:10.1016/S0168-3659(99)00015-2.
29. Kato Y, Onishi H, Machida Y. Biological characteristics of lactosaminated N-succinyl-chitosan as a liver-specific drug carrier in mice. J Control Release. 2001;70(3):295–307. doi:10.1016/S0168-3659(00)00356-4.
30. Yang KW, Li XR, Yang ZL, Li PZ, Wang F, Liu Y. Novel polyion complex micelles for liver-targeted delivery of diammonium glycyrrhizinate: in vitro and in vivo characterization. J Biomed Mater Res A. 2009;88A(1):140–8. doi:10.1002/jbm.a.31866.
31. Yuan Z-X, Sun X, Gong T, Ding H, Fu Y, Zhang Z-R. Randomly 50% N-acetylated low molecular weight chitosan as a novel renal targeting carrier. J Drug Target. 2007;15(4):269–78. doi:10.1080/10611860701289875.
32. Yang R, Yang S-G, Shim W-S, Cui F, Cheng G, Kim I-W, Kim D-D, Chung S-J, Shim C-K. Lung-specific delivery of paclitaxel by chitosan-modified PLGA nanoparticles via transient formation of microaggregates. J Pharm Sci. 2009;98(3):970–84. doi:10.1002/jps.21487.
33. Joshi N, Shirsath N, Singh A, Joshi KS, Banerjee R. Endogenous lung surfactant inspired pH responsive nanovesicle aerosols: pulmonary compatible and site-specific drug delivery in lung metastases. Sci Rep. 2014;4:7085. doi:10.1038/srep07085.
34. Ahmad Z, Pandey R, Sharma S, Khuller GK. Alginate nanoparticles as antituberculosis drug carriers: formulation development, pharmacokinetics and therapeutic potential. Indian J Chest Dis Allied Sci. 2006;48(3):171–6.

Chapter 8
Hydrogels: Stimuli Responsive to on-Demand Drug Delivery Systems

Arti Vashist, Ajeet Kaushik, Rahul Dev Jayant, Atul Vashist, Anujit Ghosal, and Madhavan Nair

Abstract Hydrogels are the three dimensional crosslinked network of polymeric materials, which have the ability to respond and adapt to the surrounding environment inside the human body. Hydrogels are the upcoming class of biomaterials, which act as excellent drug delivery systems. These materials have important characteristic feature that they respond to various external stimuli like pH, light, temperature, magnetic field, electric field and pressure. Present chapter gives a broad overview of the recent hydrogels used for drug delivery using various external stimuli. Concluding remarks and viewpoints for the future development of stimuli responsive hydrogels are addressed.

Keywords pH responsive • Hydrogels • Stimuli responsive • Drug delivery • Thermosensitive hydrogels

1 Introduction

Recent trends show that a high attention has been given on the biodegradable hydrogels for various applications including nanomedicine, cosmetics, water purification and biotechnology [1–5]. Bio-polymeric hydrogels are in high demand due to their

A. Vashist (✉) • A. Kaushik • R.D. Jayant • M. Nair (✉)
Center for Personalized Nanomedicine, Institute of Neuro immune Pharmacology,
Department of Immunology, Herbert Wertheim College of Medicine,
Florida International University, Miami, FL 33199, USA
e-mail: avashist@fiu.edu; nairm@fiu.edu

A. Vashist
Department of Biotechnology, All India Institute Medical Sciences, New Delhi, India

A. Ghosal
Department of Chemistry, School of Basic and Applied Sciences, Galgotias University,
Gautam Buddh Nagar, Uttar Pradesh India

School of Biotechnology, Jawaharlal Nehru University, New Delhi, India

A. Kaushik et al. (eds.), *Advances in Personalized Nanotherapeutics*,
https://doi.org/10.1007/978-3-319-63633-7_8

soft rubbery tissue like characteristics and the ability to be modulated in varying size and shape with the external stimuli. Presently, emphasis have been made on the development of on demand drug delivery systems using the external force like magnetic field and electric field. Hydrogels undergo sol to gel transition upon changing the temperature due to their thermal sensitivity. These stimuli responsive hydrogels are also known as smart materials [6, 7]. The basic need of hydrogels based drug delivery system arises due to the fact that existing drug delivery of the various biomolecules is quite challenging owing to their high molecular weight and exclusive three dimensional structures. The intravenous and subcutaneous route of administration have commonly been employed for the delivery of the proteins and macromolecules. The main concern for the delivery systems of the biomolecules is their route of delivery. The biomolecules are commonly prone to the proteolytic degradation and less plasma circulation time period and results to their rapid renal clearance. These conditions result in the multiple dosing of the drugs and thus induce acute toxicity and high immune response. Thus then come the need of the sustained release vehicles, which can be used to offer sustained release mechanism and controlled release of the drug [8]. Hydrogels are the three dimensional structures resulted due to the formation of crosslinked network and this crosslinking can be induced by physical entanglements, hydrogen bonds or van der Waals interactions. It also include chemical cross linkers forming covalent bonds. pH responsive hydrogels protect the drug to release in acidic environment and also in the presence of enzymes. These versatility in the hydrogels make them useful for various important application such as artificial muscles, tissue engineering [9], immobilization of various enzymes, cells [10], biosensors [11] and foremost for on-demand drug delivery [12]. Present chapter discuss the important parameters and studies carried out using various stimuli responsive hydrogels and discuss the various significant studies carried out using these systems (Fig. 8.1).

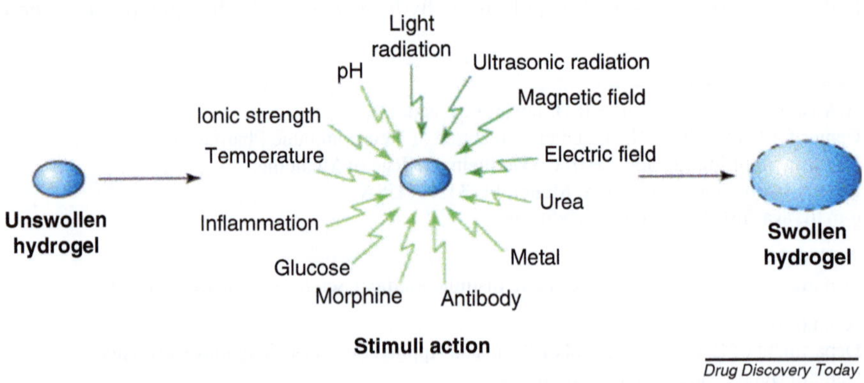

Fig. 8.1 Stimuli responsive swelling of hydrogels. Reprinted from [3], Copyright (2002), with permission from Elsevier

2 pH Responsive Hydrogels

The most common feature of the human body is the existence of various pH inside the human body ranging from acidic to basic pH. This feature provide the novel drug delivery systems to act according to the specific organ. The drug release can be made specific to the pH of the desired organ where the drug release is required. The pH responsive hydrogel systems are actively been explored for peroral drug delivery. These hydrogels are designed using polymer as a matrix material. Generally, the most studied polymers are poly(acrylamide) (PAAm), poly(acrylic acid) (PAA), poly(methacrylic acid) (PMAA), poly(diethylaminoethyl methacrylate) (PDEAEMA) and poly(dimethylaminoethyl methacrylate) (PDMAEMA) [3]. Some cationic polymers like chitosan, gelatin and dextran have also been exploited [13]. Stimuli responsive behavior is generally targeted to design functionally active materials and the introduction of more hydrophobic character is introduced. The common functional groups responsible for the pH response are the amine ($-NH_2$), carboxyl ($-COO^-$), sulphonic ($-SO_3H$), hydroxyl (–OH) etc. The polymeric amines and acids undergo protonation and becomes charged when the pH changes. As soon as these charged polymeric matrix comes in contact with the aqueous solution or physiological pH solutions the pendant groups ionized and generate an electrostatic repulsive forces which is mainly responsible for the pH responsive swelling and deswelling of the hydrogels. There occurs a change in the mesh size of the interpenetrating network. Noteworthy, with the change in the pH, the pendant group of the anionic hydrogels are unionized below and ionized above the pK_a of the polymeric backbone which results in the swelling of the hydrogel due to the large osmotic swelling force by the presence of ions. The cationic hydrogel show reversible character and swell at low pH. Thus overall the pH of the solution is an important factor for the ionization of the reactive groups.

The therapeutic drugs poses great challenges to the route they are administered such as the high administration costs, painful injections, and numerous administrations. The oral drug delivery systems having pH response are high in demand. A recent study shows that poly (l-lysine isophthalamide) (PLP), an anionic polymer has a perfect hydrogel backbone owing to its pH responsiveness and amphiphilicity. The hydrogels of these polymers were capable of loading both the hydrophobic and hydrophilic molecules of a wide range and suggests the wide application of these molecules. The major advantage highlighted by this study was that these hydrogels were capable of loading molecules larger than the 70 kDa and capable of releasing the drug in the gastrointestinal tract [14]. These systems give an added advantage over many barriers that generally are given by other routes. The pH of the stomach has a highly acidic pH 1–3.5 and exhibit harsh enzymes which have little surface area for adsorption. Thus many drug faces a limitation to get precipitated in this environment. On the other hand the small intestine has a pH range 5.5–7.5, has large surface area and the rate of absorption is very high. Thus hydrogels are excellent biomaterial, capable of retaining the drug in the acidic environment and releasing the drug in the small intestine.

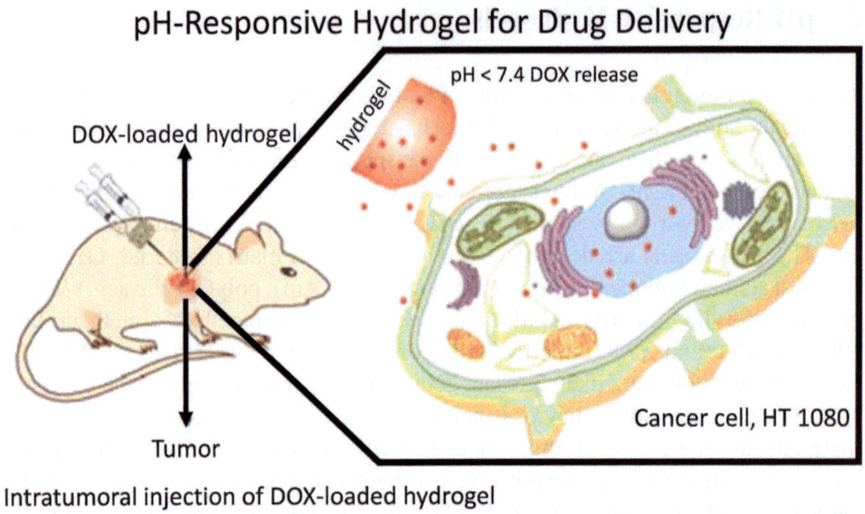

Fig. 8.2 Injectable and biodegradable pH-responsive hydrogels for localized and sustained treatment of human fibrosarcoma. Reprinted with permission from [15]. Copyright (2015) American Chemical Society

Recently, researchers are paying interest to design injectable hydrogels for the controlled release of the drugs. A recent work highlighted the use of a polyethylene glycol derivative with α,β-polyaspartylhydrazide for local cancer chemotherapy. This hydrogel was a pH responsive system, which undergo a sol–gel phase transition (Fig. 8.2). The hydrogel showed an inhibition of tumor growth when intratumoral administration in mice with human fibrosarcoma was evaluated. The tumor growth was inhibited more when the DOX-loaded hydrogel was administered. Thus these type of investigation promote the interest of researchers in developing hydrogel with injectable characteristics and high biodegradability and enhances the local chemotherapy and enhance the therapeutic effect on the human fibrosarcoma [15]. pH responsive hydrogels have been used to make biosensors and bioelectronics [16]. These hydrogels are applied by loading with the enzymes that change the pH of the local microenvironment inside the hydrogels. Commonly the enzyme glucose oxidase in developing pH responsive hydrogel systems is used that transforms glucose to gluconic acid, which lowers the local pH and affects the swelling.

3 Magnetic Fields Responsive Hydrogels

Magnetic hydrogels have significantly gained attention in last few years owing to their on demand drug release characteristics and enhanced controllability [17]. The synergism of the magnetic nanoparticles (e.g., $-Fe_2O_3$, Fe_3O_4, $CoFe_2O_4$) and the polymeric matrix result in superparamagnetic hydrogels and hold potential for drug delivery and tissue engineering. The magnetic hydrogels have been known to synthesize by various methods including blending, an in situ precipitation method or grafting-onto method. The type and the properties of the magnetic hydrogel depends on the type of hydrogel matrix, magnetic character of nanoparticles as well as their size and concentration [18]. From the last few years the magnetic hydrogels have been intensively applied for tissue engineering applications. The magnetic hydrogels as building blocks may be assembled to form complex tissue constructs with biological agents when implanted in the body. These hydrogels have attracted the in vivo cells and growth factors. The most eminent characteristic features of the magnetic scaffolds is their ability to stabilize and occupy the growth factors and bind the biological agents through the external magnetic field which simultaneously provide nutrition initial cell seeding and cell proliferation. These scaffolds give frequent response to the physical cues like mechanical stimulus by interacting between the magnetic nanoparticles and the alternating magnetic field (AMF) further control the cell biological behavior in the angiogenesis process in vitro. The other factors which can be controlled by alternating the frequency of the applied field is the stress parameter. Thus magnetic hydrogels are eligible candidate for pulsatile drug delivery and soft actuators. Further recent works shows that the mechanical stimulation induced by the magnetic forces have shown to bind the magnetic particles to the membranes of mesenchymal stem cells (MSCs) for simulation of calcium stores, upregulation of genes for bone and cartilage regeneration and mineralization. One of the interesting study by Polyak et al. [19] demonstrated the use of magnetic alginate composite scaffolds for physical simulation to cells. Elevated metabolic activity during the simulation period was observed when magnetite–impregnated alginate scaffolds are seeded with aortic endothelial cells stimulated during the first 7 days of 14 day experimental. It was also observed that the expression of proliferating cell nuclear antigen PCNA indicated the magnetically stimulated cells exhibit the lower proliferation index as compared to the non-stimulated cells. Figure 8.3 shows the SEM images which demonstrate the pore size and porous morphology of scaffolds that are not affected by the magnetic nanoparticles concentration. The study shows that the particles are homogeneously embedded in the hydrogel matrix and also indicate that the alginate- stabilized nanoparticles have fine interaction with the matrix to form the homogeneous composite materials. Recently, the concept of self-healing hydrogels have gained significant attention. They have the capability to self-repair after the damage as the biological systems does. They are known as functional materials that exist in the injectable forms and have the capacity to regenerate the integral network and sustain at the target position. One of the interesting studies showed that chitosan-Fe_3O_4 ferrofluid has been synthesized by simple method by mixing of the nanoparticles and chitosan solution which give a magnetic self-healing hydrogel by the addition of DF-PEG into the ferrofluid at room temperature in 2 min. This hydrogel showed the property of magnetism and self-healing feature. When cut into several pieces the broken pieces gather together under the influence of external magnetic field and form an integral biocompatible hydrogel again [20].

The general complication in designing a stimuli responsive gels are that they possess structural changes in shape and volume, which can kinetically hindered by

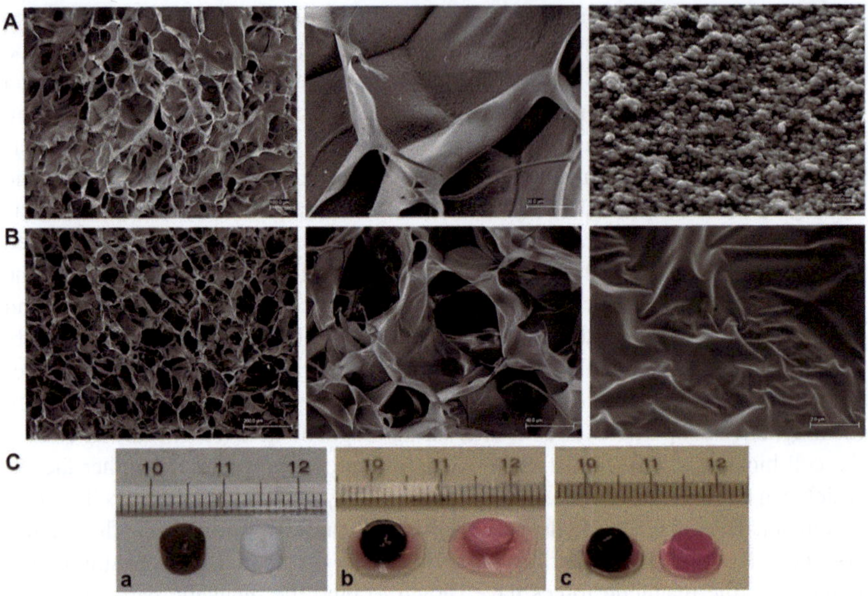

Fig. 8.3 Scaffold morphology: Scanning electron microscopy (SEM) images of the (**a**) 1.2% (w/v) MNP-alginate and (**b**) non-magnetic alginate scaffolds. (**c**) Macroscopic view of the (**a**) dry (**b**) pre-wetted for 30 min and (**c**) hydrated with culture medium for 24 h scaffolds. The black scaffold is magnetic-alginate composite. The white is control alginate scaffold without inclusion of MNPs. Reprinted from Reference [19], Copyright (2012), with permission from Elsevier

low swelling and deswelling process. Magnetic field sensitive polymeric gels have an increased response time [21]. For example, the ferrogels which is basically a chemically crosslinked polymeric network swollen by a ferrofluid. This ferrofluid is simply the term used for the magnetic fluid having colloidal dispersion of monodomain magnetic particles with a size of 10–15 nm. They have both the magnetic and elastic properties thus they are highly shape sensitive when external magnetic field is applied [22]. This property has been extensively utilized in the design of magnetically active soft polymeric actuators [23, 24].

The recent application of magnetic hydrogels are in pulsatile release systems, which are capable of depicting the release of specific peptides and hormones and achieve a prolonged drug release [25]. The study by Langer et al. [26] showed the on demand controlled release of insulin from the magnetic ethylene vinylacetate by using a low frequency oscillating magnetic field. Other fascinating study carried out by Chen et al. [21] demonstrated the fabrication of Fe_3O_4 magnetic nanoparticle PVA hydrogel with controlled drug release by switching duration time and fine tuning. The super-paramagnetic feature as well as the heating ability of the hydrogels when exposed to the external magnetic field, the magnetic hydrogels show a very quick response in the degradation, swelling as well as deformation and make them superior candidate among the stimuli responsive carriers. Thus magnetic hydrogels are potential carriers to engineer the 3D complex tissue by bottom-up assembly approaches and target the cancer region. The temperature-sensitive poly (*N*-isopropylacrylamide), PNIPA gels with the micron-sized iron and iron-oxide particles are explored for the combined hyperthermia and drug release applications. The study showed that the inducing heat to the magnetic hydrogels with varying concentration of the magnetic powder at frequency 375 kHz for magnetic field strength varying from 1.7 to 2.5 kA/m. These gel showed the collapse transition at 34 °C. The maximum temperature of 45 °C was obtained by using 2.5 wt.% Fe_3O_4 [27]. Other very remarkable studies carried out was the in situ implant formation that incorporates superparamagnetic iron oxide nanoparticles (SPIONS), which form the minimally invasive treatment for the cancer lesions by using magnetically induced local hyperthermia. This study demonstrated the use of silica microparticles to increase the flowability of the gel formulation. The use of alginate was highlighted, which induced to the formation of strong implants localizing the tumor periphery [28]. For future advancements SPIONS with combination of hydrogels are most eligible candidates for the cancer treatment.

4 Temperature-Responsive Hydrogels

The thermoresponsive hydrogels generally manage the non-linear sharp changes in the properties with the change in the temperature. The temperature act as a trigger to get the release of the drug at the desired site. The hydrogels based thermosensitive nanocarriers keep the drug stable inside the carrier and deliver the drug at the locally heated tumor site (~40–42 °C) and thus counter act the blood passage time and wash out the tumor [7]. The most commonly known polymers, which form thermosensitive hydrogels are the block copolymers having poly (ethylene glycol) (PEG) that are coupled with the hydrophobic polymer blocks. This category include the block copolymers composed of A-blocks and B-blocks arranged as ABA or BAB. The A block provides the biodegradability, biocompatibility, water solubility and the b block poly (propylene oxide) (PPO), polyesters such as (poly (D,L-lactide) (PLA), poly(D,L-lactide-co-glycolide) (PLGA), poly(ε-caprolactone) (PCL) and poly[(R)-3-hydroxybutyrate] (PHB)), polyphosphazenes, and polypeptides contributes to the hydrophobic part. These kind of hydrogel system are very important for imbibing less water soluble drugs [29]. Other important category of polymers, which are used for thermosensitive hydrogels include polymers like poly(*N*-isopropylacrylamide) (PNIPAAm) that undergo phase transition at temperature above 32 °C in an aqueous solutions. Other combination of linear and star shaped block copolymers, composed of central hydrophilic polyethylene oxide (PEO) and terminal PNIPAAm, demonstrated that these polymers form strong thermosensitive hydrogels. These materials are of great interest due to their temperature control over the in vivo studies. These systems have two primary categories viz., positively and negatively responsive systems. They are differentiated by an upper critical solution temperature (UCST) or a lower critical solution temperature (LCST), respectively.

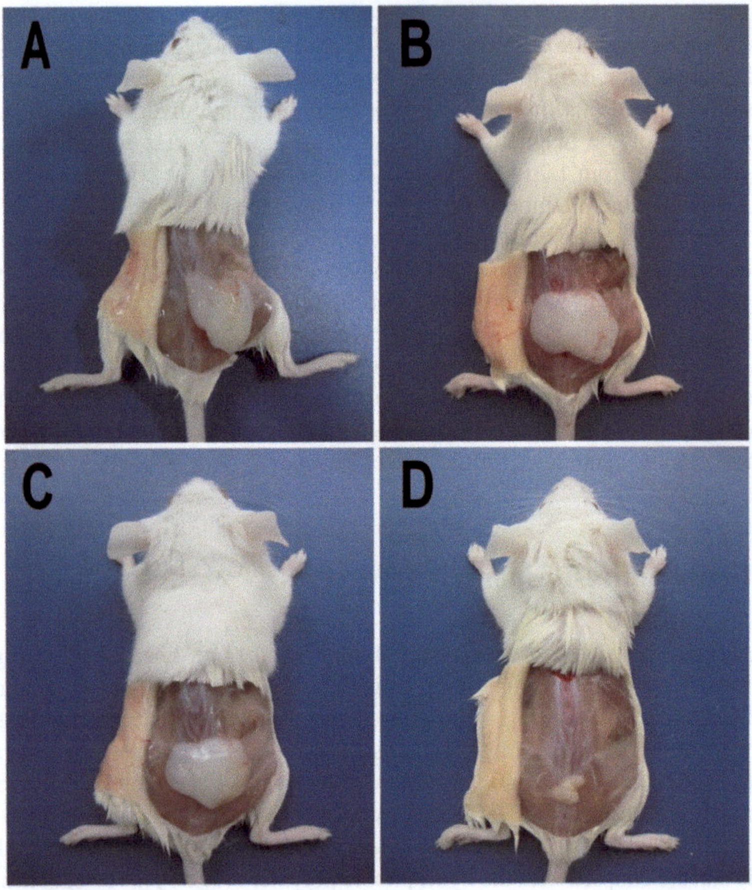

Fig. 8.4 In-vivo degradation behavior of Dex hydrogel. Reprinted from [31] with permission by Nature Publishing Group Copyright © 2015

The swelling and deswelling, which relates to the aqueous solubility describes the basis of these systems to be applied to tissue regeneration and drug delivery [30].

In recent years, the thermosensitive hydrogels have been exploited for dealing with major limitation like adhesion caused by the post-surgery. In a recent study, Dex hydrogel with dexamenthasone loaded polymeric micelles was tested for the anti-adhesion barrier as well as the controlled release of anti-adhesion drug. This hydrogel showed thermos-sensitivity and forms a non-flowable hydrogel upon subcutaneously injection in mice and also showed eventual degradation in 20 days. It was also seen that by the recurrent adhesion animal model, the adhesion alleviated significantly in hydrogel tested animal group. Figure 8.4 showed that after the subcutaneous injection the hydrogel become smaller over time and degraded. Thus this study demonstrated Dex hydrogel served as a potential anti-adhesion candidate [31]. Most importantly the very useful triblock copolymer used for the development of thermos-sensitive hydrogels is the triblock copolymer PEO–PPO–PEO trade name of Pluronic®. The increased hydrophobicity and dehydration of the PPO block with the increasing temperature result in the formation of in situ gel [32]. Researchers are working on the mechanical strength by blending PLGA with the biodegradable polymers and then applied to gene delivery, tissue engineering and drug delivery [33, 34]. Chitosan based materials are also well known for the design of the thermo-sensitive hydrogels for biomedical applications. In particular, the use of various polyols have gained attention owing to their ability to target the diseased area. The most highlighting polyol and extensively studied for in situ drug delivery is β-glycerophosphate (β-GP) [35]. These polyols maintain chitosan as a liquid and forms a shield of water around the chitosan chain in acidic media. The interactions suggested during the gelation process include the loss of electrostatic repulsion, hydrogen bonding and hydrophobic interactions between the chitosan chains [36]. The increase in pH reduce the electrostatic repulsion between the chitosan chains. Thus a high concentration of β-GP is thus used, which raises the concern of the toxicity [37]. Thus introduction of inorganic phosphate salts and added cross linkers like glyoxals, alginates and hydroxyethyl cellulose have been suggested [38]. Currently, it has been suggested that the chitosan/GP hydrogel systems need neuronal compatibility if they had to find compatibility to the nervous system thus polylysine has been allowed to diffuse away to the material and it was observed that the covalent binding of polylysine to chitosan improved the biocompatibility and neuron affinity of the system [39]. Further the thermosensitive hydrogels have also been explored for the nasal delivery, which is a preferred mode of delivery. There is an immense demand for the development of mucoadhesive viscoelastic hydrogels for nasal drug delivery. These system provide efficient system exhibiting the prolonged residence time of the active mucosal surface. Nasal formulations are in market in the form of nasal sprays or drops which are designed such that they undergo sol–gel transition at the deposition site, which directly imply that the viscosity and the rheological parameters are modulated in such a manner that result in the mucoadhesive system [40, 41]. For example, a series of *N*-trimethyl chitosan chloride polymers were synthesized using different molecular weight chitosans and designed into hydrogels using poly(ethylene glycol) and glycerophosphate. The hydrogels derived from *N*-trimethyl chitosan with a low degree of quaternisation having medium average molecular weight exhibit relatively short sol–gel conversion time. Though these hydrogels exhibited good water absorbing capacity and good mucoadhesive behavior [42]. The thermosensitive hydrogels hold great potential for the clinical application of biodegradable gels [43].

5 Pressure Sensitive Hydrogels

Pressure sensitive response of hydrogels have been explored for drug delivery in last few years [44]. It is observed that the hydrogels undergo pressure induced volume phase transition and which was conceptualized by using thermodynamic calculations on uncharged hydrogel theory. This theory explains the expansion of the hydrogels at low pressure. This effect is further explained on the basis that the total free volume of the system is reduced, which lead to the loss of entropy. The gels remain in the collapsed state and result in increase in the internal energy and the gel–gel and water–water interactions are more by the compression of the system. The study on the with poly(*N*-isopropylacrylamide) hydrogels established this prediction [45]. The other polymers that show pressure sensitivity poly(*N*-propylacrylamide), poly(*N,N*-diethylacrylamide) and poly(*N*-isopropylacrylamide) very close to their LCSTs [46]. A very interesting invention demonstrated a method for triggering release of the drug from a hydrogel to tissue at a desired location of the body using a catheter. This catheter is coated on the outer surface with hydrogel polymer which shows the capacity to incorporate a predetermined amount of drug which is inside the polymer and get released by the external condition (Pressure).

A balloon catheter is revealed in which the hydrogel is supported on the exterior surface of the balloon [47]. Thus pressure sensitive hydrogels own great potential for biomedical applications.

6 Electric Field Sensitive Hydrogels

Electric field sensitive hydrogels have shown their great potential for biomedical applications [48, 49]. An electric field is a very important stimulus that has numerous advantages such as the control over the magnitude of the current duration of electric pulses the interval between the pulses. The transdermal drug delivery have used the electric current in vivo like the electroporation and iontophoresis [50, 51]. This can also be used in the topical application. Clinically, it is in the market in USA Iontocaine®—which is a device used to deliver lignocaine by iontophoresis. The possible way of delivering the drugs is by the subcutaneous transplantation of the hydrogels which acts when the drug release is desired the electro-conducting patch is applied on the skin directly over the gel. When electric field is applied onto the skin, the electric field stimulates the hydrogel and release the drug further the patch can be removed when the required amount of drug is released. The other strategy which is suggested is the wrist watch form of the electroconductive patch and electrodes over the implant site. This eliminates the need for the surgical implantation of the hydrogel. As soon as the electric field is applied to the hydrogels they respond by swelling or bending relative to the strength of the electric field and the position of the electrodes [52] (Fig. 8.5). The bending of the hydrogels are generally observed with the mechanical devices such as valves, artificial fingers/hands/muscles, switches, "soft-actuators" and "molecular machines."

The role of electrically induced anisotropic gel was first introduced by Tanaka et al. [53] and the group suggested that the electric field produces a force on the counter ions and the immobile charged groups present on the hydrogel matrix. Figure 8.6 demonstrate the movement of ions in partially hydrolyzed polyacrylamide gels when fixed to one of the electrodes. It shows that the mobile H^+ ions tend to migrate towards the cathode and the negatively charged immobile acrylate groups

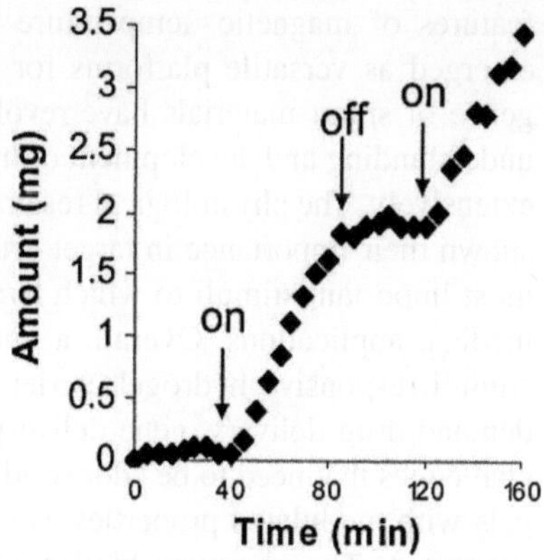

Fig. 8.5 Pulsatile drug (insulin) release profile from a chondroitin sulphate gel, when an electric field was switched "on" and "off". Reprinted from [52], Copyright (2003), with permission from Elsevier

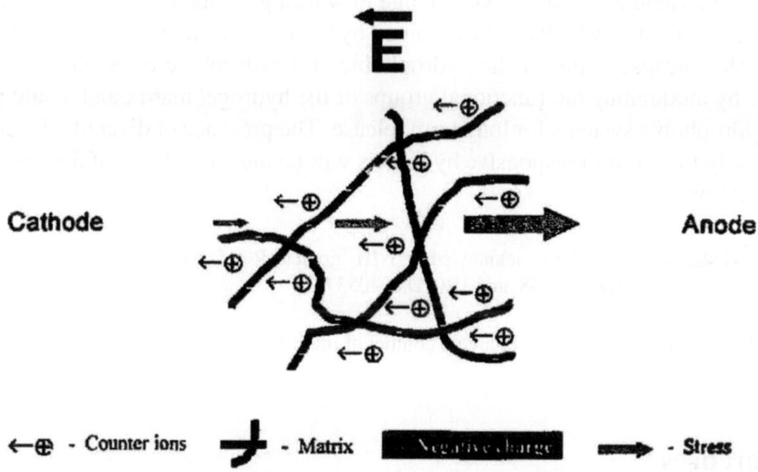

Fig. 8.6 The effect of an electric field on a polyelectrolyte gel. Positively charged counterions migrate towards the cathode, while the immobile polymeric anionic groups are attracted towards the anode. Attraction being greatest at the anode side and smallest at the cathode side, a stress gradient (greatest at the anode) is established. Reprinted from [54] Copyright (1995), with permission from Elsevier

move toward the anode. This pull creates the uniaxial stress along the hydrogel axis due to the fact that the stress is highest at the anode and lowest at the cathode thus deformation of the gel contributes to the stress gradient.

7 Conclusion and Future Prospectives

In recent years, the design and applications of stimuli responsive hydrogels have shown potential results. The innovation behind the synthesis and degradability of the hydrogel based carriers have been the primary importance. The upcoming features of magnetic, temperature and electric field responsive hydrogels have emerged as versatile platforms for biomedical applications. Moreover the emergence of smart materials have revolutionized the clinical industry. The increased understanding and development of hydrogel based biomaterials have been explored extensively. The physiological resemblance and compatibility of the hydrogels have shown their importance in target drug and gene delivery. This chapter covered the most important stimuli to which hydrogels give response and are utilized for biomedical applications. Overall, a number of studies have shown the potentials of stimuli responsive hydrogel carriers including cancer therapy, drug delivery, on-demand drug delivery, gene delivery and tissue engineering. There is still several challenges that need to be addressed for these applications. The need for 3D hydrogels with modulated properties, compatible tissue construct and specific design is suggested. The complete biodegradability of the magnetic hydrogels has to be addressed. There is a persistent challenge in working for the various types of kinetics release profiles, which can be obtained by using these stimuli responsive hydrogels. The encapsulation for the hydrophobic and hydrophilic drugs may be made easier by modulating the functional groups of the hydrogel matrix and would allow the hydrophobic systems for long term release. The presence of diversified features, it is likely that stimuli responsive hydrogels will be the eternal part of drug delivery in future biomaterials.

Acknowledgements Authors acknowledge NIH grants: RO1DA027049, R21MH 101,025, RO1DA 034547, R01DA037838, and 1R01DA040537.

Conflict of Interest Authors declare no conflict of interest.

References

1. Vashist A, Ahmad S. Hydrogels: smart materials for drug delivery. Orient J Chem. 2013;29:861–70.
2. Vashist A, Vashist A, Gupta Y, Ahmad S. Recent advances in hydrogel based drug delivery systems for the human body. J Mater Chem B. 2014;2:147–66.
3. Gupta P, Vermani K, Garg S. Hydrogels: from controlled release to pH-responsive drug delivery. Drug Discov Today. 2002;7:569–79.
4. Jing G, Wang L, Yu H, Amer WA, Zhang L. Recent progress on study of hybrid hydrogels for water treatment. Colloids Surf A Physicochem Eng Asp. 2013;416:86–94.
5. Hoare TR, Kohane DS. Hydrogels in drug delivery: progress and challenges. Polymer. 2008;49:1993–2007.
6. Uhrich KE, Cannizzaro SM, Langer RS, Shakesheff KM. Polymeric systems for controlled drug release. Chem Rev. 1999;99:3181–98.
7. Mura S, Nicolas J, Couvreur P. Stimuli-responsive nanocarriers for drug delivery. Nat Mater. 2013;12:991–1003.
8. Lin C-C, Metters AT. Hydrogels in controlled release formulations: network design and mathematical modeling. Adv Drug Deliv Rev. 2006;58:1379–408.
9. Vashist A, Ahmad S. Hydrogels in tissue engineering: scope and applications. Curr Pharm Biotechnol. 2015;16:606–20.

10. Wei Q, Xu M, Liao C, Wu Q, Liu M, Zhang Y, et al. Printable hybrid hydrogel by dual enzymatic polymerization with superactivity. Chem Sci. 2016;7:2748–52.
11. Li L, Shi Y, Pan L, Shi Y, Yu G. Rational design and applications of conducting polymer hydrogels as electrochemical biosensors. J Mater Chem B. 2015;3:2920–30.
12. Merino S, Martín C, Kostarelos K, Prato M, Vázquez E. Nanocomposite hydrogels: 3D polymer–nanoparticle synergies for on-demand drug delivery. ACS Nano. 2015;9:4686–97.
13. Tabujew I, Peneva K. Functionalization of cationic polymers for drug delivery applications. In: Samal SK, Dubruel P, editors. Cationic polymers in regenerative medicine. Cambridge: Royal Society of Chemistry; 2014.
14. Watkins KA, Chen R. pH-responsive, lysine-based hydrogels for the oral delivery of a wide size range of molecules. Int J Pharm. 2015;478:496–503.
15. Li L, Gu J, Zhang J, Xie Z, Lu Y, Shen L, et al. Injectable and biodegradable pH-responsive hydrogels for localized and sustained treatment of human fibrosarcoma. ACS Appl Mater Interfaces. 2015;7:8033–40.
16. Song HS, Kwon OS, Kim J-H, Conde J, Artzi N. 3D hydrogel scaffold doped with 2D graphene materials for biosensors and bioelectronics. Biosens Bioelectron. 2017;89:187–200.
17. Campbell S, Maitland D, Hoare T. Enhanced pulsatile drug release from injectable magnetic hydrogels with embedded thermosensitive microgels. ACS Macro Lett. 2015;4:312–6.
18. Li Y, Huang G, Zhang X, Li B, Chen Y, Lu T, et al. Magnetic hydrogels and their potential biomedical applications. Adv Funct Mater. 2013;23:660–72.
19. Sapir Y, Cohen S, Friedman G, Polyak B. The promotion of in vitro vessel-like organization of endothelial cells in magnetically responsive alginate scaffolds. Biomaterials. 2012;33:4100–9.
20. Zhang Y, Yang B, Zhang X, Xu L, Tao L, Li S, et al. A magnetic self-healing hydrogel. Chem Commun. 2012;48:9305–7.
21. Liu T-Y, Hu S-H, Liu T-Y, Liu D-M, Chen S-Y. Magnetic-sensitive behavior of intelligent ferrogels for controlled release of drug. Langmuir. 2006;22:5974–8.
22. Szabo D, Szeghy G, Zrinyi M. Shape transition of magnetic field sensitive polymer gels. Macromolecules. 1998;31:6541–8.
23. Snyder R, Nguyen V, Ramanujan R. Design parameters for magneto-elastic soft actuators. Smart Mater Struct. 2010;19:055017.
24. Fuhrer R, Athanassiou EK, Luechinger NA, Stark WJ. Crosslinking metal nanoparticles into the polymer backbone of hydrogels enables preparation of soft, magnetic field-driven actuators with muscle-like flexibility. Small. 2009;5:383–8.
25. Liu T-Y, Hu S-H, Liu D-M, Chen S-Y, Chen I-W. Biomedical nanoparticle carriers with combined thermal and magnetic responses. Nano Today. 2009;4:52–65.
26. Kost J, Wolfrum J, Langer R. Magnetically enhanced insulin release in diabetic rats. J Biomed Mater Res. 1987;21:1367–73.
27. Ang K, Venkatraman S, Ramanujan R. Magnetic PNIPA hydrogels for hyperthermia applications in cancer therapy. Mater Sci Eng C. 2007;27:347–51.
28. Le Renard P-E, Jordan O, Faes A, Petri-Fink A, Hofmann H, Ruefenacht D, et al. The in vivo performance of magnetic particle-loaded injectable, in situ gelling, carriers for the delivery of local hyperthermia. Biomaterials. 2010;31:691–705.
29. McKenzie M, Betts D, Suh A, Bui K, Kim LD, Cho H. Hydrogel-based drug delivery systems for poorly water-soluble drugs. Molecules. 2015;20:20397–408.
30. Nguyen MK, Lee DS. Injectable biodegradable hydrogels. Macromol Biosci. 2010;10:563–79.
31. Wu Q, Wang N, He T, Shang J, Li L, Song L, et al. Thermosensitive hydrogel containing dexamethasone micelles for preventing postsurgical adhesion in a repeated-injury model. Sci Rep. 2015;5:13553.
32. Fusco S, Borzacchiello A, Netti P. Perspectives on: PEO-PPO-PEO triblock copolymers and their biomedical applications. J Bioact Compat Polym. 2006;21:149–64.
33. Jeong B, Bae YH, Kim SW. In situ gelation of PEG-PLGA-PEG triblock copolymer aqueous solutions and degradation thereof. J Biomed Mater Res. 2000;50:171–7.

34. Yan S, Zhang X, Zhang K, Di H, Feng L, Li G, et al. Injectable in situ forming poly (l-glutamic acid) hydrogels for cartilage tissue engineering. J Mater Chem B. 2016;4:947–61.
35. Tahrir FG, Ganji F, Ahooyi TM. Injectable Thermosensitive chitosan/Glycerophosphate-based hydrogels for tissue engineering and drug delivery applications: a review. Recent Pat Drug Deliv Formul. 2015;9:107–20.
36. Cho J, Heuzey M-C, Bégin A, Carreau PJ. Physical gelation of chitosan in the presence of β-glycerophosphate: the effect of temperature. Biomacromolecules. 2005;6:3267–75.
37. Cho J, Heuzey M-C, Bégin A, Carreau PJ. Chitosan and glycerophosphate concentration dependence of solution behaviour and gel point using small amplitude oscillatory rheometry. Food Hydrocoll. 2006;20:936–45.
38. Zhou HY, Jiang LJ, Cao PP, Li JB, Chen XG. Glycerophosphate-based chitosan thermosensitive hydrogels and their biomedical applications. Carbohydr Polym. 2015;117:524–36.
39. Crompton K, Goud J, Bellamkonda R, Gengenbach T, Finkelstein D, Horne M, et al. Polylysine-functionalised thermoresponsive chitosan hydrogel for neural tissue engineering. Biomaterials. 2007;28:441–9.
40. Elad D, Wolf M, Keck T. Air-conditioning in the human nasal cavity. Respir Physiol Neurobiol. 2008;163:121–7.
41. Mygind N, Dahl R. Anatomy, physiology and function of the nasal cavities in health and disease. Adv Drug Deliv Rev. 1998;29:3–12.
42. Nazar H, Fatouros DG, van der Merwe SM, Bouropoulos N, Avgouropoulos G, Tsibouklis J, et al. Thermosensitive hydrogels for nasal drug delivery: the formulation and characterisation of systems based on N-trimethyl chitosan chloride. Eur J Pharm Biopharm. 2011;77:225–32.
43. Gou M, Li X, Dai M, Gong C, Wang X, Xie Y, et al. A novel injectable local hydrophobic drug delivery system: biodegradable nanoparticles in thermo-sensitive hydrogel. Int J Pharm. 2008;359:228–33.
44. Di J, Yao S, Ye Y, Cui Z, Yu J, Ghosh TK, et al. Stretch-triggered drug delivery from wearable elastomer films containing therapeutic depots. ACS Nano. 2015;9:9407–15.
45. Lee K, Cussler E, Marchetti M, McHugh M. Pressure-dependent phase transitions in hydrogels. Chem Eng Sci. 1990;45:766–7.
46. Zhong X, Wang Y-X, Wang S-C. Pressure dependence of the volume phase-transition of temperature-sensitive gels. Chem Eng Sci. 1996;51:3235–9.
47. Rosenthal A, Barry JJ, Sahatjian R. Triggered release hydrogel drug delivery system. Google Patents, 2003.
48. Yuk SH, Cho SH, Lee HB. Electric current-sensitive drug delivery systems using sodium alginate/polyacrylic acid composites. Pharm Res. 1992;09:955–7.
49. Kwon IC, Bae YH, Kim SW. Electrically credible polymer gel for controlled release of drugs. Nature. 1991;354:291.
50. Delgado-Charro MB, Guy RH. Transdermal iontophoresis for controlled drug delivery and non-invasive monitoring. STP Pharma Sci. 2001;11:404–14.
51. Vanbever R, Preat V. In vivo efficacy and safety of skin electroporation. Adv Drug Deliv Rev. 1999;35:77–88.
52. Murdan S. Electro-responsive drug delivery from hydrogels. J Control Release. 2003;92:1–17.
53. Tanaka T, Nishio I, Sun S-T, Ueno-Nishio S. Collapse of gels in an electric field. Science. 1982;218:467–9.
54. Tomer R, Dimitrijevic D, Florence AT. Electrically controlled release of macromolecules from cross-linked hyaluronic acid hydrogels. J Control Release. 1995;33:405–13.

Chapter 9
On-Demand Controlled Drug Delivery

Abhijeet Joshi, Rashmi Chaudhari, and Rahul Dev Jayant

Abstract Spatial and temporal control of drug delivery is of prime importance for establishing the therapeutic compliance of drugs for various diseases. Conventional approaches to drug delivery for temporal control of drug delivery include encapsulation, entrapment and conjugation to polymeric materials for obtaining the controlled release. Several macro, micro and nanoformulations have been researched and commercialized for producing controlled release of drugs. Apart from the control over the rate of release, a regional delivery would be highly desirable for increasing the efficacy of the drugs and reducing the undue side effects pertaining to the therapy. Some developments in region specific delivery have been utilizing physiological differences of various sites in the body. Topical formulations have been extensively explored for their region specific delivery due to ease of access of these organs like eye, ear, nasal, oral, vaginal and rectal cavities or parts of gastro-intestinal tract like mouth, stomach, intestine and colon. Several sites of the body have not been reached using drug delivery formulations to selectively deliver the drugs to particular organs due to several physiological barriers. Recent developments and approaches in material chemistry, novel polymers, and technology advancements have led to new avenues in the development of nano/micro-carriers or materials for on-demand controlled drug delivery or stimuli responsive drug delivery. On-demand drug release although complex has become possible due to materials which recognize the microenvironments and react in a dynamic way altering properties to cause release of encapsulated drugs within. The current chapter

A. Joshi (✉)
Centre for Biosciences and Bio-medical Engineering, Indian Institute of Technology Indore, Indore, Madhya Pradesh 453552, India
e-mail: abhijeet.joshi@iiti.ac.in

R. Chaudhari
Department of Biosciences and Bioengineering, Indian Institute of Technology Bombay, Mumbai 400076, India

R.D. Jayant (✉)
Center for Personalized Nanomedicine, Institute of Neuro-Immune Pharmacology, Department of Immunology, Herbert Wertheim College of Medicine, Florida International University, Miami, Florida 33199, USA
e-mail: rjayant@fiu.edu

A. Kaushik et al. (eds.), *Advances in Personalized Nanotherapeutics*,
https://doi.org/10.1007/978-3-319-63633-7_9

reviews different approaches of developing on-demand drug release using different materials and techniques.

Keywords Stimuli Responsive • On-Demand • Drug • Delivery • Controlled • Site-Specific

1 Introduction

Release of drugs can be characterized by movement of drug solutes from a vehicle into a release medium. This process appears to be very simple but is affected by multitude of factors including physico-chemical properties of drugs, materials of matrices used for vehicles, release environment and combinations of these factors. Tablets and capsules are the most primitive of the drug delivery dosage forms which function as immediate releasing mechanisms of drugs. With the advent of coating techniques like film coating, sugar coating and enteric coating, modifications in the release profile of drugs could be obtained with varying drug release times and prolonging the duration of action. A regional specificity was also obtained for delivery of drugs specifically in the intestine using enteric coated polymers. The release from the coated dosage forms is dependent on the type of coating material and the release environment which eventually brings about controlled release and site specific release. The materials used as vehicle can function as biodegradable or non-biodegradable. Drug transport from non-degradable systems can be considered as diffusion mediated governed by the Fick's law. Non degradable vehicles can be developed as a reservoir or matrix type of systems. Reservoir type systems contain an inert coating material which functions to control the rate of release. Matrix type systems contain the drug evenly distributed in the vehicle to control the rate of release. The release rate in case of reservoir systems is dependent on thickness and permeability of coating materials where as in case of matrix type systems the release rate is dependent on the concentration gradient, diffusion distance and degree of swelling. Degradable materials dissolve in a release media exposing the inner encapsulated drugs to be dissolved. Depending on the type of carrier used the drug vehicle can generate different kinds of release kinetics ranging from zero order, first order, diffusion controlled or dissolution controlled or any combination of them have been described [1].

Movement of drug follows drug release from a vehicle to the site of action, which is affected by several pharmacokinetic processes like absorption, distribution, metabolism and elimination. Bioavailability is a critical factor which is dependent on the pharmacokinetic parameters of drugs which can be understood as the rate and extent of movement of drug from the drug vehicle to site of action. The conventional systems have been employing immediate or burst releasing mechanisms which suffer with disadvantages like inefficient absorption, short time of action and local toxicity. Controlled drug delivery vehicles aim to solve these perils so that the bioavailability of drugs can be enhanced. The aim of controlled delivery has been fairly achieved in the literature but targeting, site specific release to unreachable organs and controlling release at a particular site needs further effort. In spite of several advancements in the drug delivery fields, several organs have not be extensively reached using the available drug delivery carriers and new carriers, nanomaterials or new polymeric materials need to be investigated.

2 Exogenous Stimuli for On-demand Drug Delivery

An important aspect of drug delivery is bringing about spatial and temporal control of drug delivery which is the localization of drug delivery at a particular site and controlling the rate of drug release over a period of time, respectively. Apart from the physical localization of drug delivery at a site, researchers have investigated several triggering systems which can selectively deposit molecules of interest at a particular site. Non-invasive triggering mechanisms will be a boon to healthcare as they can be formed as delivery systems without any surgical intervention. Different stimuli used for developing exogenously triggered delivery systems include temperature, magnetic or electrical field, ultrasound and light (Fig. 9.1). Most of the stimuli responsive delivery systems used in exogenous class function by increasing the temperature in the microenvironment of the drug delivery carriers. A hyperthermic effect is generated which either causes release of drug from the carriers or cause an increased temperature which itself causes cells to undergo necrosis and coagulation. Hyperthermia can be categorized into local, regional and whole body hyperthermia depending on the location and purpose of use. Local hyperthermia involves heating up only a small region of interest such as a tumor. Regional hyperthermia may involve heating up larger tissues or organs while whole body hyperthermia is heating up whole body. Whole body hyperthermia can be useful for metastatic cancer cells. A Moderate hyperthermia (41–46 °C) causes several changes in the cells

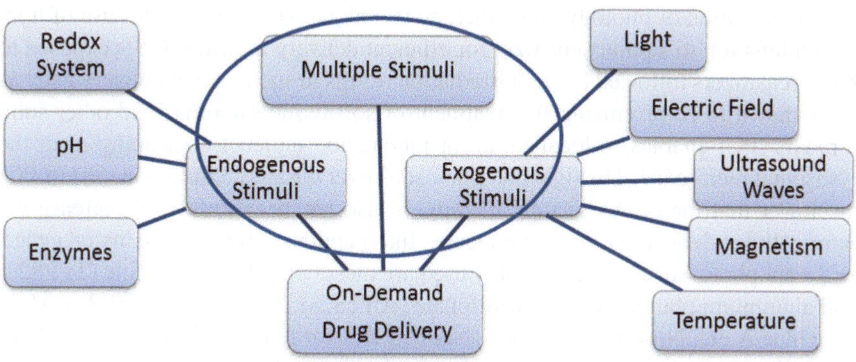

Fig. 9.1 Different mechanisms of on-demand drug delivery

which include initiation of several intra and extracellular degradation apparatuses like protein denaturation, protein folding, aggregation and DNA cross linking.

This can lead to permanent protein damage resulting in inhibition of many cellular functions. These facts can be used for treatment of localized cancers. The challenge of hyperthermic methods of therapy is to cause damage to diseased tissues and spare the normal tissues. Currently available techniques for induction of hyperthermia are ultrasound, radiofrequency, microwaves, infrared radiation, magnetically excitable thermoseeds etc. All the modes of hyperthermia can be used for developing on demand drug delivery systems.

2.1 Light Activated Drug Delivery

Light has now been known to be intriguing stimuli for on demand drug delivery. Cancer management generally encompasses use of a combination of chemotherapy and phototherapy. The main advantages of using light as a stimulus includes its simplicity, non-invasiveness and amenability of modifications in wavelength, exposure time, beam diameter and intensity providing control over the quantity, timing and location of drug released [2]. UV and visible lights due to their poor penetration and absorption by tissues are more suited for topical applications where as on the other hand for deep tissues near infra-red light (650–1000 nm) is more suitable. UV, visible or near-infra red radiations can be used as a mode for targeted imaging and photo-triggered drug delivery systems. Photo-triggered drug delivery systems are externally activated drug delivery systems which release the drug at the site and rate can be controlled based on the matrix used. Some light-responsive DDS are of a single use (i.e., the light triggers an irreversible structural change that provokes the delivery of the entire dose) while others are able to undergo reversible structural changes when cycles of light/dark are applied, behave as multi-switchable carriers (releasing the drug in a pulsatile manner).

Common modes of photon induced drug delivery involve photo-chemical triggering or photo-thermal triggering mechanism. Photochemical internalization (PCI) is a modification of photodynamic therapy technique which constitutes use of light in combination to a photosensitizer for efficient delivery of drugs. Cells exposed to photosensitizers followed by light irradiation results in such type of delivery and has been shown to be promising for treatment of soft tissue sarcomas and other solid tumours. PCI enables re-localization of therapeutic molecules escaping from the lysosomal membrane. The light and photosensitizer dosages required in case of PCI are lower that the photodynamic therapy. It also has been known to increase the intra-cellular delivery of large molecules like peptides, proteins, toxins or genes. Different photosensitizers commonly used are meso-tetraphenylporphine (TPPS2a) and aluminium phthalocyanine disulfonate (AlPcS2a), Amphinex, 3THPP, 5-ALA, Nile blue A etc. Norum et al. have described PCI for release of endocytosed bleomycin into the cytoplasm by photochemical rupture. In this study human fibrosarcoma xenograft HT1080 was used in leg muscles of athymic mice along with

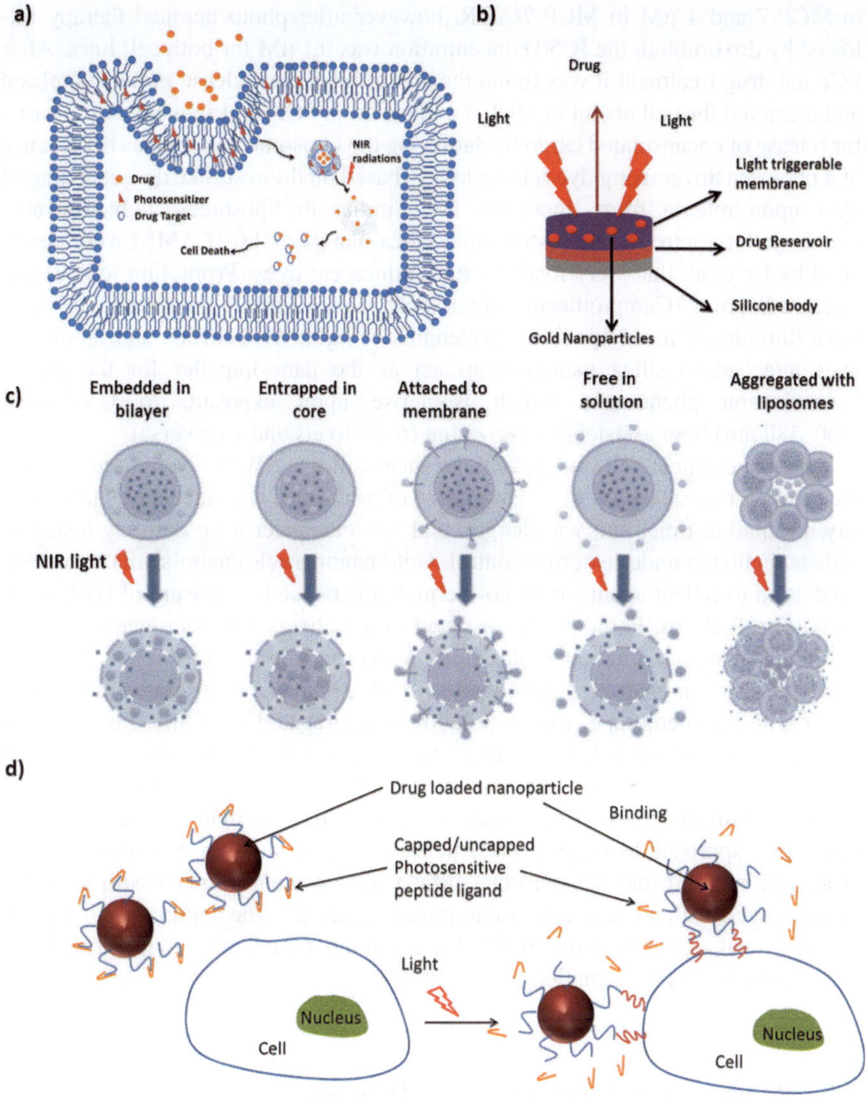

Fig. 9.2 Mechanism of (**a**) photochemical internalization using photosensitizer's (**b**, **c**) photothermal therapy using gold nanoparticles for light triggered drug delivery and (**d**) photo-triggered targeting using capped ligands conjugated to nanoparticles

di-sulfonated aluminium phthalo-cyanine. The results suggested that photochemical internalization is superior to photodynamic therapy [3]. Lou PJ et al. have described the usage of photo chemical internalization on release of doxorubicin from endocytic vesicles in MDR cells of cancer cell lines MCF-7 and MCF-7/ADR using a photosensitizer TPPS (di-sulfonated meso-tetra phyenyl porphine) and light (Fig. 9.2a). Exposure of doxorubicin alone gave the IC50 concentrations of 0.1 μM in MCF-7 and 1 μM in MCF-7/ADR, however after photochemical therapy followed by doxorubicin the IC50 concentration was 0.1 μM for both cell lines. After PCI and drug treatment it was found the doxorubicin was release in the cytoplasm and it entered the cell nuclei in MCF-7 cells without PCI [4]. Light has also be used for release of encapsulated cargo by damaging the liposome membranes in presence of a photosensitizer using dye release assays based on fluorescence de-quenching of dyes upon release from liposomes [5]. Similar to liposomes, light activated nanoimpeller-controlled mesostructured silica nanoparticles (LAMS) were developed by Lu et al. that were loaded with luminescent dyes (Propidium iodide) and anticancer drugs (Camptothecin) which were released inside the cancer cells that were illuminated using specific wavelengths of light. Azobenzene and its derivatives attached to silica nanoparticles act as the nano-impeller for the photo-isomerization phenomena which isomerise upon exposure to UV-Visible (300–380 nm) light assisted isomerization (*trans* to *cis* and vice versa).

Pancreatic cancer cell line and colon cancer cell line SW480 were used to study the uptake of particles. Intracellular release of molecules is controlled by the intensity, irradiation time, and wavelength, and the anticancer drug delivery inside of cells is regulated under external control. Gold nanoparticles/nano-shells have been used as an excellent agent for photo-thermal effects due to exposure of NIR radiations (Fig. 9.2b, c). Photothermal assisted drug delivery has been carried out by co-immobilizing gold nanoparticles along with drug carriers to form nanoformulations like gold nanoparticles embedded within the bilayer, encapsulated in the core, tethered to the membrane, free in solution or aggregated with liposomes or gold nanoshells embedded polymeric films which release the loaded drugs after light exposure [6]. Subcutaneous implantation of such a device with fast acting insulin shows a significant control over release of insulin after NIR light exposure. Another interesting approach to target nanoparticles to desired site was described by Dvir et al. who showed that nanoparticle surface was covalently linked with targeting moiety caged with a photo-removable protecting group. The caging group used in this study was a small peptide YIGSR that is required for adhesion to integrin β1 and can be removed by illumination (Fig. 9.2d) [7].

2.2 *Temperature Triggered Drug Delivery*

Temperature responsive polymers are carriers of choice for stimuli responsive drug delivery because of their volume phase transition at a particular temperature (Fig. 9.3a). Lower critical solution temperature (LCST) and upper critical solution temperature (UCST) are the parameters governing the solubility based on temperature. Polymers precipitating upon heating have a LCST and polymers which are solubilized upon heating have an UCST. Hydrogels behaving in temperature responsive manner can be characterized using lower gel transition temperature (LGTT) and upper gel transition temperature (UGTT). Typically used LCST polymers include *N*-isopropyl acrylamide (NIPAM), *N*, *N*-diethylacrylamide (DEAM),

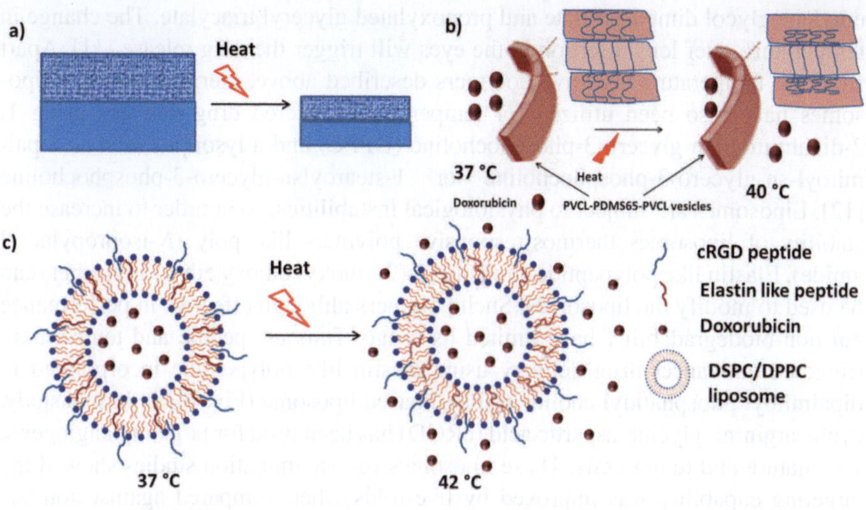

Fig. 9.3 (**a**) Volume phase transitions effects in thermos-responsive polymers in response to temperature changes, (**b**) Temperature responsive polymerosomes (**c**) Mechanism and functioning of targeted thermo-responsive liposomes

Methyl vinyl ether (MVE) *N*-vinyl caprolactum [8]. On the other hand UCST polymeric systems include combination of acrylamide and acrylic acid, PEO-b-PPO block copolymers, PEO-b-PPO-b-PEO and PEG-b-PLGA-b-PEG. Apart from the LCST other factors affecting the solubilisation of polymers is governed by molecular weight, concentration, temperature, presence of co-solvents and additives. PNIPAM co-polymers have been thoroughly investigated for temperature responsive properties.

Temperature responsive polymers based on NIPAM or *N*-viny caprolactum can act as a base for development of temperature triggered drug delivery system. Temperature responsiveness in polymerosomes developed from poly(*N*-vinylcaprolactam)n-poly(dimethylsiloxane)65-poly(*N*-vinylcaprolactam) co-polymers was described by Liu et al. for cancer chemotherapy (Fig. 9.3b). These polymersomes were described to be biocompatible, biodegradable, monodisperse, stable at room temperature, tunable size, thermally responsive and showed a high (40%) loading capacity [9]. Temperature sensitive gels using polymers are most extensively investigated for drug delivery. Hydrogels possessing swelling and deswelling properties contain insoluble polymers above LCST or UCST. Temperature triggered in-situ gels can be useful in development of ophthalmic formulations. Thermosensitive polymer pluronic F-127, HPMC-E50 LV and anti-glaucoma drug Betaxolol HCl were combined together to increase residence time of drug in the eyes for over 7 h. The polymers in the form of in situ gel showed good ocular tolerance. Such a formulation would have a great potential in glaucoma therapy [10]. Temperature triggered delivery to eye has also been described by embedding Timolol in HEMA gel which is a material for development of contact lenses. The drug release has been shown to be prolonged for 2–4 weeks using crosslinked particles of ethylene glycol dimethacrylate and propoxylated glyceryl triacylate. The change in temperature after lens insertion in the eyes will trigger the drug release [11]. Apart from the temperature sensitive polymers described above, thermoresponsive liposomes have also been utilized for temperature triggered drug delivery using 1, 2-dipalmitoyl-sn-glycero-3-phosphocholine (DPPC) and a lysolipid such as 1-palmitoyl-sn-glycero-3-phosphocholine or 1-stearoylsn-glycero-3-phosphocholine [12]. Liposomes are subject to physiological instabilities, so in order to increase the stability of liposomes thermos-responsive polymers like poly (*N*-isopropylacryl amide), Elastin like polypeptide and poly [2-(2-ethoxy) ethoxy ethyl vinyl ether) can be used to modify the liposomes. Such polymers although effective in performance but non-biodegradability have limited its usage. Tumour specific and temperature triggered anticancer drug delivery using elastin like polypeptide incorporated in dipalmitoyl phosphatidyl choline (DPPC) based liposome (Fig. 9.3c). In this study, cyclic arginine-glycine-aspartic acid (cRGD) has been used for targeting angiogenic vasculature and tumor cells. These liposomes for accumulation studies showed the targeting capability was improved by five-folds when compared against non-targeted liposomes [13]. These methods directly employ temperature dependent mechanisms to release the drugs. The change in temperature can also be generated using photothermal effects, magnetic hyperthermia and sonophoretic hyperthermia. All the other stimuli can be coupled with temperature responsive materials to obtain a temperature triggered drug delivery.

2.3 Ultrasound Triggered Drug Delivery

Ultrasound has been conventionally being used as an imaging method for soft tissues using longitudinal sound waves. In order to increase the contrast of the soft tissue imaging microbubbles have been regularly utilized. Ultrasound exposure can also be used to increase local deposition of therapeutic agents or modify vascular permeability and reduce the systemic side effects. Modifying vascular permeability, delivery of drugs in microbubbles, cavitation of skin for improving permeability of skin are some of the approaches which can be used for drug delivery. Weak and non-permeable drugs e.g. bleomycin, facilitation of the local release of chemotherapeutic agents such as doxorubicin, cisplatin and paclitaxel, promote internalization of plasmid DNA and siRNA can be improved by application of ultrasound waves. Two mechanisms (inertial cavitation and sub-cavitary oscillation) of increasing the cell permeability by ultrasound mediated microbubble delivery have been postulated. Strain resulting from the mechanical force creates gaps in the adjacent tissues, which increases the drug permeation. Microbubbles are one of the important drug delivery systems for ultrasound mediated drug delivery. They are minute air/gas filled microspheres with specific acoustic properties useful as imaging agents and drug delivery vehicles. Presence of even a single microbubble can be detected using an ultrasound. Collapse of microbubble leads to release of drugs and disappearance of hyper-echogenicity [14]. The advantage of using ultrasound and microbubbles is

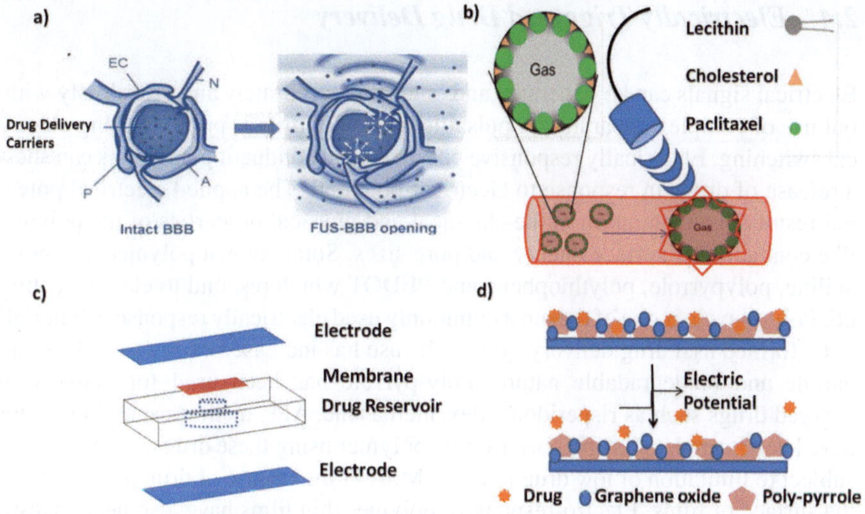

Fig. 9.4 (**a**) Focused ultrasound for blood brain barrier opening, (**b**) Paclitaxel loaded microbubbles for ultrasound mediated delivery in iliac artery restenosis, (**c**) Schematic of induction of drug release from electrically responsive materials and (**d**) Delivery of drug from Graphene-poly-pyrrole thin films by application of electrical potential

that they can be activated by ultrasound beam and that those in the path of the beam will be activated. Using ultrasound, microbubble-targeted delivery is possible for organ specific or site specific delivery. Microbubbles can be also be used for delivery of high molecular weight compounds like proteins, genes, dextran particles etc. using different pathways like macro-pinocytosis, clathrin and caveolin [15].

AFM and SEM evaluation of cavitation phenomenon has been described by Kudo et al. which suggest that the cavities can aid in improving concentrations of drugs at tumor tissues using ultrasound. Paclitaxel loaded microbubbles were investigated along with ultrasound for restenosis in rabbit iliac balloon injury model (Fig. 9.4b). It was showed that the microbubbles of paclitaxel can be visualized in the iliac artery and also be able release the encapsulated drug locally [16]. Treat et al. studied magnetic resonance guided ultrasound drug delivery using microbubbles. They showed that Dox-albumin shelled microbubbles can be taken up across BBB to a great extent with increased tissue damage (Fig. 9.4a) [17]. Jung et al. showed development of Gd(III) DOTA modified sonosensitive liposomes for delivery of DOx to a particular site, time and enhanced signal intensity in regions of accumulation. Gd (III) loaded liposomes were able to deliver DOx by ultrasound irradiation and also allow for MR imaging [18]. Polymeric carriers made of PLLA with varying capsule shell thickness can be triggered both in vitro and in vivo using different ultrasonic pressures for release of drugs and dyes like Evans blue. Nanocomposite fibers and silica nanotubes when tested for ultrasound responsive drug release showed differences in release profile when model drugs like (fluorescein and rhodamine) or Ibuprofen [19, 20].

2.4 Electrically Triggered Drug Delivery

Electrical signals can be generated and controlled accurately and effortlessly without use of complex apparatus. A pulsatile release of drug is possible using electrical switching. Electrically responsive polymers and conducting polymers can show a release of drugs in response to electrical potential. The applied electrical potential results in an alteration in the chemical and physical properties of the polymer like conductivity, surface energy and pore sizes. Some known polymers are polyaniline, polypyrrole, polythiophene and PEDOT which respond to electrical stimuli. Poly-pyrrole is one of the most commonly used electrically responsive materials to be formed as a drug delivery system. Its use has increased owing to its biocompatible and biodegradable nature. Poly-pyrrole has been used for delivery of charged drugs such as risperidone, dexamethasone, ATP, methotrexate. Thin films have been formulated using poly-pyrrole polymer using these drugs. Thin films are subject to limitation of low drug loading. Most of the release of drug happens from the surface of films. Electro-responsive polymer thin films have also been investigated for wound healing applications using antibiotic, analgesic or anti-inflammatory drugs (Fig. 9.4c) [21]. Use of nanoparticles can enhance the surface drug release owing to higher surface to volume ratio of nanoparticles in comparison to films. Use of electrically responsive polymers for drug release is influenced by several parameters like current, voltage, time, pH, ionic strength, temperature and concentration. Samanta et al. described the loading of fluorescein (water-soluble), piroxicam (hydrophobic), and insulin (high molecular weight polypeptide) [22]. The nanoparticles can contain large numbers of ionisable groups confers responsiveness by shrinking and swelling of the polymers. Poly-anionic polymers like hyaluronic acid, hydrolysed polyacrylamide, alginate, chondroitin sulphate and agarose have been useful to developing electro-responsive hydrogels. Gel swelling, de-swelling and erosion are the mechanisms used for electrically inducing the drug release. Poly (sodium 4 vinyl benzene sulfonate) (PSS) is a polyelectrolyte which is ionisable and can be used for in vitro triggering and increase the release of antiepileptic drug (Phenytoin sodium) release from hydrogel nanoparticles [23]. The main limitation of electro responsive drug delivery is the increase in the temperature due to resistance of carriers after exposure to electric field. This causes issues with reproducibility of drug release during pulsatile stimuli which can be overcome by choosing materials with good electrical properties. Nanocomposites can also been developed for improving the electrical responsiveness like Graphene oxide-conducting polymer (poly-pyrrole), Graphene-poly-methyl methacrylate (PMMA) and polymer-multi-wall carbon nanotubes. Such a nanocomposite can cause linear and a good temporal control of drug release (Fig. 9.4d) [24–26]. Apart from the electrically responsive polymers, drug delivery using microchips and MEMS devices have also been performed which are wirelessly controlled multi-reservoir systems are also gaining popularity. Microchips and MEMS devices are miniaturized micro-fabricated systems with an externally controllable circuitry. Such an implantable system on a chip with eight drug reservoirs was described by

Huang et al. which was made of silicon drug container, PDMS layer and a wireless controller for actuation wherein drug release happens based on electro-thermal mechanisms [27, 28].

2.5 Magnetism Triggered Drug Delivery

Magnetic materials are ideal candidates in the development of advanced multifunctional systems considering their properties of targeting, negative contrast for MRI, magnetic hyperthermia and field dependent controlled drug delivery. Several magnetic materials including metallic nanoparticles of Mn, Fe, Co, Ni, Zn, Gd, Mg and their respective oxides can serve as agents for above applications. Hyperthermia, targeting and field triggered controlled delivery are the applications which can be considered under magnetic activation. Several ferrites like cobalt ferrite ($CoFe_2O_4$), manganese ferrite ($MnFe_2O_4$), nickel ferrite ($NiFe_2O_4$), lithium ferrite ($Li_{0.5}Fe_{2.5}O_4$), mixed ferrites of nickel–zinc–copper ($Ni_{0.65}Zn_{0.35}Cu_{0.1}Fe_{1.9}O_4$) and cobalt–nickel ferrite ($Co_xNi_{(1-x)}Fe_2O_4$) are known to be magnetic in nature and can be used to produce changes in temperature. However, most common biologically relevant magnetic material is the use of iron based magnetic nanoparticles owing to its lack of toxicity, excellent biocompatibility and biodegradation potential to maintain iron homeostasis in cells.

Hyperthermia induced by magnetic materials is the most commonly used process investigated for killing cancer cells. The process involves distribution of MNPs in tumor tissues followed by application of alternating magnetic field which generates heat. The heat generation in MNPs is attributed to relaxation and hysteresis loss. Varying the power of electromagnetic pulses can optimize changes in temperature. The effectiveness of hyperthermia is dependent on type of magnetic materials used, particle size, size distribution, magnetic properties such as magnetic moment, magnetic anisotropy, viscosity of fluid, temperature changes generated at the targeted site of action, duration of exposure and type of cancer tissues in question. However, unavoidable heating of healthy tissues and variabilities of heat generation lead to side effects or recurrence of tumor growth. Magnetic hyperthermia for cancer tissues was first investigated in 1957 using magnetic nanoparticles (MNP). MNP are of sizes below 10 nm and can easily pass through complex physiological geometries like blood brain barrier and tumors. Targeting coupled with magnetic hyperthermia using ligands such as antibodies can selectively heat the tumor tissues and spare the normal healthy tissues. Deger et al. tested a combination of hyperthermia with 3D conformal radiotherapy using Co–Pd thermoseeds for the treatment of prostate cancer and the intraprostatic temperatures achieved were about 42–46 °C with no side effects [29].

Due to agglomerating properties of magnetic materials they need to be stabilized using biocompatible polymeric materials like dextran, poly-vinyl alcohol, lauric acid, liposomes, polyethylene glycol (PEG), PEG-PMMA, PLA, Pluronic F127, PEO-PPO-PEO, chitosan etc. Polymer coating also provide amenability to conjugate

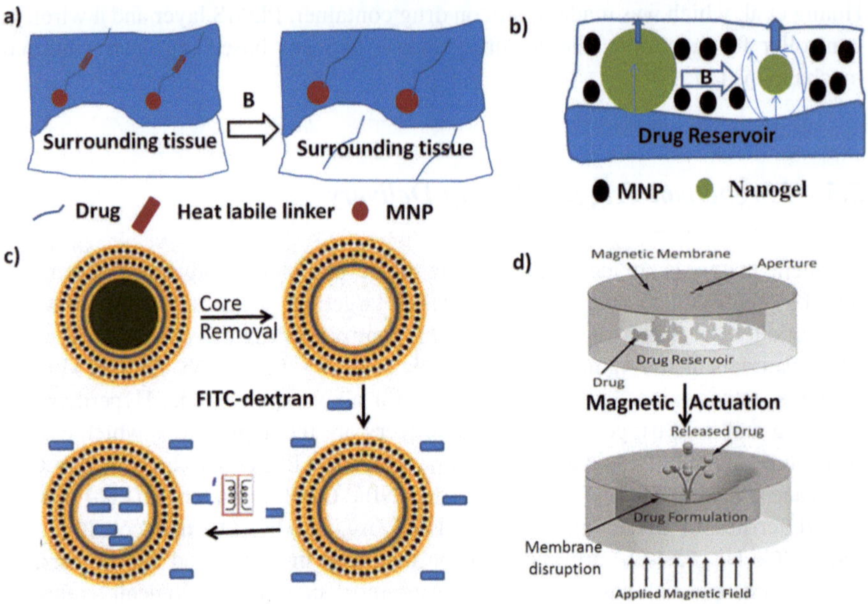

Fig. 9.5 (**a**) MNP linked to drugs using a heat labile DNA linker, (**b**) De-swelling of nanogels due to MNP hyperthermia over drug reservoir, (**c**) Loading and release of FITC dextran using LBL assembled MNP nanoparticles and (**d**) MEMS based drug releasing matrix in response to magnetic actuation

or immobilize drugs within/on the carriers. Several formulations containing magnetic materials and polymers have been developed till date including liposomes, resealed erythrocytes, emulsions, contraceptive systems, infusion pumps, implants which bring significant advances for triggering and targeting drugs to desired site. Drug loaded magnetic nanoparticles can be localized in tumors using an external magnetic field for possible drug release. Hyperthermia induced drug release can occur due to breaking of bonds or due to enhanced permeability of the polymers. An example of bond breaking due to application of alternating magnetic field is a use of heat labile marker for release of fluorescein-labeled 18 bp single stranded DNA (a model antisense therapeutic) in a model tumor near the posterior mammary fat pad of mice (Fig. 9.5a) [30].

Another mechanism of hyperthermia induced drug release can be obtained by formulations containing polymers, hydrogels and thermos-responsive polymers. Temperature sensitive polymers like PNIPAM microgels along with magnetic nanoparticles can tune the encapsulated drug release. On-demand “on–off” release of sodium fluorescein over multiple magnetic cycles was successfully achieved using PNIPAM based nanogels and MNPs (Fig. 9.5b) [31]. The earliest reports on MNP induced hyperthermia for controlled release occurred in 2005, which showed creation of nanopores in layer-by-layer (LBL) self-assembled carriers due to heat generation. Polymer coated MNP lead to formation of core-shell structures that can be beneficial to modify physical and biological properties for drug delivery systems.

Core-shell structures can be useful in generating multi-functionality particles because core and shell can have different properties. Shell can provide protection to sensitive molecules encapsulated in core of nanoparticles. It is observed that the heat released can be increased to an extent of four to five times by using core-shell type structures. In an earliest report, ferromagnetic gold coated cobalt nanoparticles (3 nm) embedded in polymeric capsules developed using LBL self-assembly technique. Microcapsules developed by LBL self-assembly (4 bilayers poly (sodium styrene sulfonate)/poly (allyl amine hydrochloride) were up-to 5 μm diameter microcapsules had wall structures consisting of 4 bilayers. Oscillating magnetic fields of 100–300 Hz resulted in distortion of capsule walls and release of macromolecules like FITC-dextran (Fig. 9.5c) [32]. Conjugation of urokinase and dextran coated iron oxide nanoparticles were evaluated for targeted thrombolysis which showed possibility of usage in the treatment of cardiovascular diseases [33]. PNIPAM and PNIPAM-MAA composites of iron oxide nanoparticles are commonly developed by radical polymerization can provide magnetic targeting provides simultaneous magnetic hyperthermia and resultant drug release from a thermoresponsive polymer [34, 35]. Fe_3O_4 nanoparticles coated with PEG formed from alkaline co-precipitation method of ferrous chloride and ferric chloride hexahydrate were developed by Xu et al. These nanoparticles were later modified with 3-Aminopropyltriethoxysilane (3-APTES) and lysozyme [36]. Dual-purpose chitosan and polyethyleneimine (PEI) coated magnetic micelles (CP-MAG-micelles) that can deliver nucleic acid-based therapeutic agents and also provide magnetic resonance imaging (MRI) have been developed by Wang et al. Certain MEMS devices are also reported for magnetic actuation of drug delivery. Docetaxel on demand release was investigated for diabetic retinopathy using a MEMS device for about 35 days. The device has been fabricated from elastic magnetic PDMS membrane which had a laser drilled aperture. Magnetic field based actuation was used which caused a drug release at a rate of 171 ng per actuation (Fig. 9.5d) [37].

3 Endogenous Stimuli for On-demand Drug Delivery

Endogenous stimuli can be considered as the most closely associated triggers in biological systems. Drug delivery systems responding accurately to the endogenous stimuli can function in a better manner in comparison to external stimuli responsive systems. This is because of the fact that additional extraneous variations are needed to optimize for exogenously administered systems. For e.g. in case of light triggered drug delivery using NIR light, the penetration depth, wavelength of light, flux variations, skin thickness need to be considered in addition to the physiological variations. Similarly in case of magnetic targeting, type of magnetic materials, magnetic field, temperature changes due to alternating field etc. are the external variables apart from the internal physiological variables. In case of endogenous stimuli responsive delivery systems physiological stimuli like pH, redox states, and presence of enzymes are used as triggers for the release of drugs (Fig. 9.1).

Spatial control of drug release can very well be obtained using internal stimuli without avoiding toxicity to healthy tissues. The following section will discuss approaches and drug delivery mechanisms used for such internal triggers for controlled drug release.

3.1 pH Triggered Drug Delivery

Differences in pH of physiological and pathological conditions like inflammation, infection and cancer can be explored as an endogenous triggering mechanism. pH based releasing systems are the most common forms of triggered or stimuli responsive delivery systems. It would be highly desirable if drugs can be released in response to physiological needs, locations and times. pH dependent release systems have already been explored for orally administered drugs. Preferential release of drug in a particular absorption window can improve the overall bioavailability. Enteric coating to formulations has been used for preferential release of drugs in the intestine and protection from gastric pH conditions. In case of cancer, pH of extracellular tissue lies in the range of 6.5–7.2 and is considered slightly acidic than normal tissue pH. The pH in the lysosomes is also acidic in nature, which can be explored for cell internalization of carriers. The pH, surface charge, density of low density lipoprotein receptors are some differences in normal and cancer tissues. Different pH responsive nano-formulations that are investigated include polymeric nanoparticles, polymer-drug conjugates, micelles, liposomes, dendrimers etc. All pH sensitive polymers have acidic (COOH, SO_3H) and basic (e.g. ammonium) functional groups so that they can change their morphology in the environment. Polymers with acidic functionalities will be un-swollen in acidic pH and will swell in alkaline pH and vice versa in case of basic polymer functionalities. Swelling or de-swelling induced by ionization of polymer backbone in a particular pH media is generally used for establishing the drug release.

Most commonly used pH sensitive polymers for oral applications include derivatives of acrylic acid and celluloses. Poly acrylic acid (PAA) becomes ionized at high pH whereas poly (*N*, *N*-diethyl aminoethyl methacrylate) (PDEAEM) becomes ionized in at a low pH. The swelling and deswelling of polymers can be controlled by using neutral co-monomers such as 2-hydroxyl ethyl methacrylate, methyl methacrylate, maleic anhydride etc. Hydrogels of poly-methacrylic acid (PMA) grafted with PEG can generate unique pH sensitive properties wherein at acidic pH, protons of COOH groups of PMA hydrogen with ether oxygen of PEG resulting in gel shrinking [38]. Hydrogels can be crosslinked with azo-aromatic cross-linkers in addition to carboxylic acid based crosslinkers so that drug can be released in a colon specific manner (Fig. 9.6a). Lysosomal release of genes can be achieved using amphoteric poly (amido amine) and poly L-histidine polymers due to endo-osmolytic activity at lower pH helping in internalization of genes in the cells [39]. Hydrogel microspheres of methylene-bis acrylamide/methacrylic acid loaded with

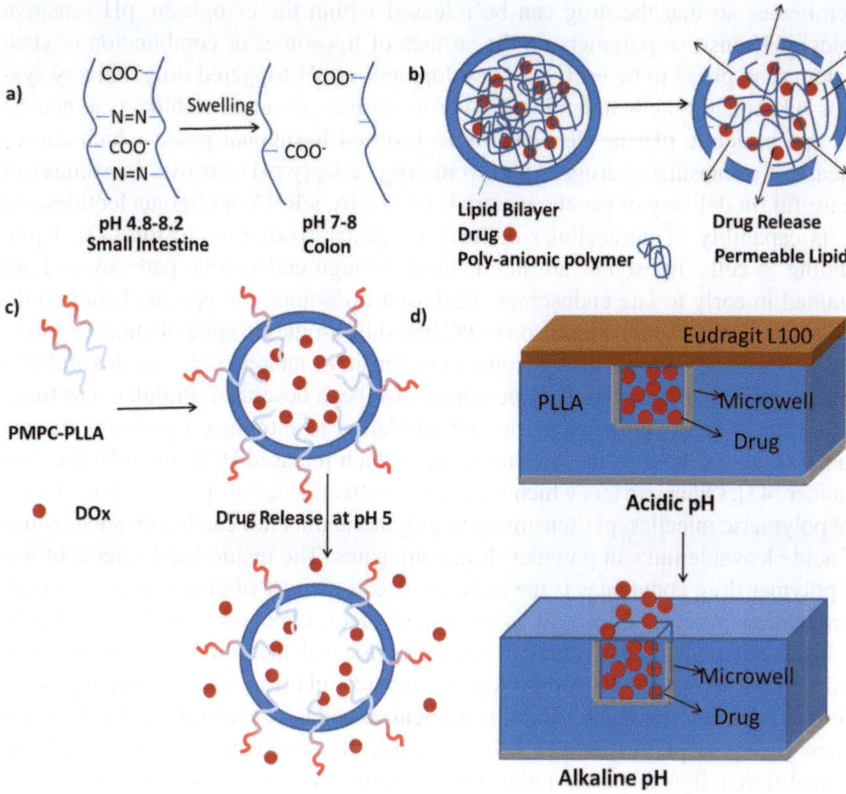

Fig. 9.6 (**a**) Hydrogels based on diazo mediated crosslinking for colon specific drug release (**b**) Lipid permeability of drug encapsulated poly-anionic polymer for intestinal delivery (**c**) Polymerosomes for pH-triggered drug release of Dox and (**d**) PLLA microwells for intestinal delivery of drugs

doxorubicin and coated with lipid bilayer can be disrupted by electroporation so that encapsulated drug can be released (Fig. 9.6b).

pH sensitive polymers such as Eudragits (E, L, S) and Cellulose acetate phthalate (CAP) etc. can dissolve at different pH values and hence has be used for oral formulations. A tunable release can be obtained using pH sensitive microspheres developed using these polymers like Eudragit E-100, Hydroxy-propyl methyl cellulose phthalate and HPMC acetate succinate etc [40]. An application of Eudragit polymers is to coat Polymer microwells fabricated using poly-L-lactic acid for oral drug delivery using amorphous sodium salt of furosemide as a model drug. Drug was filled in microwells (diameter: depth = 240:100 micron) using screen printing technique followed by Eudragit L100 coating (Fig. 9.6d). These microwells showed a good pH sensitivity for oral drug delivery in pH 1.6 and pH 6.5 [41]. pH sensitive liposomes have been also showing promise as they are stable at physiological pH and get destabilized under acidic conditions. The lipid membrane can fuse with membranes so that the drug can be released within the cytoplasm. pH sensitive lipids, pH sensitive polymers on the surface of liposomes or combination of such systems can prove to be useful for development of pH triggered drug delivery systems. Phosphatidyl ethanolamine (PE) or its derivatives act as stabilizers at neutral pH and at acidic pH the PE converts to inverted hexagonal phase which causes release of encapsulated drugs. Apart from drug delivery pH sensitive liposomes can be useful for delivery of genetic materials like DNA, siRNA or oligonucleotides due to its capability of intracellular delivery by destabilization or fusion [42]. Upon binding to cells, liposomes are internalized through endocytotic pathway and are retained in early to late endosomes. PE based liposomes can provide better endosomal drug delivery in comparison to PC based liposomes in spite of almost equivalent cell uptake. Similar to liposomes pH sensitive nanoparticle for doxorubicin delivery using pullulan like polymers have also been described. Pullulan was functionalized to carboxymethylization and amidation to introduce hydrazide groups. Sub 100 nm particles of these were formed which released DOX in a pH sensitive manner [43]. Other carriers which have shown effectiveness as pH sensitive carriers are polymeric micelles. pH sensitivity in polymeric micelles can be obtained by use of acid cleavable links in polymer-drug conjugates. The major disadvantage of use of polymer drug conjugates is the maintenance of activity of drug during the entire conjugation process. Lee et al. investigated block copolymer of PEG and poly (L-histidine) to fabricate pH sensitive micelles in which histidine residue function as hydrophobic at neutral pH and protonated at acidic pH (Fig. 9.6c). Variable protonation of histidine residues results in critical micellar concentration which is ten times greater at pH 5 than pH 7.4. These micelles can further be functionalized using different ligands for obtaining active targeting. Poly(L-histidine)-b-PEG along with PLLA-b-PEG and adriamycin was studied for extracellular tumor targeting. The formulation showed a transition at pH 6.6 where Adriamycin is released from micelles [44]. Dendritic polymers are synthetic, highly branched, macromolecular nanomaterials as drug carriers with well-defined molecular architecture. Doxorubicin can be linked to 2, 2 bis-(hydroxymethyl) propanoic acid via a pH sensitive linkage to obtain a stimuli responsive drug release. The hydrolysable linkages like amide, or hydrazine, acetal, with polymer may provide the necessary pH sensitivity to dendritic carriers [45, 46].

3.2 Enzyme Triggered Drug Delivery

Enzymes provide necessary specificity for drug release by interaction to specific substrates. Physiologically particular enzymes may be present at a particular site, which will provide the desired localization of therapy. Enzyme triggered delivery requires carriers which degrade or change morphology upon exposure to enzymes. Prodrugs are classical forms of enzyme-triggered therapy wherein enzyme metabolizes prodrugs in to active drugs inside the body. For e.g. Levodopa is a prodrug for dopamine and is converted by dopa-decarboxylase to active drug dopamine.

The prodrugs are developed in order to surpass the pharmacokinetic perils of active drugs. Azo-reductase enzyme has been used for triggering release of sulfasalazine used in ulcerative colitis and inflammatory bowel disease. Enzymes used in such type of release system commonly present in the body, can be classified as hydrolases and oxido-reductases. Typical classes of enzymes used by researchers are phospholipases, proteases, matrix metalloproteinase, phosphodiesterases, caspase, phosphatases, amidase, urease, trypsin, hyaluronidase and glucose oxidase. Among the hydrolases, phospholipases are the most investigated for development of enzyme-triggered delivery systems. They have been used in strategies requiring cleavage of lipid conjugated prodrugs. Similarly matrix metalloproteinase (MMP) are useful in overcoming interstitial fluid pressure and enhancing uptake of nanoparticles in solid tumours.

3.2.1 Phospholipase

Phospholipases are enzymes involved in hydrolysis of phospholipids into fatty acids and smaller units. They are classified into different types (A, B, C, D) based on the type of reaction they catalyse. Phospholipase A2 which is an important component of venoms of snake and bees are known to act on lecithins and hydrolyse them to fatty acid and lyso-lecithin. Enzyme activated lipid hydrolysis by phospholipase has been reported by Pignatello et al., Kaasgaard et al. and Zerouga et al. for methotrexate conjugated to lipoamino acids, lipid membranes of liposomes and docosahexanoic acid (DHA) linked to phosphatidyl choline, respectively. Methotrexate linked DHA-Phosphatidyl choline was hydrolysed by phospholipase A2 causing a dose dependent reduction of murine leukaemic cell growth [47]. Owing to its lipid hydrolysing capability liposomes/micelles would be considered as best candidates for development of drug delivery systems. Further to that phospholipase A2 activity was greater, when anionic phosopholipids were used in comparison to the Zwitterionic lipids. Phospholipase A2 has also been found to be overexpressed in infections, inflammatory diseases and cancerous tissues including prostrate, colon, breast, pancreas etc. making it suitable for therapeutic intervention for triggered delivery of anti-cancer drugs. PEGylated lipids were described to be enhancing the activity of this enzyme as indicated by Andresen et al. who used 1,2-distearoyl-sn-glycero-phosphoethanol amine-*N*-methoxy (PEG)-2000 (DSPE-PEG 2000). Authors indicated that rapid release of cisplatin and doxorubicin was possible using these lipids and were shown to enhance cytotoxicity of Colo 205 colon cancer cells and significantly improved anticancer efficacy in breast cancer xenograft (MT-3) model in mice [47, 48]. Another example of lipase triggered release of drugs from MSNP was proposed by Kim et al. In this work cyclodexrins were used as gatekeepers attached to nanoparticles via a lipase sensitive linker. Cleaving of lipase sensitive groups liberates the cyclodextrin groups making available release of molecules entrapped in pores. The cleavable group was also found to be sensitive to α-amylase, making this application useful for acute pancreatitis cases [49].

3.2.2 Proteases

Proteases are involved in cleavage of larger peptides into smaller fragments by hydrolysis of peptide bonds. MMP are thought to play an important role in cancer (invasion, angiogenesis and metastasis). Apart from cancer, MMP are found to be important in rheumatoid arthritis, osteoarthritis and auto-immune skin disorders. Among several proteases matrix metalloproteinases have been used by several researchers for getting triggered release of drugs. Fukumara et al. proposed MMP-2 responsive 100 nm nanoparticle aggregates which can further be broken down into 10 nm particles. 10 nm particles were found to be penetrating deep in tumor tissues in spite of interstitial fluid pressure due to enhanced permeation and retention [50]. Kim et al. reviewed protease responsive drug delivery responsive DOx-peptide coated magnetic silica nanoparticle conjugate for intracellular delivery in vitro [51]. Surgical removal of orthotopic and metastatic breast tumors was performed by matrix metalloproteinase targeted fluorophores by Chi et al. [52]. Liposomes developed from lipopeptides which are sensitive to MMP-9 and induce liposome destabilization can cause triggered drug release. Similar research of Harashima et al., PEG masked liposomes with configuration of PEG-peptide-DOPE having a lipidic core which is degraded by MMP-2 by destabilization-fusion and aggregation. The results also indicated that the liposomes of this configuration were less stable in circulation as against PEGylated liposomes. Macromolecular carriers like Gly-Phe-Leu-Gly can be used to program drug release from the nanocarriers by the lysosomal enzyme cathepsin B which is only present in intra-cellular compartment. The encapsulated drug exerts its effect only after internalization in the cells. Such an approach improves the therapeutic efficacy and reduces the toxic effects. This peptide sequence has been investigated for DOx linked to HPMA nanoparticles through aminoglutethimide (AGM) which acts as an aromatase inhibitor [53].

3.2.3 Glycosidase

Glycosidases are enzymes which hydrolyze carbohydrates to yield small sugars. α-amylase is highly expressed in tumour environment making it suitable to design a drug delivery system containing sugar molecules coated on the nanocarriers. A successful proof of this concept was showed by linking dextran with phospholipase via succinoylation. Lactose and starch derivatives coupled to MSNP with fluorophore trapped inside the pores when treated with pancreatin or β-D-galactosidase cause release of dye. DOx loaded and starch modified MSNP showed a great impact on cell survival upon release of drug [54]. Dzamukova showed selective delivery of anticancer drug doxorubicin and Brilliant green using halloysite nanotube carriers (50 nm diameter). Dextrin end stoppers were used to prevent release of Brilliant green; however when intercellular glycosyl hydrolases degrade the dextrin stoppers the entrapped molecules can be liberated. Rate of internalization of nanotubes was dependent on the growth rates of tumour tissues and hence were very much suitable for malignant cancer tissues [55].

3.2.4 Oxido-reductase

Glucose oxidase is the most commonly used oxido-reductase enzyme researched for glucose responsive drug delivery. Use of glucose oxidase could be used for feedback controlled release of anti-diabetic molecules. Glucose oxidase upon reaction with glucose forms gluconic acid and hydrogen peroxide. Gluconic acid production causes a drop in pH and based on these facts pH sensitive carriers or inherent pH based solubility of drugs can be used for anti-diabetic drug release in response to in vivo glucose changes. Solubility changes due to change in pH along with enzymes can prove to be very promising as triggered drug delivery system. Polymers like ethylene/vinyl acetate copolymer have been commonly employed for immobilization of insulin and glucose oxidase.

3.2.5 Other Enzymes

Several other enzymes like phosphatase, β-lactamase, Amidases and urease have been used to develop enzyme triggered solutions. Dephosphorylation activity of phosphatase was used by Davis and Szoka to catalyse conversion of lamellar liposomes to hexagonal phase liposomes which are loaded with plasmid DNA for transfection applications. Phosphatase triggered co-assembly of ICG doped nanofibers have been developed for tumour targeting to exploit the NIR, photoacoustic and photothermal properties of ICG. Phosphatase assisted delivery increased the uptake of ICG upto 25 folds than free ICG [56]. β-lactamase sensitive delivery of β-LEAP linked with photosensitizer (EtNBS) and cephalosporin linker was described by Zheng et al. The photosensitizers remain quenched in intact structure but when cleaved by β-lactamase (which is produced by resistant bacterial population only) can be used for fluorescence imaging. This strategy is an interesting attempt to target resistant bacteria and distinguishes between human and bacterial cells. Amidase and urease mediated delivery of $[Ru(bipy)_3]^{2+}$ dye was investigated using nanoscopic silica mesoporous particles which are capped with enzyme hydrolysable bulky organic moieties containing amide and urea linkages. Simultaneous treatment with both amidase and urease showed a synergistic effect showing complete release of the molecules entrapped in it. The release of $[Ru\ (bpy)_3]^{2+}$ dye was analysed spectro-fluorimetrically [57].

3.3 Redox Triggered Drug Delivery

The basis of redox systems lies in the conversion of disulphide bonds to –SH bonds in response to Glutathione (GSH). Glutathione concentrations are found to be different in extracellular/intracellular compartments and tumour/healthy tissues. Delivery systems containing disulphide bond as linker in polymer chains or incorporated in the hydrophobic backbone of micelles or liposomes can bring about redox sensitivity in drug delivery systems with variations in glutathione concentrations. Glutathione sensitive cross linkers incorporated in carriers core or shell of micelles is an important method of achieving redox triggered drug release [58]. Many delivery systems have been investigated using such an approach e.g. micelles, capped mesoporous materials, dendrimer-drug conjugates, liposomes of quinone-lipid conjugate and nanogels. Oxidative environment due to the presence of reactive oxygen species in inflammatory tissues can also be used as trigger for drug release. A thioketal based nanoparticles formulated with reactive oxygen sensitive polymer showed a capability of TNF-iRNA delivery at the sites of intestinal inflammation providing gene silencing after oral administration [59].

4 Multiple Stimuli Responsive and Feed-back Based Drug Delivery

Previous sections have summarized single stimuli based delivery systems for site specific and controlled drug delivery. Improvement in sensitivity of drug release can be achieved by developing systems which are responsive to more than one stimuli applied sequentially or simultaneously.

Combination of dual or multiple stimuli like temperature and pH responsive polymers can exhibit dual responsive behaviours and can lead to interesting biomedical applications. An example of such dual responsive drug delivery was performed by Chilkoti et al., who have designed doxorubicin-polypeptide conjugate for cancer therapy. The first stimuli is temperature which uses LCST behaviour of the polymer. The polymer undergoes a phase transition at higher temperature and becomes insoluble after reaching the tumor target site. The second stimuli occurs when the carriers reach the tumor tissues where the extracellular pH is towards acidic side. Lower pH causes detachment of DOx due to solubilisation of acid labile linker [65]. Similarly, coexistence of pH and oxidative environment in pathological conditions like cancers can be employed. Conjugation of antisense bcl2 oligonucleotides and doxorubicin to form PEG with acid cleavable and redox reducible linkers can lead to development of such pH and redox triggers. Table 9.1 summarizes some reports for multiple stimuli responsive drug delivery systems.

In the body several hormones and proteins are tightly controlled to maintain a balance by a feedback system called homeostasis. It would be highly desirable if drugs are delivered by judging the magnitude of signal caused by the disease causing the right amount of drug release. Such systems are called feed-back based drug delivery systems and have been investigated for management of diseases like diabetes. Glucose responsive drug delivery systems can be used for diabetes management wherein phenyl boronic acid and its derivatives which use swelling as the trigger when glucose reacts with phenyl boronic acid units. PBA swelling has been used to activate release of insulin from poly (ethylene glycol) b-(polyacrylic acid –co-acrylamimdophenyl boronic acid) based micelles and poly(ethylene glycol)-b-poly(styrene boroxole) polymersomes [66]. These approaches required

Table 9.1 Summary of multi-stimuli responsive drug delivery carriers

Stimuli	Polymer	Drug	References
Temperature/pH/Redox	PNIPAAm-SS-P(THP-protected HEMA) micelles	Nile Red	[60]
Tempearture/pH/Magnetic	P(NIPAAm-co-MAA) coated magnetic MSNs	Doxorubicin	[61]
pH/Redox/Magnetic	Fe(II) loaded PMAA crosslinked by *N,N*-methylene-bisacrylamide and *N,N*-bis(acryloyl)-cystamine	Daunorubicin hydrochloride	[62]
Temperature/Redox/Guest molecule	Vesicles based on hosteguest complex formation between C4AS and MVC12	Doxorubicin	[63]
Temperature/pH/Guest molecule	Cucurbit(8)uril micelles, methylviologene-functionalize PNIPAAm and naphthalene-terminated PDMAEMA	Doxorubicin	[64]

high glucose concentrations for generating sufficient amount of insulin release. Recently, glucose responsive drug delivery systems have been researched which involve pH as a trigger due to formation of gluconic acid during a *Glucose oxidase* catalytic reaction. For example pH dependent loosening of multiple layers of mesoporous silica nanoparticles has shown to achieve a sustained insulin release. Implantable systems based on such approach will not only provide controlled release but also could be useful to bring about feedback based control in glucose levels in management of diabetes. Despite several attempts clinically suitable feedback based drug delivery systems are yet to be achieved.

5 Conclusions and Future Prospects

The most common peril of conventional delivery systems is the poor bioavailability at the site of action. Most of the drugs used in cancer chemotherapy reach to an extent of about 5% at the site of action. Significant advances in the fields of materials synthesis, nanotechnology and drug delivery have led to novel on demand drug releasing concepts. Structural heterogeneity of targets and bioavailability problems of drugs together are detrimental for drug targeting and site specific release. Although enhanced permeation and retention effect in cancers has led to research reports on passive targeting however it is highly unlikely that they will clear the clinical trials because of inherent physiological variabilities. In this aspect stimulus responsive or on demand drug delivery systems may appear superior to other systems. The wide range of stimuli which can trigger the drug release at the right place and time, along with the kinds of advanced materials allowing flexibility can greatly improve the attempts in designing delivery systems of improved performance. Several studies showing in vitro proof of concepts have been described in this chapter; however attempts need to be made for establishing in vivo studies and validation. Complexities in synthetic procedures, scale up issues, toxicities can hamper their translation in to clinics. Studies pertaining to scale up and toxicity issues of these advanced materials are still sought before testing these formulations at the clinics. A critical analysis of benefit/risk need to be calculated for therapy of diseases based on side effects, issues of biocompatibility and degradation. As a general rule, the simpler and easier the development of a system is, the better its chances of reaching the clinic. We have shown in this chapter, immense progress in materials chemistry and drug delivery leading to design of smart stimuli-responsive concepts using nano-engineered systems. In future the focus needs to shift towards clinically relevant approaches.

Acknowledgements Rahul Dev Jayant would like to acknowledge financial support from Herbert Wertheim College of Medicine Pilot funding grant (# 800008542); The Campbell Foundation grant (# 800008886); and other support from Center for Personalized Nanomedicine (CPNM) and Institute of Neuro-Immune Pharmacology (INIP) from Department of Immunology, Florida International University (FIU). Abhijeet Joshi acknowledges the INSPIRE Faculty award and fellowship provided by Department of Science and Technology, Government of India. Authors would like acknowledge financial support from DBT and SERB.

References

1. Fu Y, Kao WJ. Drug release kinetics and transport mechanisms of non-degradable and degradable polymeric delivery systems. Expert Opin Drug Deliv. 2010;7(4):429–44. doi:10.1517/17425241003602259.
2. Miranda D, Lovell JF. Mechanisms of light-induced liposome permeabilization. Bioeng Transl Med. 2016;1(3):267–76. doi:10.1002/btm2.10032.
3. Norum OJ, Gaustad JV, Angell-Petersen E, Rofstad EK, Peng Q, Giercksky KE, Berg K. Photochemical internalization of bleomycin is superior to photodynamic therapy due to the therapeutic effect in the tumor periphery. Photochem Photobiol. 2009;85(3):740–9. doi:10.1111/j.1751-1097.2008.00477.x.
4. Lou PJ, Lai PS, Shieh MJ, Macrobert AJ, Berg K, Bown SG. Reversal of doxorubicin resistance in breast cancer cells by photochemical internalization. Int J Cancer. 2006;119(11):2692–8. doi:10.1002/ijc.22098.
5. Pashkovskaya A, Kotova E, Zorlu Y, Dumoulin F, Ahsen V, Agapov I, Antonenko Y. Light-triggered liposomal release: membrane permeabilization by photodynamic action. Langmuir. 2010;26(8):5726–33. doi:10.1021/la903867a.
6. Timko BP, Arruebo M, Shankarappa SA, McAlvin JB, Okonkwo OS, Mizrahi B, Stefanescu CF, Gomez L, Zhu J, Zhu A, Santamaria J, Langer R, Kohane DS. Near-infrared–actuated devices for remotely controlled drug delivery. Proc Natl Acad Sci U S A. 2014;111(4):1349–54. doi:10.1073/pnas.1322651111.
7. Rai P, Mallidi S, Zheng X, Rahmanzadeh R, Mir Y, Elrington S, Khurshid A, Hasan T. Development and applications of photo-triggered theranostic agents. Adv Drug Delivery Rev. 2010;62(11):1094–124. doi:10.1016/j.addr.2010.09.002.
8. Schmaljohann D. Thermo- and pH-responsive polymers in drug delivery. Adv Drug Delivery Rev. 2006;58(15):1655–70. doi:10.1016/j.addr.2006.09.020.

9. Liu F, Kozlovskaya V, Medipelli S, Xue B, Ahmad F, Saeed M, Cropek D, Kharlampieva E. Temperature-sensitive polymersomes for controlled delivery of anticancer drugs. Chem Mater. 2015;27(23):7945–56. doi:10.1021/acs.chemmater.5b03048.
10. Geethalakshmi A, Karki R, Sagi P, Jha SK, Venkatesh DP. Temperature triggered in-situ gelling system for betaxolol in glaucoma. J Appl Pharm Sci. 2013;3(2):153–9.
11. Jung HJ, Chauhan A. Temperature sensitive contact lenses for triggered ophthalmic drug delivery. Biomaterials. 2012;33(7):2289–300. doi:10.1016/j.biomaterials.2011.10.076.
12. Needham D, Anyarambhatla G, Kong G, Dewhirst MW. A new temperature-sensitive liposome for use with mild hyperthermia: characterization and testing in a human tumor xenograft model. Cancer Res. 2000;60(5):1197–201.
13. Kim MS, Lee D-W, Park K, Park S-J, Choi E-J, Park ES, Kim HR. Temperature-triggered tumor-specific delivery of anticancer agents by cRGD-conjugated thermosensitive liposomes. Colloids Surf, B. 2014;116:17–25. doi:10.1016/j.colsurfb.2013.12.045.
14. Deckers R, Moonen CTW. Ultrasound triggered, image guided, local drug delivery. J Controlled Release. 2010;148(1):25–33. doi:10.1016/j.jconrel.2010.07.117.
15. Meijering BD, Juffermans LJ, van Wamel A, Henning RH, Zuhorn IS, Emmer M, Versteilen AM, Paulus WJ, van Gilst WH, Kooiman K, de Jong N, Musters RJ, Deelman LE, Kamp O. Ultrasound and microbubble-targeted delivery of macromolecules is regulated by induction of endocytosis and pore formation. Circ Res. 2009;104(5):679–87. doi:10.1161/circresaha.108.183806.
16. Zhu X, Guo J, He C, Geng H, Yu G, Li J, Zheng H, Ji X, Yan F. Ultrasound triggered image-guided drug delivery to inhibit vascular reconstruction via paclitaxel-loaded microbubbles. Sci Rep. 2016;6:21683. doi:10.1038/srep21683.
17. Treat LH, McDannold N, Vykhodtseva N, Zhang Y, Tam K, Hynynen K. Targeted delivery of doxorubicin to the rat brain at therapeutic levels using MRI-guided focused ultrasound. Int J Cancer. 2007;121(4):901–7. doi:10.1002/ijc.22732.
18. Jung SH, Na K, Lee SA, Cho SH, Seong H, Shin BC. Gd(III)-DOTA-modified sonosensitive liposomes for ultrasound-triggered release and MR imaging. Nanoscale Res Lett. 2012;7(1):462. doi:10.1186/1556-276x-7-462.
19. Song B, Wu C, Chang J. Ultrasound-triggered dual-drug release from poly(lactic-co-glycolic acid)/mesoporous silica nanoparticles electrospun composite fibers. Regenerative Biomater. 2015;2(4):229–37. doi:10.1093/rb/rbv019.
20. Kapoor S, Bhattacharyya AJ. Ultrasound-triggered controlled drug delivery and biosensing using silica nanotubes. J Phys Chem C. 2009;113(17):7155–63. doi:10.1021/jp9000863.
21. Justin GA, Zhu S, Nicholson TR, Maskrod J, Mbugua J, Chase M, Jung Jh, Mercado RML. On-demand controlled release of anti-inflammatory and analgesic drugs from conducting polymer films to aid in wound healing. In: 2012 Annual International Conference of the IEEE Engineering in Medicine and Biology Society, Aug. 28 2012-Sept. 1 2012 2012. pp 1206–1209. doi:10.1109/EMBC.2012.6346153.
22. Samanta D, Hosseini-Nassab N, Zare RN. Electroresponsive nanoparticles for drug delivery on demand. Nanoscale. 2016;8(17):9310–7. doi:10.1039/C6NR01884J.
23. Ying X, Wang Y, Liang J, Yue J, Xu C, Lu L, Xu Z, Gao J, Du Y, Chen Z. Angiopep-conjugated electro-responsive hydrogel nanoparticles: therapeutic potential for epilepsy. Angew Chem Int Ed. 2014;53(46):12436–40. doi:10.1002/anie.201403846.
24. Weaver CL, LaRosa JM, Luo X, Cui XT. Electrically controlled drug delivery from graphene oxide nanocomposite films. ACS Nano. 2014;8(2):1834–43. doi:10.1021/nn406223e.
25. Servant A, Leon V, Jasim D, Methven L, Limousin P, Fernandez-Pacheco EV, Prato M, Kostarelos K. Graphene-based electroresponsive scaffolds as polymeric implants for on-demand drug delivery. Adv Healthcare Mater. 2014;3(8):1334–43. doi:10.1002/adhm.201400016.
26. Servant A, Methven L, Williams RP, Kostarelos K. Electroresponsive polymer–carbon nanotube hydrogel hybrids for pulsatile drug delivery in vivo. Adv Healthcare Mater. 2013;2(6):806–11. doi:10.1002/adhm.201200193.

27. Liu Y, Servant A, Guy OJ, Al-Jamal KT, Williams PR, Hawkins KM, Kostarelos K. An electric-field responsive microsystem for controllable miniaturised drug delivery applications. Sensors Actuators B Chem. 2012;175:100–5. doi:10.1016/j.snb.2011.12.069.
28. Huang Y-J, Liao H-H, Huang P-L, Wang T, Yang Y-J, Wang Y-H, Lu S-S. An implantable release-on-demand CMOS drug delivery SoC using electrothermal activation technique. J Emerg Technol Comput Syst. 2012;8(2):1–22. doi:10.1145/2180878.2180884.
29. Deger S, Boehmer D, Türk I, Roigas J, Budach V, Loening SA. Interstitial hyperthermia using self-regulating thermoseeds combined with conformal radiation therapy. Eur Urol. 2002;42(2):147–53. doi:10.1016/S0302-2838(02)00277-4.
30. Derfus AM, von Maltzahn G, Harris TJ, Duza T, Vecchio KS, Ruoslahti E, Bhatia SN. Remotely triggered release from magnetic nanoparticles. Adv Mater. 2007;19(22):3932–6. doi:10.1002/adma.200700091.
31. Hoare T, Santamaria J, Goya GF, Irusta S, Lin D, Lau S, Padera R, Langer R, Kohane DS. A magnetically triggered composite membrane for on-demand drug delivery. Nano Lett. 2009;9(10):3651–7. doi:10.1021/nl9018935.
32. Lu Z, Prouty MD, Guo Z, Golub VO, Kumar CSSR, Lvov YM. Magnetic switch of permeability for polyelectrolyte microcapsules embedded with Co@Au nanoparticles. Langmuir. 2005;21(5):2042–50. doi:10.1021/la047629q.
33. Bi F, Zhang J, Su Y, Tang Y-C, Liu J-N. Chemical conjugation of urokinase to magnetic nanoparticles for targeted thrombolysis. Biomaterials. 2009;30(28):5125–30. doi:10.1016/j.biomaterials.2009.06.006.
34. Purushotham S, Ramanujan RV. Thermoresponsive magnetic composite nanomaterials for multimodal cancer therapy. Acta Biomater. 2010;6(2):502–10. doi:10.1016/j.actbio.2009.07.004.
35. Akbarzadeh A, Zarghami N, Mikaeili H, Asgari D, Goganian AM, Khiabani HK, Samiei M, Davaran S. Synthesis, characterization, and in vitro evaluation of novel polymer-coated magnetic nanoparticles for controlled delivery of doxorubicin. Nanotechnol Sci Appl. 2012;5:13–25. doi:10.2147/NSA.S24328.
36. Xu L, Kim M-J, Kim K-D, Choa Y-H, Kim H-T. Surface modified Fe3O4 nanoparticles as a protein delivery vehicle. Colloids Surf A Physicochem Eng Asp. 2009;350(1–3):8–12. doi:10.1016/j.colsurfa.2009.08.022.
37. Pirmoradi FN, Jackson JK, Burt HM, Chiao M. On-demand controlled release of docetaxel from a battery-less MEMS drug delivery device. Lab Chip. 2011;11(16):2744–52. doi:10.1039/C1LC20134D.
38. Peppas NA, Klier J. Controlled release by using poly(methacrylic acid-g-ethylene glycol) hydrogels. J Controlled Release. 1991;16(1):203–14. doi:10.1016/0168-3659(91)90044-E.
39. You Han B, Kun N (2005) pH-sensitive polymers for drug delivery. In: Polymeric drug delivery systems. Drugs and the pharmaceutical sciences. Informa Healthcare, pp 129–194. doi:10.1201/9780849348129.ch3
40. Yang M, Cui F, You B, Wang L, Yue P, Kawashima Y. A novel pH-dependent gradient-release delivery system for nitrendipine: II. Investigations of the factors affecting the release behaviors of the system. Int J Pharm. 2004;286(1–2):99–109. doi:10.1016/j.ijpharm.2004.08.007.
41. Nielsen LH, Nagstrup J, Gordon S, Keller SS, Østergaard J, Rades T, Müllertz A, Boisen A. pH-triggered drug release from biodegradable microwells for oral drug delivery. Biomed Microdevices. 2015;17(3):55. doi:10.1007/s10544-015-9958-5.
42. Balamuralidhara V, Pramodkumar TM, Srujana N, Venkatesh MP, Vishal Gupta NV, Krishna KL, Gangadharappa HV. pH sensitive drug delivery systems: a review. Am J Drug Discov Develop. 2011;1:24–48.
43. Lu D-x, X-t W, Liang J, X-d Z, Z-w G, Y-j F. Novel pH-sensitive drug delivery system based on natural polysaccharide for doxorubicin release. Chin J Polym Sci. 2008;26(03):369–74. doi:10.1142/S0256767908003023.
44. Lee ES, Na K, Bae YH. Polymeric micelle for tumor pH and folate-mediated targeting. J Controlled Release. 2003;91(1–2):103–13. doi:10.1016/S0168-3659(03)00239-6.

45. Gillies ER, Jonsson TB, Fréchet JMJ. Stimuli-responsive supramolecular assemblies of linear-dendritic copolymers. J Am Chem Soc. 2004;126(38):11936–43. doi:10.1021/ja0463738.
46. Ihre HR, Padilla De Jesús OL, Szoka FC, Fréchet JMJ. Polyester dendritic systems for drug delivery applications: design, synthesis, and characterization. Bioconjugate Chem. 2002;13(3):443–52. doi:10.1021/bc010102u.
47. Davidsen J, Vermehren C, Frokjaer S, Mouritsen OG, Jørgensen K. Drug delivery by phospholipase A2 degradable liposomes. Int J Pharm. 2001;214(1–2):67–9. doi:10.1016/S0378-5173(00)00634-7.
48. Anderson JL. Lipoprotein-associated phospholipase A_2: an independent predictor of coronary artery disease events in primary and secondary prevention. Am J Cardiol. 2008;101(12):S23–33. doi:10.1016/j.amjcard.2008.04.015.
49. Park C, Kim H, Kim S, Kim C. Enzyme responsive nanocontainers with cyclodextrin gatekeepers and synergistic effects in release of guests. J Am Chem Soc. 2009;131(46):16614–5. doi:10.1021/ja9061085.
50. Wong C, Stylianopoulos T, Cui J, Martin J, Chauhan VP, Jiang W, Popović Z, Jain RK, Bawendi MG, Fukumura D. Multistage nanoparticle delivery system for deep penetration into tumor tissue. Proc Natl Acad Sci U S A. 2011;108(6):2426–31.
51. Kim CS, Duncan B, Creran B, Rotello VM. Triggered nanoparticles as therapeutics. NanoToday. 2013;8(4):439–47. doi:10.1016/j.nantod.2013.07.004.
52. Chi C, Zhang Q, Mao Y, Kou D, Qiu J, Ye J, Wang J, Wang Z, Du Y, Tian J. Increased precision of orthotopic and metastatic breast cancer surgery guided by matrix metalloproteinase-activatable near-infrared fluorescence probes. Sci Rep. 2015;5:14197. doi:10.1038/srep14197.
53. Vicent MJ, Greco F, Nicholson RI, Paul A, Griffiths PC, Duncan R. Polymer therapeutics designed for a combination therapy of hormone-dependent cancer. Angew Chem. 2005;117(26):4129–34. doi:10.1002/ange.200462960.
54. Bernardos A, Mondragon L, Aznar E, Marcos MD, Martinez-Manez R, Sancenon F, Soto J, Barat JM, Perez-Paya E, Guillem C, Amoros P. Enzyme-responsive intracellular controlled release using nanometric silica mesoporous supports capped with “saccharides”. ACS Nano. 2010;4(11):6353–68. doi:10.1021/nn101499d.
55. Dzamukova MR, Naumenko EA, Lvov YM, Fakhrullin RF. Enzyme-activated intracellular drug delivery with tubule clay nanoformulation. Sci Rep. 2015;5:10560. doi:10.1038/srep10560.
56. Huang P, Gao Y, Lin J, Hu H, Liao H-S, Yan X, Tang Y, Jin A, Song J, Niu G, Zhang G, Horkay F, Chen X. Tumor-specific formation of enzyme-instructed supramolecular self-assemblies as cancer theranostics. ACS Nano. 2015;9(10):9517–27. doi:10.1021/acsnano.5b03874.
57. Agostini A, Mondragón L, Coll C, Aznar E, Marcos MD, Martínez-Máñez R, Sancenón F, Soto J, Pérez-Payá E, Amorós P. Dual enzyme-triggered controlled release on capped nanometric silica mesoporous supports. Chemistry Open. 2012;1(1):17–20. doi:10.1002/open.201200003.
58. Koo AN, Lee HJ, Kim SE, Chang JH, Park C, Kim C, Park JH, Lee SC. Disulfide-cross-linked PEG-poly(amino acid)s copolymer micelles for glutathione-mediated intracellular drug delivery. Chem Commun. 2008;48:6570–2. doi:10.1039/B815918A.
59. Wilson DS, Dalmasso G, Wang L, Sitaraman SV, Merlin D, Murthy N. Orally delivered thioketal-nanoparticles loaded with TNFα-siRNA target inflammation and inhibit gene expression in the intestines. Nat Mater. 2010;9(11):923–8. doi:10.1038/nmat2859.
60. Klaikherd A, Nagamani C, Thayumanavan S. Multi-stimuli sensitive amphiphilic block copolymer assemblies. J Am Chem Soc. 2009;131(13):4830–8. doi:10.1021/ja809475a.
61. Chang B, Sha X, Guo J, Jiao Y, Wang C, Yang W. Thermo and pH dual responsive, polymer shell coated, magnetic mesoporous silica nanoparticles for controlled drug release. J Mater Chem. 2011;21(25):9239–47. doi:10.1039/C1JM10631G.
62. Bilalis P, Chatzipavlidis A, Tziveleka L-A, Boukos N, Kordas G. Nanodesigned magnetic polymer containers for dual stimuli actuated drug controlled release and magnetic hyperthermia mediation. J Mater Chem. 2012;22(27):13451–4. doi:10.1039/C2JM31392H.

63. Wang K, Guo D-S, Wang X, Liu Y. Multistimuli responsive supramolecular vesicles based on the recognition of p-sulfonatocalixarene and its controllable release of doxorubicin. ACS Nano. 2011;5(4):2880–94. doi:10.1021/nn1034873.
64. Loh XJ, del Barrio J, Toh PPC, Lee T-C, Jiao D, Rauwald U, Appel EA, Scherman OA. Triply triggered doxorubicin release from supramolecular nanocontainers. Biomacromolecules. 2012;13(1):84–91. doi:10.1021/bm201588m.
65. Furgeson DY, Dreher MR, Chilkoti A. Structural optimization of a "smart" doxorubicin–polypeptide conjugate for thermally targeted delivery to solid tumors. J Controlled Release. 2006;110(2):362–9. doi:10.1016/j.jconrel.2005.10.006.
66. Wang B, Ma R, Liu G, Li Y, Liu X, An Y, Shi L. Glucose-responsive micelles from self-assembly of poly(ethylene glycol)-b-poly(acrylic acid-co-acrylamidophenylboronic acid) and the controlled release of insulin. Langmuir. 2009;25(21):12522–8. doi:10.1021/la901776a.

Chapter 10
CNS Drug Delivery for Diseases Eradication: An Overview

Anil Kumar Kalvala, Priyanka Giri, Jyothirmai Kaligatla, Wahid Khan, and Ashutosh Kumar

Abstract Central nervous system (CNS) diseases lead to severe disability and account for considerable amount of health burden globally making them one of the most devastating diseases affecting the mankind. Though new drugs to eradicate CNS diseases are much warranted, drug development in this field is going at snail's pace due to potential limitation of low success rate in clinical trials. Therapeutic delivery to the brain offers superior challenge to scientific community because of the presence of blood brain barrier (BBB), the lipophilic barrier which hampers the entry of foreign substances into brain and drug efflux transporters which prevents drug accumulation in CSF (Cerebro spinal fluid). Further it is assumed that BBB permeability and brain pharmacokinetics alters with pathological condition which further complicates the drug delivery. In view of this altered physiology, scientists have designed some pharmaceutically active formulations of drug molecules which can lodge in brain because of their altered lipophilicity, selective permeability or ligand based targeting. Such delivery options enhance the efficient delivery of active therapeutic molecules to the target of interest without showing any side effects. Such site directed target delivery also ensures improved brain pharmacokinetics of the compounds. This book chapter offers information about myriad range of CNS diseases; unmet needs in their management and several novel methods of CNS directed delivery options for therapeutics to alleviate those nervous disorders with maximum efficacy. In particular by realizing the emergence of nano technology and nano medicine, we have discussed nano technology based formulations and their potential importance for CNS delivery in detail.

A.K. Kalvala • P. Giri • J. Kaligatla • A. Kumar (✉)
Department of Pharmacology and Toxicology, National Institute of Pharmaceutical Education and Research Hyderabad (NIPER Hyderabad), Balanagar, Telangana 500037, India
e-mail: ashutosh@niperhyd.ac.in

W. Khan
Department of Pharmaceutics, National Institute of Pharmaceutical Education and Research Hyderabad (NIPER Hyderabad), Balanagar, Telangana 500037, India

A. Kaushik et al. (eds.), *Advances in Personalized Nanotherapeutics*,
https://doi.org/10.1007/978-3-319-63633-7_10

Keywords CNS • BBB • Blood-CSF barrier • Nanoparticles • Nanocarriers • Polymers

1 Background

World health organization (WHO) defined health as "a state of complete physical, mental and social well-being and not merely the absence of disease or infirmity". Brain disorders are often accompanied by perturbations in mental state of health. Mental disturbances comprises of social, cultural, environmental, economic and political factors of individual; and resulted in global burden of nearly 16.3 million US dollars in between 2011 and 2013 [1]. As per WHO, suicide is the second most cause of death in young individuals, which can be attributed to CNS disorders, for instance people suffering with schizophrenia and depression were shown 60 and 40% more susceptibility to premature deaths respectively compared with total deaths in young individuals [2]. Globally, it was estimated that approximately 85% of the people from low to middle income countries and 35–50% of people from high income countries are deprived of proper treatment facilities for CNS [3]. This may be due to lack of resources, availability of high cost and low effective drugs, and limited clinical success of CNS drugs. Accumulating literature showed annual expenditure of less than 0.25 US dollars per person on mental health in developing countries [4]. An average time for conducting clinical trials for CNS interventions is about 8.8–12.6 years, which cost around 1.3 billion US dollars, a dramatically higher cost compared to diabetes and cardio vascular trials [5, 6] . In spite of the huge investment, a mere 7% of success rate has been reported in clinical trials for CNS interventions, considerably very low comparative to other trials [5]. The complexity of CNS system is the major obstacle in the drug development, which limits the drug availability in brain through the presence of blood brain barrier (BBB) or blood-cerebrospinal fluid (blood-CSF) barrier or blood-tumor barrier. Even if drug molecules are able to cross these membranes and available in the CNS, non-specificity in their actions restrict their utility for particular indication. In addition, it is very difficult in functional assessment of neuronal diseases because of the involvement of subjective measures such as cognition and physical function which varies from patients to investigators. Further, limited biomarker development in the field of brain diseases equally impacting the treatment failure in clinical trials [7].

1.1 Anatomical Barriers to CNS Drugs Delivery

Currently many of the potential drugs and pharmaceutical agents like antibiotics, antineoplastic agents, neuropeptides and neuroprotectants against neurodegenerative diseases were proved to be effective in vitro studies but failed to mimic the same in vivo studies [8]. This is because of the presence of BBB, not only as an anatomical barrier but also as transport and metabolic barrier (Fig. 10.1). However, to perform normal physiological functioning of brain, transport of various biological mediators including therapeutic agents across biological membranes is highly indispensable. Thus, the BBB is regarded as the bottleneck for the successful development of the drugs that specifically target the CNS. Therefore, various strategies to enhance the drug delivery to CNS were developed based on BBB breaching. To

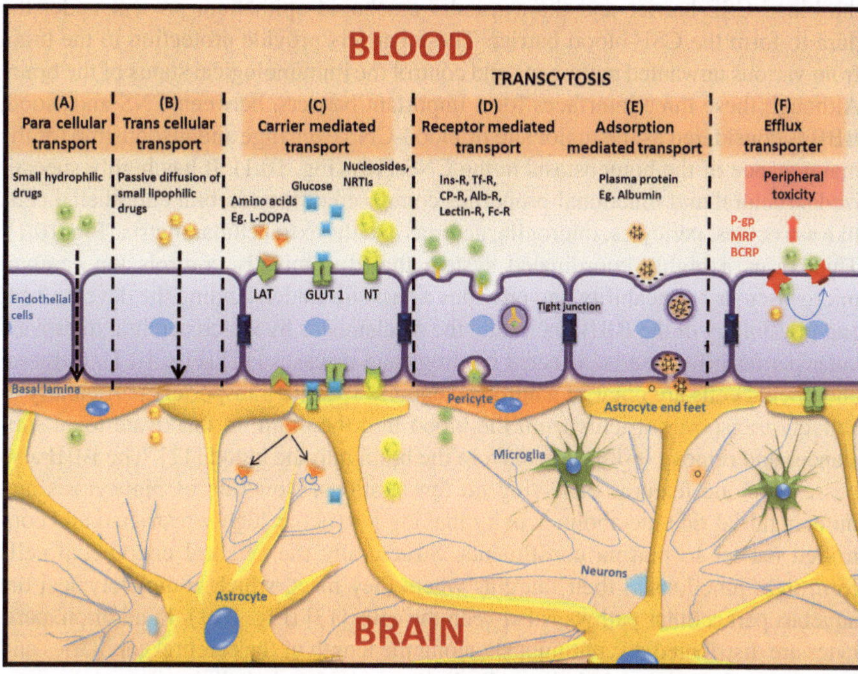

Fig. 10.1 Illustration of the BBB and the possible mechanism and challenges that can be approached for CNS drug delivery. BBB largely comprises the endothelial cells forming the capillary bed and are held together by tight junctions. The brain facing side of endothelial cells are surrounded by a layer of basal lamina, pericyte cells, feet end of astrocytes and microglias to limit the passage of xenobiotics through the BBB. However certain transport mechanisms are present to facilitate the movement of molecules across the BBB. (A) Paracellular transport of small hydrophilic drugs across the aqueous pores. (B) Transcellular transport of small lipophilic drugs through the endothelial cells. (C) Carrier mediated transport of agents like amino acid, glucose, nucleoside via respective carriers. (D) Receptor mediated transport involve transport of proteins and peptides by their binding to respective receptors followed by internalization of receptors (transferrin binds to transferrin receptor). (E) Adsorption mediated transport involves adsorption of cationic drugs to the negative surface of endothelial cells thereby promoting Clatherin mediated pit formation followed by endocytosis. (F) Efflux transport: transport of proteins such as P-gp, MRP efflux out drugs from endothelial cells back to the blood contributing to the factor of peripheral toxicity. *L-DOPA* Levodopa, *NRIs* Nucleoside reverse transcriptase inhibitor, *LAT* L-amino acid transporter, *GLUT 1* Glucose Transporter 1, *NT* Nucleoside Transporter, *Ins-R* Insulin Receptor, *Tf-R* Transferrin Receptor, *CP-R* Ceruloplasmin Receptor, *Alb-R* Albumin Receptor, *P-gp* P-glycoprotein, *MRP* Multidrug resistance associated protein, *BCRP* Breast cancer resistance protein

develop a successful therapy to target CNS, firstly the structural and functional characteristics have to be understood. Here we have discussed various barriers of CNS and their influence on the penetration of drugs. The first study to demonstrate the existence of BBB was done by Ehrlich in 1885 [9]. There are three main interfaces in the brain which protect the neurons by preventing the entry of foreign molecules and helps in maintenance of water homeostasis: the blood-brain interface, blood-CSF interface and the CSF-blood interface. The cerebral endothelium forms the largest interface i.e. BBB, while the epithelial cells of the choroid plexus form the blood-CSF barrier and the avascular arachnoid epithelium present under the dura to form the CSF-blood barrier. These barriers provide protection to the brain from various unwanted pathogens and control the immunological status of the brain. Although these three interfaces form important barriers, between CNS and blood, BBB is considered as the major site of blood-CNS exchange and responsible for the maintenance of the homeostasis in the CNS [10] (Fig. 10.1). It has highly specialized structural and functional properties composed of the cerebral endothelial cells like astrocytes, pericytes, microglia, neurons and the extracellular matrix (Fig. 10.1). This forms a highly coordinated system that dynamically controls the cerebral microvascular permeability and provides a basis for understanding the development and physiology of the BBB, including the mechanisms by which cerebral microvascular permeability can be affected by drugs and diseases [11] (Fig. 10.1). Cerebral endothelial cells are polar in nature and express specific transporters particularly to actively transport nutrients from the blood into the brain, to inactivate toxic substances and remove (efflux) them from the brain into the blood [12]. The BBB also possess unique features of having no fenestrations, slow rate of pinocytosis and mitochondrial rich metabolic activity making it to act as highly restricted and controlled barrier to plasma constituents. Structurally the cerebral endothelial cells form tight junctions at their margins where they meet, which completely seal the aqueous paracellular pathways between the cells [13] (Fig. 10.1). In addition, pericytes are distributed discontinuously along the length of the cerebral capillaries and partially enclose the endothelium. Both the cerebral endothelial cells and the pericytes are surrounded by, and contribute to, the same extracellular matrix rich in laminins 8 and 10. The foot processes of astrocytes also form a network that fully encloses the capillaries (Fig. 10.1). Axons from the neurons surround the cells that contain neuropeptides and neurotransmitters. Microglial cells are present as immunocompetent cells of the brain which are derived from systemic circulating monocytes and macrophages [14] (Fig. 10.1). These tight junctions formed by various endothelial cells and other neuronal cells, in particular, the polar expressions of transport proteins are responsible for inducing BBB properties and differentiation in the cerebral endothelial cells. Though BBB forms such a restricted and controllable environment, vital solutes cross the BBB either by diffusion or transporters [15]. Leukocytes may cross the BBB adjacent to, or by modifying, the tight junctions to provide immunocompetence. Solutes may diffuse passively due to their more lipid-solubility properties through the cell membrane and cross the endothelium. Active efflux carriers may interrupt some of these passively passing solutes and pump them out of the endothelial cell [16] (Fig. 10.1). Many essential polar molecules such as glucose, amino acids, and nucleosides may be transported through carrier-mediated influx into the CNS [17] (Fig. 10.1). Receptor-mediated transcytosis can carry macromolecules such as peptides and proteins across the cerebral endothelium. Absorptive mediated transcytosis appears to be induced non-specifically by negatively charged macromolecules and is also useful in transportation across the BBB. Tight junction modulation relaxes the junctions and totally or partially opens the paracellular aqueous diffusional pathway [18] (Fig. 10.1). Due to these properties of BBB, the treatment of CNS-related disorders or diseases; drug delivery to the CNS at therapeutically active concentration has become critical step for many promising drug candidates.

Treatment for several neurological diseases generally follows invasive or non-invasive routes of administration. The invasive techniques include neurological direct injections or structural disruption of BBB by using ultrasound. Neurological direct injections include intracerebral injection/infusion, where the drug is directly injected into brain parenchyma or intraventricularly or intrathecally into CSF and intranasal delivery, where nasally administered drug gets access to the olfactory nerves and they get transported by retrograde axonal cytoplasmic flow back into the CNS. For successful delivery and the proper distribution of drugs at targeted location in CNS, disruption of the BBB is essential. This is induced by pulsed ultrasound in the presence of preformed gas bubbles [19]. It is operated very cautiously because over pitch sound may damage brain tissues. Other neurosurgical delivery strategies that are available include BBB disruption by osmotic imbalance (using mannitol) or by using vasoactive compounds, intraventricular drug infusion, and intracerebral implants [20]. Other methods which provide broad spectrum of action and deliver a wide range of drugs to CNS are pharmacological and physiological methods which are relatively safe and non-invasive. The physiological strategies are designed based on the endogenous transport mechanisms by using either carrier-mediated transport of nutrients like glucose transporter (GLUT1), efflux transporters like P-glycoprotein (P-gp) or receptor-mediated transport of peptides like lipoprotein receptor-related proteins 1 (LRP1) and 2 (LRP2) [21]. These physiological strategies are proved as better and potential delivery methods, because of wider safety cover provided by drug transport. But it is not suitable for all kind of neurotherapeutics. In pharmacological methods, lipid carrier or liposomes are used for drug delivery. To achieve the objective of enhanced targeted drug delivery to CNS, the conventional pharmacological methods can be improved by using liposomes, colloidal drug carriers, micelles, chimeric peptide technology and importantly nanotechnology [22].

Nanoparticles are colloidal delivery systems which have been identified for their significant potential in treatment of CNS diseases. The reason for introduction of nanoparticles in medical field is, they possess some unique characteristics compared to conventional drugs. This includes low molecular size (100 nm) that provides huge surface area which enhances absorption, site specific delivery that overcomes the problems of conventional drugs whose penetration is limited by BBB. Surface-modified nanoparticles are advantageous in terms of site-specific delivery, reducing peripheral toxicity and causes the control release kinetics [23] (Fig. 10.2).

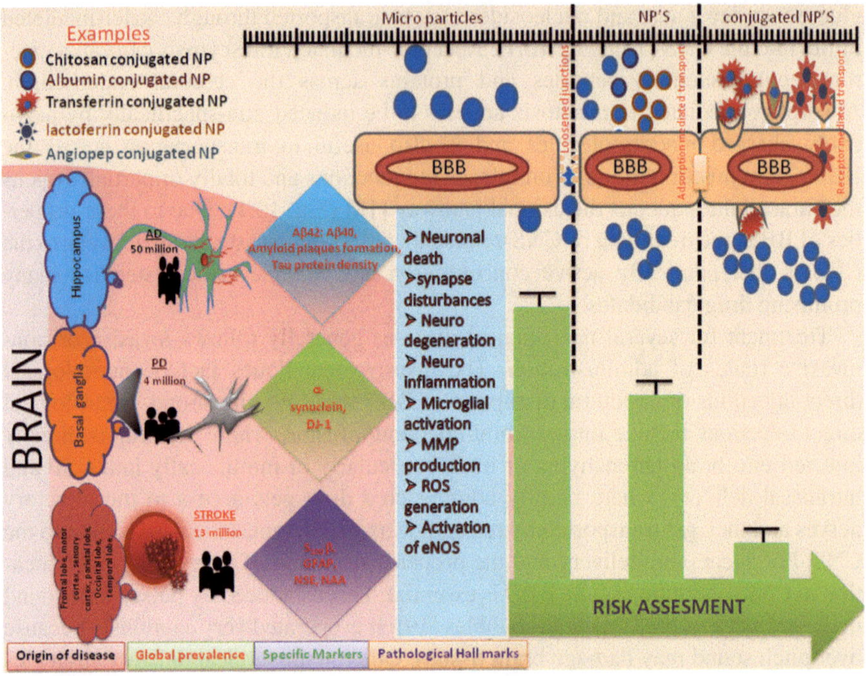

Fig. 10.2 Blood brain barrier (BBB) transport mechanisms of nano drug formulations and simultaneous risk assessment of the CNS diseases with Nanoparticles (NP's) and conjugated NP's of the drug that cross the BBB via different transport mechanisms. NP's conjugated with natural polymers like chitosan and albumin follows adsorption mediated transcytosis. Majority of NP's currently developed were conjugated with ligands like transferrin, angiopep-2, lactoferrin, COG 133, CRM 197 and ox26 to cross the BBB via receptor mediated transport system. In this figure, the global prevalence, origin of the CNS disease, disease specific markers and pathological hall marks of the CNS diseases were represented. Further, disease risk assessment with different drug formulations was represented by drawing an arbitary histogram. *AD* Alzheimers disease, *GFAP* Glial fibrillary acidic protein, *MMP* Matrix metalloproteases, *NAA* N-acetyl aspartate, *nNOS* Neuronal nitric oxide synthase, *NSE* Neuron specific enolase, *ROS* Reactive oxygen species

Surface modification of nanoparticles could be achieved by these two approaches; one of them is to modify the surface of the drug-loaded nanoparticle by conjugating it with a targeting ligand that is particularly specific to the brain and another approach is modification of nanoparticles with surfactants (Fig. 10.2).

2 CNS Diseases: Treatment and Unmet Needs

2.1 Alzheimer's Disease (AD)

In United States officially AD declared as sixth leading cause of death, and postulated it as the only disease among the top ten causes of death in America that cannot be prevented, cured or even slowed [24]. In 2013, over 84,000 Americans died because of AD and it is estimated that 700,000 Americans will die in 2016. The cost of Medicare in 2016 is estimated to be 236 billion US dollars for AD patients and expected to project one trillion US dollars in 2050 unless something contributed in this AD research [24]. Globally, 50 million individuals are suffering from AD and in 2050 it is estimated that 120 million people will be affected with this disease [25]. From past 30 years AD research is surrounding around the deposition of Aβ aggregates followed by formation of neurofibrillary tangles on and in the hippocampus neurons respectively. AD development involves long prodromal/pre symptomatic phase followed by symptomatic phase. Approximately, it takes 20 years to develop the clinical symptoms of AD [25]. In this period, maximum damage to the brain prevails and becomes prominent challenge in reversing the disease. In fact, from past 20 years drugs were developed, validated, and progressed to some extent in preclinical-clinical trials, yet no drug is currently available in the market for the treatment of AD [25] except few drugs providing symptomatic relief. So far some of monoclonal antibodies like aducanumab which is under phase 3 trials [26], active immunization with AN1792 [27] and bapineuzumab [28], solanezumab anti Aβ antibodies in which the latter completed phase 3 clinical trials [29] were showing reduction in Aβ burden and provided symptomatic relief of AD. Although these drugs are in positive progress, experts demonstrated these drugs may be associated with high cost that will inflate financial burden. Apart from this, drug cost also affected by cost of clinical trials, current expenditure for phase three clinical trials in AD is about 350–500 million US dollars if they involve multiple amyloid scans and other imaging studies [25]. Besides, myth in identification of biomarkers is one more barrier for failure of drugs in clinical trials. Available biomarkers such as amyloid imaging, imaging of CSF, Aβ42:Aβ40 ratios, imaging volumetric change magnetic resonance imaging (MRI) or the recent tau imaging studies were not precise and have many obligations in identifying the prodromal phase of AD [30] (Fig. 10.2). Moreover, these biological markers identification techniques are costly unlike any other diagnostic/biomarkers tools. In recent advances, blood exosomes have been found as biological markers in the AD patients which is under validation studies [31]. In spite of it, the other challenges faced in treatment of AD are the drug delivery aspects. In recent times, many natural compounds, antibodies have shown efficacy in treating AD in vitro but failed in in vivo correlation of the same due to low reachability of drug in to the brain. In order to increase their penetration and bioavailability, these drugs if encapsulated in nanoparticles (surface modified with ligand) can effectively cross the BBB and produces the efficient therapeutic effect (Fig. 10.2). Example, an anti-oxidant Tempol loaded into poly lactic-co-glycolic acid (PLGA) nanoparticles, surface modified with transferrin, favors receptor mediated endocytosis across endothelium of brain capillaries and enhances the brain uptake [32] (Table 10.1). Further, Bapineuzumab and Solanezumab, when conjugated with nanoparticles were tried against amyloid β and were effective in solubilizing plaque [33]. Trimethylated chitosan-modified co-enzyme Q10 loaded PLGA nanoparticles were found to have enhanced neuroprotective activity in an in-vivo model of AD [34] (Table 10.1).

Table 10.1 Summary of nanotechnology based drug formulations applied in the treatment of various CNS diseases/disorders

Sr. No.	Disease/disorder	Drug	Target	Formulation	Mechanistic indication	References
1.	AD	Tempol	Antioxidant	PLGA nanoparticles	Transferrin mediated transcytosis	[32]
		Q10	Antioxidant	TMC modified nanoparticles	Adsorption mediated transcytosis	[33]
2.	PD	Levodopa	Dopamine agonist	PLGA nanoparticles	Promotes Sustained release there by decreases the akinetic effects of levodopa when administered alone.	[42]
		Dopamine	D1 and D2 receptors of basal ganglia	Cellulose acetate phthalate nanoparticles	Adsorption mediated transport	[42]
3.	Depression	Venlafaxin	SSRI andSNRI	Alginate nanoparticles	Paracellular transport	[144]
		ZnO	Increases neuronal excitability	Nano-sized ZnO	Transported through loosened junctions of BBB	[68]
4.	Stroke	t-PA	Fibrinolytic	Polystyrene latex nanoparticles	Adsorption mediated transport	[23]
		SOD	Antioxidant	PLGA nanoparticles	Adsorption mediated transport	[72]
5.	Glioma	Doxorubicin	DNA and Topoisomerase II	Lactoferrin modified procationic liposome	Lactoferrin receptor mediated transcytosis	[82]
		Anti-miR-21	miR-21	Cholorotxin conjugated nanoparticle	Metalloproteinase-2 receptor mediated transcytosis	[145]
		NK012 (prodrug of SN-38)	Topoisomerase I	PEG-PLGA polymeric nanomicelle	Adsorption and diffusion mediated transcytosis	[146]
		Paclitaxel	Tubulin	Paclitaxel loaded c(RGDyK)- PEG-PLA	Integrin receptor mediated transcytosis	[147]
		Fe(0)	ROS liberation	Mesoporous Silica nanoparticle	Passage through glioma disrupted BBB	[87]

Abbreviations: *BBB* Blood brain barrier, *PEG* Polyethylene glycol, *PLGA* Poly (lactic-co-glycolic acid), *ROS* Reactive oxygen species, *TMC* Trimethylated chitosan, *t-PA* Tissue plasminogen activator, *ZnO* Zinc oxide

2.2 *Parkinson's Disease (PD)*

In the year 2007, based on funding resources, progress in research, FDA approvals, and technologies adopted to eradicate global burden of PD, WHO projected the deaths caused by PD is about 0.2% (~1.75 millions) in 2015 and 0.23 % (~2.38 millions) in 2030 of total world's death population [35]. In 2015, September 29th Parkinson's disease foundation reported that more than four million people worldwide are suffering from PD [36]. The projected statistics by WHO and reported epidemiology of PD by Parkinson's disease foundation were almost comparable which represents there is no success towards eradication of PD in past few years. The cost of patient care for PD disease is about 14 billion dollars per year in US alone [37]. This explains the therapeutic unmet need of PD. From last 30 years much advancement in the development of drugs to relieve symptoms of Parkinsonism has been addressed yet there is no single drug has been found to modify the disease progression [37]. The pathology of disease remained on neurochemical basis of dopamine depletion and genetic variations. Thus, dopamine analogues, agonists, degradative inhibitors of dopamine and reuptake inhibitors were under usage to treat symptomatic disease from decades [38]. Apart from this, researchers put forth their efforts in identifying the disease modifying agents on genetic and neurochemical basis which have shown potential effects in preclinical studies but yet to be of value in clinical trials [39]. The possible explanations of strategic failure in drug development of PD treatment are attributable to poor understanding of pathophysiology. Incomplete establishment of links between the targets and poor understanding of downstream signaling pathways making the process of drug development more complicated [37]. Improper animal and cellular models in screening interventions are a major translational barrier in PD drug development programs [40]. Additionally, anatomical barriers like BBB, CSF-blood barrier, also contribute to translational barriers against the development of PD symptomatic or disease modifying agents. This led to the development of formulations to reduce the attrition due to these anatomical barriers in translational medicine. For example, Pillay has showed that, the inclusion of dopamine loaded cellulose acetate phthalate nanoparticles into a binary cross linked alginate facilitated local dopamine delivery in rat model [41]. Intra peritoneal (I.P.) administration of dopamine loaded polyvinyl Pyrrolidone (PVP) or polyvinyl acrylic acid (PAA) improved the striatal dopamine levels and catalepsy scores in reserpine treated rats [42] (Table 10.1). Levodopa was encapsulated with Benserazide in PLGA nanoparticles has successfully abolished levodopa induced dyskinesia in rats [42] (Table 10.1). Although these studies have shown some promise in improving the bioavailability of dopaminergic agonists, more pre-clinical validations are yet to be done before being applied in clinical settings.

2.3 Schizophrenia

WHO projected Schizophrenia is one among the top ten diseases which is reason for major economic burden worldwide that affected more in late adolescence and in early adult hood populations [43]. In 2005, 0.5–0.8 % of world's population diagnosed with schizophrenia [44] and now the disease occupied 1% among total world's population [45]. Schizophrenia is characterized with positive (hallucinations and delusions; loss of reality, unhealthy individuals), negative (healthy, loss of motivation, disorganized speech, social withdrawal) and cognitive symptoms (loss of remembrance and reutilization of already known facts) which affects 80–90% of unemployment in patients unlike other brain diseases [46]. In England, epidemiological survey on schizophrenia estimated a loss of 11.8 billion British pounds per year which includes direct expenditure on medical care and indirect cost on unemployment [47]. This disease is unique in variations unlike other brain diseases, because there are no clear boundaries to fragment the disease state. Schizophrenia has been reported for its sudden attack, slow progression, frequent relapse and remittance, with unpredictable combination of clinical symptoms in patients [48–50]. Recent advancements in the schizophrenia research has been focused on dopamine dysfunction and glutamate toxicity [51]. Growing research evidence suggested that schizophrenia affects various regions of the brain especially prefrontal cortex, hippocampus, and reduction in grey and white matter, temporal lobe, thalamus and caudate nucleus [52–54]. Loss of synapses between the neurons centered as unifying theory in affected regions of brain [55]. Further, researchers identified involvement of dopamine and glutamate in constructing synapses [56, 57]. In spite of it, growing research evidenced involvement of more than 110 genes (polygenic and pleotropic risk) in the development of schizophrenia [58]. Yet, no significant advancement in the clinical research of this disease has been made which is mainly due to lack of specific biomarkers, no good preclinical models which mimics human health, uncertainty and complexity of targets involved with indefinite downstream pathways and anatomical barriers of translational medicine [58, 59]. Serendipitous discovery of chlorpromazine during 1950s and synthesis of clozapine in late 1960 brought new insights in drug development of treating schizophrenia [60]. Current drugs on extensive usage are D2 receptors antagonists which were classified into first to third generation of drugs [60]. These drugs have limited therapeutic efficacy towards positive symptoms and no effect on negative and cognitive symptoms [61, 62]. Indeed, research efforts to develop other than D2 antagonists were sidetracked because of its limited clinical utility due to severe side effects or refractory response in treated patients. Although drugs have approved for the treatment, they still suffer from poor penetrability and are known to precipitate many peripheral side effects, and hence their long term clinical utilization were limited. For instance amisulpride and sulpride have limited BBB penetration [63]. In this context, experts opined the future drug development strategies will be seen to fall on either of following categories (1) identification of appropriate targets and discovery of drugs towards novel targets (2) modifications of existing drugs to minimize their side effect profiles or making more efficacious drugs (3) focusing on adjuvant therapy to potentiate the therapeutic action of the potent drugs [60].

2.4 Depression

It is a most common widely spread psychiatric disorder that has been rising across the world. It is characterized by sadness, suicidal tendency, lack of interest and altered mood. WHO stated that it has become fourth leading cause for disability worldwide and it has been prophesied that it would be in second position by 2020 [64]. It is known that depression is mainly due to decreased monoamine function, the use of selective serotonin reuptake inhibitors (SSRIs) and serotonin noradrenaline reuptake inhibitor (SNRIs) has become the prime targets. Even though there are many drug treatments for depression, but still they are unmet with clinical needs due to minimal efficacy and increased side effects. Other reasons include poorly understood pathophysiology, lack of established biomarkers, failure in drug exposure to CNS target due to partial penetration, failure in clinical trial validation, high metabolism of drugs and drug-drug interactions [5]. These are not only specific to depression but also include other CNS disorders. In depression, SSRIs and SNRIs are considered as first line drugs that were launched in 1990s, and still contemplated as first line therapy. In order to decrease their side effects and improve efficacy of drugs, combinational therapies and other novel approaches such as neurokinin (NK1 peptidergic), neurotrophins, modulation of cholinergic and gabanergic transmission have been targeted [65]. Fluoxetine and paroxetine, (SSRIs) are selective, while vidalazone is combined SSRI and 5HT1A partial agonist. Vidalazone has undergone phase 2 clinical trial in 2000 and FDA has granted for market on 2011. It increases the extracellular 5-HT (5-hydroxy tryptamine) levels, apart from this it showed severe adverse effects on gastrointestinal system like diarrhea and nausea [66]. Other drug is venlafaxine (VLF), SSRI and SNRI, is available as oral tablet. But it has limitations like short t1/2, side effects like tachycardia, fatigue and hypertension. Indeed, VLF loaded with alginate nanoparticles by intranasal from nose to brain has increased the bioavailability, reduced side effects and improved risk/benefit ratio [67]. Another drug for management of depression is agomelatine, a melatonin (MT) agonist and 5-HT2c antagonist. It reduces the symptoms of depression and enhances the sleep quality. It has approved in some of regions globally in 1990s but, after a phase 3 clinical trial held in U.S, it is was discontinued due to severe hepatotoxicity [66]. A nano ZnO would play a potential role in depression by increasing neuronal excitability and release of neurotransmitter. This nano ZnO enhances the behavioral and cognitive functions in mice model. But it causes genotoxicity through oxidative stress and lipid peroxidation [68] (Table 10.1) . More investigations are needed for better understanding of neuro-effect of nanoparticles that provide basis for new therapeutic targets in treatment of depression.

2.5 Stroke

It is a third leading cause of death in U.S supported by evidence that about one-third of stroke survivors have become permanently disabled and in 2010 worldwide prevalence has raised to 33 million [69] (Fig. 10.2). It is estimated that 16 million people have suffered from first-ever stroke and it is expected that this may increase to 7.8 million deaths by 2030 [70]. Stroke may be ischemic or hemorrhagic that causes disability and death across the world, caused by inadequate blood flow to brain resulted from ruptured blood vessel due to trauma and other pathological states. Inspite of many approaches for management of stroke, its treatment remains one of the major challenges in current medicine. These approaches include cerebral reperfusion, neuroprotection and improving in current drug delivery aspects. Cerebral reperfusion mainly focuses to restore the brain blood flow and this can be achieved through thrombolytic drugs. t-PA (tissue Plasminogen Activator) is a FDA approved drug that has licensed in 1996 for the treatment of stroke. But it is benefited only in selected patients who produce stroke symptoms within 3 h of onset. t-PA has not been approved in European countries because of its limited success, (which does not show satisfactory effect in after 6 h onset of stroke in their Acute Stroke Studies (ASS) [71]. Apart from this it also has non-specificity and other hemorrhagic complications. In order to increase the specificity of t-PA , it was tried with polystyrene latex nanoparticles modified with antifibrin antibody, which makes t-PA specific to clot [23] (Table 10.1). Stroke promotes various pathobiochemical processes and leads to neuronal cell death. These death promoting mechanisms includes oxidative stress, excitotoxicity and neuroinflammation. Based on these, many neuroprotective approaches have been developed. Even though they produced effective outcomes in animal models but none of them were effective in patients [71]. Oxidative stress can be diminished by antioxidant enzymes like Super oxide dismutase (SOD), catalase and glutathione peroxidase. But these enzymes have short $t_{1/2}$, and are not taken up by neuronal cells even though they cross BBB. So these enzymes mainly SOD, encapsulated with PLGA nanoparticles showed significant neuroprotective effects because it increases the bioavailability, due to active release of enzyme, and intracellular delivery of SOD in brain [72] (Table 10.1). Other targets include glutamate receptor antagonists like NMDA, AMPK antagonists in which some are not successful (selfotel, eliprodil). A phase 3 clinical trial with YM872 is still ongoing [71].

2.6 Glioma

Glioma is a common term for primary brain tumors originating from the glial tissues resulting in significant morbidity and mortality. According to the CBTRUS (Central Brain Tumor Registry of the United States) report 2015, the world wide incidence rate of primary malignant brain and CNS tumors estimated in 2008–2012 when age-adjusted patients was 3.4 per 100,000 with mortality rate in US is estimated of 4.3 per 100,000 [73]. These statistics does not refer specifically to Glioma incidences but since Glioma constitutes 80% of the primary malignant brain tumors, their incidence may affect the statistics. This global burden of Glioma represents group of tumors, namely neuroepithelial tumors of CNS like astrocytomas, oligodendrogliomas, ependymomas and mixed neuronal Gliomas characterized by International Classification of Diseases and Oncology, Version 3 (ICD-O-3) [74]. Moreover, WHO graded various Gliomas from I to IV according to the malignant behavior and histologic site in adult which include astrocytomas (Grade I–IV), oligodendrogliomas (grade II–III) and oilgoastrocytoma (grade II–III) [74]. However, the most common and most severe form of Glioma is Glioma multiforme also referred as glioblastoma (grade IV) with survival time of approximately 12 months. Various studies indicate environmental factors (radiation, chemical exposure), genetic aggregation (gene polymorphism), immunological responses (Asthma, RA (Rheumatoid Arthritis), allergy) are responsible for the prognosis of Glioma [75]. In spite of these attempts to study the cause of Glioma, the results were inconclusive and draws the hope more towards identifying its molecular pathology.

Classical tumor molecular and cytogenetic studies showed that development of Glioma multiforme is majorly through dysregulated pathway by progression of primary glioblastoma along with secondary glioblastoma [76] marked by p53 mutation (p53 protein accumulation) and EGFR gene overexpression respectively [77, 78]. Several strategies targeting EGFR include monoclonal anti-EGFR (cetuximab, panitumumumab, nimouzumab), alkylating agent (temezolomide), EGFR inhibitors (erlotinib, gefitinib, lapatinib), EGFR specific siRNA therapy [79]. Phase I clinical trials for evaluating safety and efficacy of mebendazole (NCT01837862) in pediatric Glioma is currently in progress. Additionally, besides targeting EGFR, RAS proves to be a prime target in Glioma patients. Recently, phase I/II trial (NCT00050986) was conducted to evaluate the safety and efficacy of Zarnestra (tipifarnib) in combination with temezolomide.

Although many targets and therapeutics have been highlighted in the treatment of Gliomas, certain factors still remain the point of concern. These factors include resistance of tumor towards single drug therapy and inability of therapeutics to cross BBB thereby limiting their efficacy and also promoting peripheral toxicity [80]. However, in an attempt to enhance the therapeutic efficacy, various drug combinations are currently in the clinical trials. Recently dose-escalation study of XL765 (SAR245409) in combination with Temezolomide with and without radiation in subjects with Malignant Glioma have completed phase II clinical trial (NCT00704080). The aforementioned study proposes the adjuvant therapy of XL765; a PI3K and mTOR inhibitor thought to potentiate the efficacy of temezolomide. Of note growing interest in research evidenced that application of nanotechnology and exploring its uptake mechanism via BBB can potentially pave a way for brain targeted drug delivery enabling those drug which cannot transverse the BBB to not only reach the site of action but also to prevent the side effects of peripheral toxicity [81]. Huali Chen et al. developed doxorubicin loaded liposomes formulated of neutral or negative charged cholesterol and surface modified with lactoferrin [82] (Table 10.1). Lactoferrin is a mammalian cationic iron-binding glycoprotein belonging to the transferrin family which binds to lactoferrin (transferrin) receptors present only on BBB [83]. These modified liposomes with its surface attached to lactoferrin binds to the lactoferrin receptors, converts to cationic liposome and accordingly promote uptake in the brain by transferrin receptor as well as adsorption mediated transcytosis. Another tumor targeted ligand nanoparticles for drug delivery in brain is depicted in chlorotoxin conjugated nanoparticles where the nanoparticles are adsorbed with chlorotoxin; a toxin binding specifically to metalloproteinase [84]. Metalloproteinase is an enzyme which cleaves extracellular matrix components and is significantly found to be overexpressed in Glioma [85]. Therefore, ligand conjugated nanoparticles can prove to be novel target specific drug delivery in Glioma owing to transportation of drug and its controlled release [86]. Moreover, an in vivo study conducted on glioblastoma has shown that zero valent ferrous ions which are impregnated in mesoporous silica nanocarriers enhanced ferrous delivery into the tumor tissues selectively. The plausible mechanism for ferrous cytotoxicity was proposed to be production of ROS thereby reducing the tumor size [87] (Table 10.1).

3 Drug Delivery Aspects of CNS Diseases: Focus on Nanotherapeutics

The BBB has always been the primary gating mechanism and a challenge for delivering therapeutics to the brain. The drug delivery aspects of CNS diseases and disorders can be broadly carried out by

3.1 Invasive Drug Delivery

Invasive drug delivery approaches involve physically entering the CNS by disruption of the BBB, or injecting a therapeutic agent directly into the target site within the CNS. Various strategies have been proposed to circumvent the BBB.

3.1.1 Direct Injection

An invasive approach which bypasses the BBB by directly injecting the drug, by drilling a hole in the head and injecting drugs intracerebrally or intracerebroventricularly. Direct injection of a drug into the CNS has the advantage of infusing the therapeutic agent directly into the target site, but it is largely diffusion dependent, meaning that treatment area may remain small, often within a few millimeters of the injection site [88]. One of the invasive drug delivery for localized action in brain is convection enhanced drug delivery (CED). CED uses direct intracerebral injection of the drug and positive hydrostatic pressure during drug infusion. In CED, an external pressure gradient is established, typically through a syringe pump, and agents are infused continuously into the brain tissue via bulk fluid flow [89]. Experimental data have shown that the amount of drugs delivered to the target site matches the amount that was infused, implying that with CED gives relative control and uniform distribution of the drug throughout the target region, which increases delivery efficiency [90]. In a preclinical study carried out by Murad et al. CED techniques were used to deliver tumor-targeted cytotoxin interleukin 13 bound to Pseudomonas exotoxin (IL13-PE), which is a drug of choice in treating certain malignant brain tumors. However, injection of drug surpassing the BBB raises certain questions regarding the convenience and acceptability [91].

3.1.2 BBB Disruption

Modulation of vascular permeability involves disruption of the cells in the BBB, thus allowing macromolecular drugs to leak into the CNS. Osmotic BBB disruption has been a mainstay for increasing permeability of the BBB, usually using high concentrations of mannitol introduced into the carotid artery but involves problems related to target specificity. Other techniques of BBB modulation are by Optical modulation of vascular permeability [92] and Ultrasound-mediated blood–brain barrier disruption (USMD) [93].A major limitation of the former technique is that it currently is able to only penetrate approximately 1 mm into the tissue [94]. Drawback of USMD is that, although USMD has the ability to deliver high concentrations of unmodified drugs into the tissues within the brain, it may cause toxic substances such as bacteria antibodies to leak into the CNS while the BBB is disrupted [95].

3.2 Non-Invasive Drug Delivery

Non-invasive techniques of delivery may be of a chemical or biological nature. The non-invasive drug delivery aspect involves alteration in the property of the drug like prodrug or lipophilic derivation, involving chemical drug delivery or carrier-mediated drug delivery and/or specialized biological drug delivery like Nanoparticles [96].

3.2.1 Chemical Aspect of CNS Drug Delivery

Prodrug Mediated Drug Delivery

A method to circumvent the problem of penetration of macromolecules has been to chemically modify the molecule called as prodrug. Prodrugs are bio reversible derivatives of drug molecules and that undergo an enzymatic or chemical change in vivo to release the active parent drug. Three main strategies are involved while designing a prodrug: lipidisation, carrier mediated prodrug transport (CMT), and receptor-mediated prodrug transport (RMT) [96]. Lipidisation is the chemical modification of a drug to mask polar or charged substituents with non-polar groups to convert a water-soluble drug to a lipophilic prodrug. The process allows for the prodrug to passively diffuse through the BBB and then subsequently various enzymes such as esterases or oxidases may cleave off the protecting groups. DP-155 is a lipidised prodrug of indomethacin connected to lecithin via a 5-carbon length linker which under action of enzyme PLA2 (Phospholipase A2) in the brain releases indomethacin thus reducing the problem of GI irritability and produce CNS effect. These effects included reduced AD-related pathology in in vitro and in vivo models, as well as analgesia [97]. CMT and RMT both utilize endogenous transporters to carry the drug into the brain by disguising the drug as a substrate for the specific transporter. The difference between the two, is that CMT is associated with transporters of small molecule substrates such as amino acids or sugars whereas RMT is associated with larger substrates such as insulin and transferrin [98]. One of the example of using prodrugs via a carrier mediated transport is use of L-DOPA, a precursor of dopamine in Parkinson's disease which targets LAT. LAT (L amino acid transporters) can transport not only naturally occurring biologics but also amino acid related products like L-Dopa [99]. Dopamine cannot cross BBB therefore, Levodopa when administered acts as a substrate for LAT and cross the BBB. Eventually in brain, levodopa decarboxylates into dopamine [100].

3.2.2 Biological/Formulation Aspect of CNS Drug Delivery

Nanomedicine

Nanotechnology can deliver individual molecules that are normally poorly distributed to target tissue may be easily loaded into a nanocarrier, which can produce higher therapeutic concentrations than would previously have been achieved in specific targets until the therapeutic effect is achieved; then the material breaks down and is eliminated. The use of nanotechnology eliminates the BBB, owing to the small size of the nano-delivery system enough to travel through the normal transport mechanisms of BBB and also pass directly through the two cell membranes before being released in the brain. Nanoparticles made of bio resorbable elements can be used to deliver drugs to a specific position within the body or to particular types of tissue scattered throughout the body giving rise to various formulation and drug delivery system.

Fullerenes

Fullerenes (Fig. 10.3a) are carbon cage molecules considered to be three-dimensional analogues of benzene nanoparticles which are typically around 7 nm in diameter and formulated as nanocarrier for water soluble drugs unable to pass the BBB. The most abundant form of fullerenes is Buckminster fullerene (C_{60}) with 60 carbon atoms also known as buckyballs C60 [101].Apart from the buckyballs C60 use for encapsulation of biologic agent, buckyballs itself is a potential antioxidant [102] and has neuroprotective activity [103]. Numerous works describe the remarkable antioxidant and neuroprotective effects of water soluble fullerene derivatives including its use to attenuate the cortical infarction in rat brain [104]. Besides, malonic acid C60 derivate are capable of eliminating both superoxide anion and H_2O_2, and were effective inhibitors of lipid peroxidation (LPO) in the nervous tissue [105]. Toxicity depends strongly on whether any molecules are attached to the surface.

Polymer Nanoparticles

Many polymers like poly(ethyleneimines), polyalkynoacrylates, polymethyledine maloniate and polyesters are used to fabricate polymers nanoparticles (Fig. 10.3b) [106]. The nanocarriers can be in the range of 1–30 nm thus allowing passage of the drugs through blood brain barrier. These are stable and allow high loading of drugs, providing control of drug release and are relatively safe in use. These

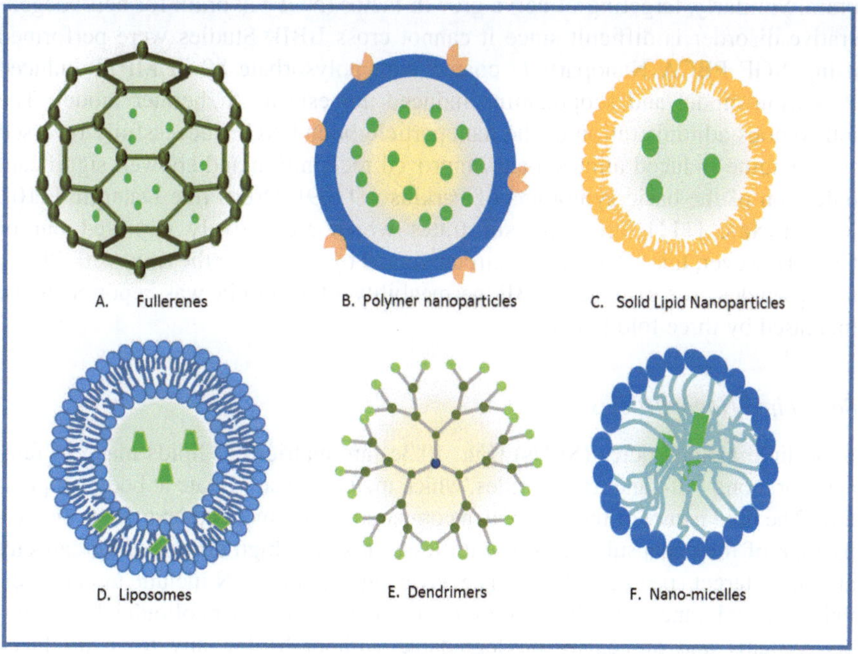

Fig. 10.3 Different aspect of drug delivery for CNS diseases and disorders: (**a**) Fullerenes: Spheroidal molecule of carbon having 60 or more atoms, enclosing the drug molecules. (**b**) Polymer nanoparticles: Hydrophobic drug enclosed in hydrophobic polymer (blue) and surface modified to increase site specificity. (**c**) Solid lipid nanoparticles; involved in drug delivery of lipophilic drug. (**d**) Liposomes; for drug delivery of both hydrophilic (in the core) and/or hydrophobic drug (between the hydrophobic part of the bilayer). (**e**) Dendrimers: Tree like branched structure with distinct surface functional groups attached to core to enhance solubility, release and absorption of the drug candidate. (**f**) Nano-micelles: Monolayer micelles for carrying non-polar drugs in the hydrophobic core of micelles

nanoparticles not only are medium of crossing BBB and protect the drug but also trigger timed release of the agent after the bindings of ligands on nanocapsules with the target tissue [106]. The process of polymeric nanoparticles to CNS depends on the ability for receptor mediated transcytosis and adsorptive mediated mediated transcytosis through BBB [107]. Polybutylcyanoacrylate (PBCA) nanoparticles were first polymer based nanoparticle system used to deliver drugs to CNS. In one study, PBCA polymers were created with polysorbate 80 and loaded with Loperamide to produce water soluble opioid agonist to deliver across the BBB. The analgesic effect after i.v injection was used to indicate drug transport through blood brain barrier. The experiment obtained results concluding: (1) Coating of polysorbate-80 to the PBCA polymer enhances the penetration of nanoparticles across the BBB as compared to formulation of simple polymer nanoparticles. (2) The polysorbate 80 coated with PBCA nanoparticles enable the transfer of water soluble Loperamide to the brain producing analgesia [108]. Local delivery of drugs loaded in nanoparticles can be done by convection enhanced delivery leading to distribution of therapeutics over large volumes in brain. Similarly, targeting of nerve growth factor (NGF) in brain for neurodegenerative disorder is difficult since it cannot cross BBB. Studies were performed using NGF PBCA Nanoparticle coated with polysorbate 80 in MPTP-induced Parkinson model and scopolamine induced amnesia as Alzheimer model. The intravenous administration of the nanoparticle-bound NGF successfully reversed scopolamine-induced amnesia and improved recognition and showed significant reduction of the basic symptoms of Parkinson [109]. Drugs like Dalargin [110], Methotrexate [111], are P-gp substrates which are actively exported out of CNS. However, this effect can be circumvented by delivering the drug with PBCA nanoparticles, for example, BBB permeability of Dalargin was reported to be increased by three fold [112].

Solid Lipid Nanoparticles

Solid lipid nanoparticles (SLNs) (Fig. 10.3c) are matrices of lipids made of fatty acids or mono-, di-, and triglycerides, which maintain a solid state at body temperature. The lipid nature of the SLN when compared to polymer Nanoparticles, enables passage of the encapsulated drug with lesser toxicity, high drug loading capacity and more target specificity [113]. The advantages of the SLN include target specificity, host tolerance, high bioavailability as compared to other colloidal drug delivery systems and are easily biodegradable without having any toxic products. Improved brain accumulation of several compounds like paclitaxel [114] and doxorubicin [115] for glioblastoma and brain malignancy was demonstrated when the drug was delivered using SLN, lipid nanocapsules or their PEGylated forms. SLN are less efficient for encapsulation of hydrophilic compounds, and may require heating for preparation, which can degrade heat labile drugs. The encapsulation efficiency of SLN for hydrophilic compounds can be improved by including an amphiphilic polymer to the lipids to form complexes with the charged, water-soluble drug molecules [116]. Surface modification by adsorbing proteins like apoE on the

SLN along with use of surfactant like p80 showed to enhance the target-site specificity [117].

Magnetic Nanoparticles

Magnetic nanoparticles, although may contain other elements, are often iron oxides. Iron oxide nanoparticles in applications are superparamagnetic which have single magnetic domain. When an external magnetic field is applied to the superparamagnetic nanoparticles, they tend to align along the magnetic field, leading to a net magnetization [118]. Magnetic nanoparticles have been studied for its therapeutic use in conditions like gliomas [119]. Poly-[aniline-co-N-(1-one-butyric acid) aniline] coated Fe_3O_4 particles with 1,3-bis(2-chloroethyl)-1-nitrosourea (BCNU) could be concentrated at targeted sites in vitro and in vivo using an externally applied magnet. When applied to brain tumors, magnetic targeting was found to increase the concentration and retention of bound-BCNU-3 [120]. Also studies showed that surface covalently conjugating lactoferrin (Lf) to a poly(ethylene glycol) (PEG)-coated modification like Fe_3O_4 showed to facilitate the transport of the nanoparticles across the blood-brain barrier (BBB) by receptor-mediated transcytosis via the Lf receptor present cerebral endothelial cells [121]. Though the use of magnetic Nanoparticles can provide enhanced drug delivery for neurological disease but there are chances of embolization due the formation of magnetic nanoparticle aggregates in the capillaries [122].

Liposomes

Liposomes (Fig. 10.3d) are the nano-carriers subjected to deliver different categories of drugs to improve their physio-chemical characteristics that offer wide range of therapeutic benefits to humans in combating life threatening diseases. Owing to biological inertness, versatility of conjugation with different ligand molecules, improvement in targeting efficacy, enhancing drug stability, decreasing drug toxicity made the liposomes as unique nano-carriers for brain targeting delivery [123]. Indeed, recent advances in research suggested the application of brain delivery by ligand conjugated liposomes is because of receptor, adsorptive or transporter mediated transcytosis [124]. For instance, GSH-PEG liposomal formulation of opioid peptide DAMGO doubled the concentration of drug in the brain. The plausible mechanisms for this study were suggested as receptor and transporter mediated endocytosis through BBB [125]. Often, α- mangostein, proved as a potent compound for treatment of Alzheimer's disease and its therapeutic utility was limited to its poor efficiency in crossing BBB. Further, transferrin, a glycoprotein conjugated with α- mangostein liposomes were improved in CNS delivery that is evident by correlating in vitro and in vivo experiments [126]. Nevertheless, mannose labeled liposomes was most efficiently incorporated in the brain compared to galactose and fructose conjugated liposomes [123]. This makes attractive strategies in delivering small molecules via liposomes. Many studies showed the drug entrapment in liposomes with different ligands increasing the CNS delivery. Particularly, one comparative study reported the entrapment of drugs in R17217 (OX26) anti mouse antibody of transferring receptor conjugated liposome have higher rates of delivering the drugs in CNS; crossing BBB, compared to Holo transferin, COG133, angiopep-2 and CRM 197 ligands conjugated liposomes [127]. Moreover, the pharmacokinetics and tissue distribution largely depends upon the nature of coupling ligand with liposomes.

Dendrimers

Unlike liposomes, dendrimers (Fig. 10.3e) compose with large repetitive branching chains of distinct structure containing inside core, repeated internal layers and external surface. The branching chains depends upon the number of repeated reactions (generations) also reflects the size of dendrimer [128]. Research states that less than 11.7–11.9 nm size dendrimers are able to cross the pores of BBB [129]. This limited size usage can be modulated to higher size range to cross the BBB by conjugating its surface with carrier ligand molecules like transferrin, angiopep-2, OX26 along with pegylation. The branching surface of a dendrimer offers a wide range of conjugation with drug, carrier ligand, and polymer ligation unlike other Nano delivery aspects. These features serve dendrimers as superior nano-carriers for CNS delivery of drugs. Till date, poly (amindoamine) or PAMAM, poly (propylenamine), triazine, poly-L-lysine dendrimers are highly explored in biological systems [130–132]. They have applications in gene delivery, nanotherapeutics, blood substitute studies, and diagnostics. Regarding CNS delivery aspects, drugs like methotrexate, doxorubicin, paclitaxel, docetaxel, siRNA, monoclonal antibodies like Cetuximab were entrapped in PMAM dendrimers using ligand conjugating agents like transferrin, T 7 peptide, angiopep-2, D-glucosamine, polysorbate 80 etc. [130, 133]. Moreover, these delivery aspects involved dual targeting mechanisms that deliver the bioactive compound targeting BBB for brain tumors treatment. Nevertheless, the major limitation of these dendrimers is attributable to their non-biodegradable nature which appears in lysosomes.

Nano Micelles

Micelles between 5 and 100 nm size ranges are considered as Nano micelles (Fig. 10.3f). These micelles are aggregates of amphiphilic molecules that accomplish to form a stable complex with exposing hydrophilic ends outside and hydrophobic area towards the core which makes a structured micelle in hydrophilic environment. Perhaps, the size of the micelle mostly depends upon the type of amphiphile, critical micellar concentration and Kraft point [134]. There are unstable conventional detergents to highly stable amphiphilic polymers which are currently available in the market for micellar formulations [135]. In fact, highly insoluble drugs like Tamoxifen, Paclitaxel, Campothecin, Porfyrin were made bioavailable in the micellar formulations [136]. As research advanced the current development of Nano micellar formulations with different compositions of polymers to make it with required hydrophobicity and to required size are achieved for CNS drug delivery.

For instance, transferrin-conjugated docetaxel (DTX)-loaded-alpha-tocopherol polyethylene glycol 1000 succinate (vitamin E TPGS or TPGS) micelles for targeted brain cancer therapy showed size range of 20 nm, 85% of drug loading efficiency and efficient drug transport through the BBB, with 50% of drug release in 24 h of time that evidenced through conducting in vitro and in vivo experiment [137]. Further, loading of paclitaxel in transferrin-conjugated polyphospho ester micelles and in Transferrin-Modified c [RGDfK] hybrid Micelle increased brain concentrations of paclitaxel [138]. Apart from this, polymer mediated micelles offered degree of freedom in conjugating with BBB carrier ligands like angiopep-2, COG133, CRM 197 and OX26 for delivery of insoluble bioactive compounds across CNS barriers to treat various CNS diseases and disorders.

4 Current Advancement in Nanotherapeutics: Perspective on CNS Diseases

Natural discoveries are still more beautified by inventing the new technologies; one such is nanotechnology in drug development process. The number of patents and new concepts developed often forecasts nanotherapeutics as a gold standard treatment for many diseases and disorders across the globe. Of note, research contributed by Ting Xu is an example for forecasting the above statement. Recently, she was awarded patent for developing 3HM technology for treatment of glioblastoma multiform (octopus cancer). This 3HM technology represents coiled-coil 3-helix micelles which are at only 20 nm in size and had good penetration through BBB and selectively accumulates in octopus tumors. They found almost 2 times release of bioactive compounds through 3 HM technology in the tumors as compared to FDA approved formulations [139]. Justin Hanes Patented for nonadhesive particles coated larger than 110 nm was coated with PEG and further making it neutral showed high penetration into the brain (patent no: US20130183244A1). Nevertheless, significant progresses have been made for targeted drug delivery of nano-therapeutics in clinical practice for both prevention and eradication of CNS diseases. Copaxone® (NDA 020622) marketed by Teva Pharmaceuticals was approved by USFDA in 1996 after the phase III clinical trials against relapsing form of Multiple Sclerosis. The active ingredient being Glatiramer acetate is a heterogenous mixture of immunogenic polypeptide containing four amino acids, namely **L**-glutamic acid, **L**-lysine, **L**-alanine and **L**-tyrosine in colloidal suspension. Similarly, nanotechnology has been used for targeting chemotherapeutic drugs in brain for many brain related cancers. One of the example is Opaxio® (Paclitaxel polyglumex) approved by USFDA in 2012 against glioblastoma. The formulation comprises of Paclitaxel covalently linked to solid nanoparticles composed of polyglutamate which releases the drug in the brain after the enzymatic hydrolysis of the polyglutamate nanoparticles. Recently, USFDA have approved NanoTherm™ therapy by MagForce to treat glioblastoma (Patent No. EP 1871423 B1). The NanoTherm™ therapy involves the drug delivery with the help of magnetic nanoparticles in which Aminosilane-coated supermagnetic iron oxide nanoparticles are targeted to solid tumors of glioblastomas followed by thermal ablations of tumours performed by heating the nanoparticles in alternative magnetic field. Another use of nanotechnology is involved in the marketed form Dipirivan® (NDA 019627) which is oil in water nanoemulsion of Propofol designed to increase the drug solubility and bioavailability to produce enhanced analgesic activity. Although nanotechnology has appeared in late 1970s, the term 'nanotherapeutics in CNS' has recently been the field of interest therefore not many CNS drug delivery systems pass the clinical trials as compared to cancer nanotherapy indicating that more work is needed to be done to overcome the challenges in CNS drug discovery. Perhaps, this persistent research in the field of Nanotechnology brings new advances in the development of high efficacious drugs in eradicating CNS diseases of future generations.

5 Future Directions

Increased incidence of neurodegenerative conditions and cognitive decline in elderly patients has attracted large number of scientists to study the novel therapeutics to alleviate such conditions [140]. Several nanoparticles based drug delivery systems have been designed and tested for their efficacy in preclinical models of disease. In most of times the preclinical efficacy of such delivery systems is far more superior than their respective naive formulations/drugs [141]. Though elevated treatment outcome is observed in the case of nanotherapeutics, the therapeutic concentrations achieved at the targeted site are far less from ideal concentration [142]. Further, increased surface of nanoparticles and non-specific distribution of ligand receptors in other tissues increases the chances of toxicity potentially limiting their application [143]. Hence, future research should account for altered BBB characteristics and more brain selective ligand based targeting options to ensure optimal delivery with minimal side effects.

References

1. Stefan LR. Psychological training and counseling for the increase of personnel well-being and flight safety. Rev Air Force Acad. 2015;3(30):149. doi:10.19062/1842-9238.2015.13.3.26.
2. Organization WH. The world health report 2001: mental health: new understanding, new hope. Geneva: World Health Organization; 2001.
3. Saxena S, Funk M, Chisholm D. Comprehensive mental health action plan 2013–2020. East Mediterr Health J. 2015;21:461.
4. Lu Y. Rural-urban migration and health: evidence from longitudinal data in Indonesia. Soc Sci Med. 2010;70(3):412–9.
5. Pangalos MN, Schechter LE, Hurko O. Drug development for CNS disorders: strategies for balancing risk and reducing attrition. Nat Rev Drug Discov. 2007;6(7):521–32.

6. DiMasi JA, Grabowski HG. The cost of biopharmaceutical R&D: is biotech different? Manag Decis Econ. 2007;28(4-5):469–79.
7. De Smaele E, Ferretti E, Gulino A. MicroRNAs as biomarkers for CNS cancer and other disorders. Brain Res. 2010;1338:100–11.
8. Bicker J, Alves G, Fortuna A, Falcao A. Blood-brain barrier models and their relevance for a successful development of CNS drug delivery systems: a review. Eur J Pharm Biopharm. 2014;87(3):409–32.
9. Hawkins BT, Davis TP. The blood–brain barrier/neurovascular unit in health and disease. Pharmacol Rev. 2005;57(2):173–85.
10. Upadhyay RK. Drug delivery systems, CNS protection, and the blood brain barrier. Biomed Res Int. 2014;2014:1–37.
11. Nagpal K, Singh SK, Mishra DN. Drug targeting to brain: a systematic approach to study the factors, parameters and approaches for prediction of permeability of drugs across BBB. Expert Opin Drug Deliv. 2013;10(7):927–55.
12. Gabathuler R. Approaches to transport therapeutic drugs across the blood–brain barrier to treat brain diseases. Neurobiol Dis. 2010;37(1):48–57.
13. Petty MA, Lo EH. Junctional complexes of the blood–brain barrier: permeability changes in neuroinflammation. Prog Neurobiol. 2002;68(5):311–23.
14. Banks WA. From blood-brain barrier to blood-brain interface: new opportunities for CNS drug deliveryNat Rev Drug Discov, vol. 15; 2016. p. 275–92.
15. Redzic Z. Molecular biology of the blood-brain and the blood-cerebrospinal fluid barriers: similarities and differences. Fluids Barriers CNS. 2011;8(1):3.
16. Sanchez-Covarrubias L, Slosky LM, Thompson BJ, Davis TP, Ronaldson PT. Transporters at CNS barrier sites: obstacles or opportunities for drug delivery? Curr Pharm Des. 2014;20(10):1422–49.
17. Abbott NJ, Patabendige AAK, Dolman DEM, Yusof SR, Begley DJ. Structure and function of the blood-brain barrier. Neurobiol Dis. 2010;37(1):13–25.
18. Genes-Hernandez LI. Development of a microfluidic based microvascular model: towards a complete blood brain barrier (BBB) mimic. 2008. ProQuest
19. Sheikov N, McDannold N, Vykhodtseva N, Jolesz F, Hynynen K. Cellular mechanisms of the blood-brain barrier opening induced by ultrasound in presence of microbubbles. Ultrasound Med Biol. 2004;30(7):979–89.
20. Pardridge WM. Drug delivery to the brain. J Cereb Blood Flow Metab. 1997;17(7):713–31.
21. Pardridge WM. Blood-brain barrier drug targeting: the future of brain drug development. Mol Interv. 2003;3(2):90.
22. Benita S. Microencapsulation: methods and industrial applications. Boca Raton: CRC; 2005.
23. Kanwar JR, Sriramoju B, Kanwar RK. Neurological disorders and therapeutics targeted to surmount the blood–brain barrier. Int J Nanomedicine. 2012;7:3259.
24. Alzheimer's Association. 2016 Alzheimer's disease facts and figures. Alzheimers Dement. 2016;12(4):459–509.
25. Golde TE. Overcoming translational barriers impeding development of Alzheimer's disease modifying therapies. J Neurochem. 2016;139:224–36.
26. Underwood E. Alzheimer's amyloid theory gets modest boost. Science. 2015;349(6247):464.
27. Holmes C, Boche D, Wilkinson D, Yadegarfar G, Hopkins V, Bayer A, Jones R, Bullock R, Love S, Neal J. Long term effects of Ab42 immunization in Alzheimer's disease: immune response, plaque removal and clinical function. Lancet. 2008;372:216–23.
28. Liu E, Schmidt ME, Margolin R, Sperling R, Koeppe R, Mason NS, Klunk WE, Mathis CA, Salloway S, Fox NC. Amyloid-β 11C-PiB-PET imaging results from 2 randomized bapineuzumab phase 3 AD trials. Neurology. 2015;85(8):692–700.
29. Siemers ER, Friedrich S, Dean RA, Gonzales CR, Farlow MR, Paul SM, DeMattos RB. Safety and changes in plasma and cerebrospinal fluid amyloid β after a single administration of an amyloid β monoclonal antibody in subjects with Alzheimer disease. Clin Neuropharmacol. 2010;33(2):67–73.

30. Blennow K, Hampel H, Weiner M, Zetterberg H. Cerebrospinal fluid and plasma biomarkers in Alzheimer disease. Nat Rev Neurol. 2010;6(3):131–44.
31. Goetzl EJ, Boxer A, Schwartz JB, Abner EL, Petersen RC, Miller BL, Kapogiannis D. Altered lysosomal proteins in neural-derived plasma exosomes in preclinical Alzheimer disease. Neurology. 2015;85(1):40–7.
32. Carroll RT, Bhatia D, Geldenhuys W, Bhatia R, Miladore N, Bishayee A, Sutariya V. Brain-targeted delivery of Tempol-loaded nanoparticles for neurological disorders. J Drug Target. 2010;18(9):665–74.
33. Sahni JK, Doggui S, Ali J, Baboota S, Dao L, Ramassamy C. Neurotherapeutic applications of nanoparticles in Alzheimer's disease. J Control Release. 2011;152(2):208–31.
34. Wang ZH, Wang ZY, Sun CS, Wang CY, Jiang TY, Wang SL. Trimethylated chitosan-conjugated PLGA nanoparticles for the delivery of drugs to the brain. Biomaterials. 2010;31(5):908–15.
35. World Health Organization. Global burden of neurological disorders: estimates and projections. In: Neurological disorders: public health challenges. Geneva: WHO Press; 2007. p. 27–39.
36. Gordon R, Kelley MD. Parkinson's disease incidence and prevalence; 2000. Parkinsons Disease Foundation
37. van der Brug MP, Singleton A, Gasser T, Lewis PA. Parkinson's disease: from human genetics to clinical trials. Sci Transl Med. 2015;7(305):205–20.
38. Sanchez-Mut JV, Heyn H, Vidal E, Moran S, Sayols S, Delgado-Morales R, Schultz MD, Ansoleaga B, Garcia-Esparcia P, Pons-Espinal M. Human DNA methylomes of neurodegenerative diseases show common epigenomic patterns. Transl Psychiatry. 2016;6(1):e718.
39. Atlas D. DopAmide: novel, water-soluble, slow-release l-dihydroxyphenylalanine (l-DOPA) precursor moderates l-DOPA conversion to dopamine and generates a sustained level of dopamine at dopaminergic neurons. CNS Neurosci Ther. 2016;22:461–7.
40. Shen T, Pu J, Si X, Ye R, Zhang B. An update on potential therapeutic strategies of Parkinson's disease based on pathogenic mechanisms. Expert review of neurotherapeutics (just-accepted). Expert Rev Neurother. 2016;16(6):711–22.
41. Pillay V, Choonara YE, Sibeko B, Harilall S-L, Pillay S, Modi G, Iyuke SE, Naidoo D. Polymeric pharmaceutical dosage form in sustained release; 2009. Google Patents
42. Gendelman HE, Anantharam V, Bronich T, Ghaisas S, Jin H, Kanthasamy AG, Liu X, McMillan J, Mosley RL, Narasimhan B. Nanoneuromedicines for degenerative, inflammatory, and infectious nervous system diseases. Nanomedicine. 2015;11(3):751–67.
43. Mathers C, Fat DM, Boerma JT. The global burden of disease: 2004 update. Geneva: World Health Organization; 2008.
44. Carpenter WT, Koenig JI. The evolution of drug development in schizophrenia: past issues and future opportunities. Neuropsychopharmacology. 2008;33(9):2061–79.
45. Bernard A, Fischer M, Robert W, Buchanan M. Schizophrenia: epidemiology and pathogenesis, vol. 2016. Alphen: Wolters Kluwer; 2016.
46. Modini M, Tan L, Brinchmann B, Wang M-J, Killackey E, Glozier N, Mykletun A, Harvey SB. Supported employment for people with severe mental illness: systematic review and meta-analysis of the international evidence. Br J Psychiatry. 2016;115:165092.
47. Commission S. The abandoned illness: a report from the Schizophrenia Commission. London: Rethink Mental Illness; 2012.
48. Lieberman JA, Perkins D, Belger A, Chakos M, Jarskog F, Boteva K, Gilmore J. The early stages of schizophrenia: speculations on pathogenesis, pathophysiology, and therapeutic approaches. Biol Psychiatry. 2001;50(11):884–97.
49. Addington J, Heinssen R. Prediction and prevention of psychosis in youth at clinical high risk. Annu Rev Clin Psychol. 2012;8:269–89.
50. Lewandowski K, Cohen B, Öngur D. Evolution of neuropsychological dysfunction during the course of schizophrenia and bipolar disorder. Psychol Med. 2011;41(02):225–41.
51. Moghaddam B, Javitt D. From revolution to evolution: the glutamate hypothesis of schizophrenia and its implication for treatment. Neuropsychopharmacology. 2012;37(1):4–15.

52. Meyer-Lindenberg A, Tost H. Neuroimaging and plasticity in schizophrenia. Restor Neurol Neurosci. 2014;32(1):119–27.
53. Haijma SV, Van Haren N, Cahn W, Koolschijn PCM, Pol HEH, Kahn RS. Brain volumes in schizophrenia: a meta-analysis in over 18 000 subjects. Schizophr Bull. 2013;39(5):1129–38.
54. Barch DM, Ceaser A. Cognition in schizophrenia: core psychological and neural mechanisms. Trends Cogn Sci. 2012;16(1):27–34.
55. Ting JT, Peça J, Feng G. Functional consequences of mutations in postsynaptic scaffolding proteins and relevance to psychiatric disorders. Annu Rev Neurosci. 2012;35:49–71.
56. Ripke S, Neale BM, Corvin A, Walters JT, Farh K-H, Holmans PA, Lee P, Bulik-Sullivan B, Collier DA, Huang H. Biological insights from 108 schizophrenia-associated genetic loci. Nature. 2014;511(7510):421.
57. Fromer M, Pocklington AJ, Kavanagh DH, Williams HJ, Dwyer S, Gormley P, Georgieva L, Rees E, Palta P, Ruderfer DM. De novo mutations in schizophrenia implicate synaptic networks. Nature. 2014;506(7487):179–84.
58. Cardiff U, Owen MJ, Sawa A, Mortensen PB. Lancet. 2016;388:86–97.
59. Bhaskar S, Tian F, Stoeger T, Kreyling W, de la Fuente JM, Grazú V, Borm P, Estrada G, Ntziachristos V, Razansky D. Multifunctional nanocarriers for diagnostics, drug delivery and targeted treatment across blood-brain barrier: perspectives on tracking and neuroimaging. Part Fibre Toxicol. 2010;7(1):1.
60. Miyamoto S, Miyake N, Jarskog L, Fleischhacker W, Lieberman J. Pharmacological treatment of schizophrenia: a critical review of the pharmacology and clinical effects of current and future therapeutic agents. Mol Psychiatry. 2012;17(12):1206–27.
61. Green MF. What are the functional consequences of neurocognitive deficits in schizophrenia? Am J Psychiatry. 1996;153(3):321.
62. Rosenheck RA, Krystal JH, Lew R, Barnett PG, Fiore L, Valley D, Thwin SS, Vertrees JE, Liang MH. Long-acting risperidone and oral antipsychotics in unstable schizophrenia. N Engl J Med. 2011;364(9):842–51.
63. Pereira JNDS, Tadjerpisheh S, Abed MA, Saadatmand AR, Weksler B, Romero IA, Couraud P-O, Brockmöller J, Tzvetkov MV. The poorly membrane permeable antipsychotic drugs amisulpride and sulpiride are substrates of the organic cation transporters from the SLC22 family. AAPS J. 2014;16(6):1247–58.
64. Kessler RC, Bromet EJ. The epidemiology of depression across cultures. Annu Rev Public Health. 2013;34:119.
65. Schechter LE, Ring RH, Beyer CE, Hughes ZA, Khawaja X, Malberg JE, Rosenzweig-Lipson S. Innovative approaches for the development of antidepressant drugs: current and future strategies. NeuroRx. 2005;2(4):590–611.
66. Dale E, Bang-Andersen B, Sánchez C. Emerging mechanisms and treatments for depression beyond SSRIs and SNRIs. Biochem Pharmacol. 2015;95(2):81–97.
67. Haque S, Md S, Sahni JK, Ali J, Baboota S. Development and evaluation of brain targeted intranasal alginate nanoparticles for treatment of depression. J Psychiatr Res. 2014;48(1):1–12.
68. Xie Y, Wang Y, Zhang T, Ren G, Yang Z. Effects of nanoparticle zinc oxide on spatial cognition and synaptic plasticity in mice with depressive-like behaviors. J Biomed Sci. 2012;19(1):1.
69. American Heart Association. Heart disease and stroke statistics—at-a-glance. 2015
70. Mukherjee D, Patil CG. Epidemiology and the global burden of stroke. World Neurosurg. 2011;76(6):S85–90.
71. Jordán J, Ikuta I, Garcia-Garcia J, Calleja S, Segura T. Stroke pathophysiology: management challenges and new treatment advances. J Physiol Biochem. 2007;63(3):261–77.
72. Yun X, Maximov VD, Yu J, Vertegel AA, Kindy MS. Nanoparticles for targeted delivery of antioxidant enzymes to the brain after cerebral ischemia and reperfusion injury. J Cereb Blood Flow Metab. 2013;33(4):583–92.
73. Ostrom QT, Gittleman H, Fulop J, Liu M, Blanda R, Kromer C, Wolinsky Y, Kruchko C, Barnholtz-Sloan JS. CBTRUS statistical report: primary brain and central nervous system tumors diagnosed in the United States in 2008–2012. Neuro-Oncology. 2015;17(suppl 4):iv1–iv62.

74. Louis DN, Ohgaki H, Wiestler OD, Cavenee WK, Burger PC, Jouvet A, Scheithauer BW, Kleihues P. The 2007 WHO classification of tumours of the central nervous system. Acta Neuropathol. 2007;114(2):97–109.
75. Ostrom QT, Bauchet L, Davis FG, Deltour I, Fisher JL, Langer CE, Pekmezci M, Schwartzbaum JA, Turner MC, Walsh KM. The epidemiology of glioma in adults: a "state of the science" review. Neuro Oncol. 2014;16:896–913.
76. Ohgaki H, Kleihues P. Genetic pathways to primary and secondary glioblastoma. Am J Pathol. 2007;170(5):1445–53.
77. Mao H, LeBrun DG, Yang J, Zhu VF, Li M. Deregulated signaling pathways in glioblastoma multiforme: molecular mechanisms and therapeutic targets. Cancer Investig. 2012;30(1):48–56.
78. Schwartzbaum JA, Fisher JL, Aldape KD, Wrensch M. Epidemiology and molecular pathology of glioma. Nat Clin Pract Neurol. 2006;2(9):494–503.
79. Taylor TE, Furnari FB, Cavenee WK. Targeting EGFR for treatment of glioblastoma: molecular basis to overcome resistance. Curr Cancer Drug Targets. 2012;12(3):197–209.
80. Huse JT, Holland EC. Targeting brain cancer: advances in the molecular pathology of malignant glioma and medulloblastoma. Nat Rev Cancer. 2010;10(5):319–31.
81. Ferrari M. Cancer nanotechnology: opportunities and challenges. Nat Rev Cancer. 2005;5(3):161–71.
82. Chen H, Qin Y, Zhang Q, Jiang W, Tang L, Liu J, He Q. Lactoferrin modified doxorubicin-loaded procationic liposomes for the treatment of gliomas. Eur J Pharm Sci. 2011;44(1):164–73.
83. Fillebeen C, Descamps L, Dehouck M-P, Fenart L, Benaïssa M, Spik G, Cecchelli R, Pierce A. Receptor-mediated transcytosis of lactoferrin through the blood-brain barrier. J Biol Chem. 1999;274(11):7011–7.
84. Fu Y, An N, Li K, Zheng Y, Liang A. Chlorotoxin-conjugated nanoparticles as potential glioma-targeted drugs. J Neuro-Oncol. 2012;107(3):457–62.
85. Choe G, Park JK, Jouben-Steele L, Kremen TJ, Liau LM, Vinters HV, Cloughesy TF, Mischel PS. Active matrix metalloproteinase 9 expression is associated with primary glioblastoma subtype. Clin Cancer Res. 2002;8(9):2894–901.
86. Gao H, Yang Z, Zhang S, Cao S, Shen S, Pang Z, Jiang X. Ligand modified nanoparticles increases cell uptake, alters endocytosis and elevates glioma distribution and internalization. Sci Rep. 2013;3:2534.
87. Shevtsov M, Parr M, Ryzhov V, Zemtsova E, Arbenin AY, Ponomareva A, Smirnov V, Multhoff G. Zero-valent Fe confined mesoporous silica nanocarriers (Fe (0)@ MCM-41) for targeting experimental orthotopic glioma in rats. Sci Rep. 2016;6:29247.
88. Pathan SA, Iqbal Z, Zaidi S, Talegaonkar S, Vohra D, Jain GK, Azeem A, Jain N, Lalani JR, Khar RK. CNS drug delivery systems: novel approaches. Recent Pat Drug Deliv Formul. 2009;3(1):71–89.
89. Raghavan R, Brady ML, Rodríguez-Ponce MI, Hartlep A, Pedain C, Sampson JH. Convection-enhanced delivery of therapeutics for brain disease, and its optimization. Neurosurg Focus. 2006;20(4):E12.
90. Allard E, Passirani C, Benoit J-P. Convection-enhanced delivery of nanocarriers for the treatment of brain tumors. Biomaterials. 2009;30(12):2302–18.
91. Ay I, Francis JW, Brown RH. VEGF increases blood–brain barrier permeability to Evans blue dye and tetanus toxin fragment C but not adeno-associated virus in ALS mice. Brain Res. 2008;1234:198–205.
92. Choi M, Ku T, Chong K, Yoon J, Choi C. Minimally invasive molecular delivery into the brain using optical modulation of vascular permeability. Proc Natl Acad Sci. 2011;108(22):9256–61.
93. Aryal M, Arvanitis CD, Alexander PM, McDannold N. Ultrasound-mediated blood–brain barrier disruption for targeted drug delivery in the central nervous system. Adv Drug Deliv Rev. 2014;72:94–109.
94. Andresen V, Alexander S, Heupel W-M, Hirschberg M, Hoffman RM, Friedl P. Infrared multiphoton microscopy: subcellular-resolved deep tissue imaging. Curr Opin Biotechnol. 2009;20(1):54–62.

95. Treat LH, McDannold N, Vykhodtseva N, Zhang Y, Tam K, Hynynen K. Targeted delivery of doxorubicin to the rat brain at therapeutic levels using MRI-guided focused ultrasound. Int J Cancer. 2007;121(4):901–7.
96. Stockwell J, Abdi N, Lu X, Maheshwari O, Taghibiglou C. Novel central nervous system drug delivery systems. Chem Biol Drug Des. 2014;83(5):507–20.
97. Dvir E, Elman A, Simmons D, Shapiro I, Duvdevani R, Dahan A, Hoffman A, Friedman JE. DP-155, a lecithin derivative of indomethacin, is a novel nonsteroidal antiinflammatory drug for analgesia and Alzheimer's disease therapy. CNS Drug Rev. 2007;13(2):260–77.
98. Vytla D, Combs-Bachmann RE, Hussey AM, McCarron ST, McCarthy DS, Chambers JJ. Prodrug approaches to reduce hyperexcitation in the CNS. Adv Drug Deliv Rev. 2012;64(7):666–85.
99. Uchino H, Kanai Y, Kim DK, Wempe MF, Chairoungdua A, Morimoto E, Anders M, Endou H. Transport of amino acid-related compounds mediated by L-type amino acid transporter 1 (LAT1): insights into the mechanisms of substrate recognition. Mol Pharmacol. 2002;61(4):729–37.
100. Salameh F, Karaman D, Mecca G, Scrano L, Bufo SA, Karaman R. Prodrugs targeting the central nervous system (CNS). World J Pharm Pharm Sci. 2015;4(8):208–37.
101. Thakral S, Mehta R. Fullerenes: an introduction and overview of their biological properties. Indian J Pharm Sci. 2006;68(1):13–6.
102. Tykhomyrov AA, Nedzvetsky VS, Klochkov VK, Andrievsky GV. Nanostructures of hydrated C 60 fullerene (C 60 HyFn) protect rat brain against alcohol impact and attenuate behavioral impairments of alcoholized animals. Toxicology. 2008;246(2):158–65.
103. Dugan LL, Turetsky DM, Du C, Lobner D, Wheeler M, Almli CR, Shen CK-F, Luh T-Y, Choi DW, Lin T-S. Carboxyfullerenes as neuroprotective agents. Proc Natl Acad Sci. 1997;94(17):9434–9.
104. Lin AM-Y, Fang S-F, Lin S-Z, Chou C-K, Luh T-Y, Ho L-T. Local carboxyfullerene protects cortical infarction in rat brain. Neurosci Res. 2002;43(4):317–21.
105. Yin J-J, Lao F, Fu PP, Wamer WG, Zhao Y, Wang PC, Qiu Y, Sun B, Xing G, Dong J. The scavenging of reactive oxygen species and the potential for cell protection by functionalized fullerene materials. Biomaterials. 2009;30(4):611–21.
106. Wong HL, Wu XY, Bendayan R. Nanotechnological advances for the delivery of CNS therapeutics. Adv Drug Deliv Rev. 2012;64(7):686–700.
107. Patel T, Zhou J, Piepmeier JM, Saltzman WM. Polymeric nanoparticles for drug delivery to the central nervous system. Adv Drug Deliv Rev. 2012;64(7):701–5.
108. Alyautdin RN, Petrov VE, Langer K, Berthold A, Kharkevich DA, Kreuter J. Delivery of loperamide across the blood-brain barrier with polysorbate 80-coated polybutylcyanoacrylate nanoparticles. Pharm Res. 1997;14(3):325–8.
109. Kurakhmaeva KB, Djindjikhashvili IA, Petrov VE, Balabanyan VU, Voronina TA, Trofimov SS, Kreuter J, Gelperina S, Begley D, Alyautdin RN. Brain targeting of nerve growth factor using poly (butyl cyanoacrylate) nanoparticles. J Drug Target. 2009;17(8):564–74.
110. Aliautdin R, Petrov V, Ivanov A, Kreuter J, Kharkevich D. Transport of the hexapeptide dalargin across the hemato-encephalic barrier into the brain using polymer nanoparticles. Eksp Klin Farmakol. 1995;59(3):57–60.
111. Seo D-H, Jeong Y-I, Kim D-G, Jang M-J, Jang M-K, Nah J-W. Methotrexate-incorporated polymeric nanoparticles of methoxy poly (ethylene glycol)-grafted chitosan. Colloids Surf B Biointerfaces. 2009;69(2):157–63.
112. Dwibhashyam V, Nagappa A. Strategies for enhanced drug delivery to the central nervous system. Indian J Pharm Sci. 2008;70(2):145.
113. Blasi P, Giovagnoli S, Schoubben A, Ricci M, Rossi C. Solid lipid nanoparticles for targeted brain drug delivery. Adv Drug Deliv Rev. 2007;59(6):454–77.
114. Chirio D, Gallarate M, Peira E, Battaglia L, Muntoni E, Riganti C, Biasibetti E, Capucchio MT, Valazza A, Panciani P. Positive-charged solid lipid nanoparticles as paclitaxel drug delivery system in glioblastoma treatment. Eur J Pharm Biopharm. 2014;88(3):746–58.

115. Battaglia L, Gallarate M, Peira E, Chirio D, Muntoni E, Biasibetti E, Capucchio MT, Valazza A, Panciani PP, Lanotte M. Solid lipid nanoparticles for potential doxorubicin delivery in glioblastoma treatment: preliminary in vitro studies. J Pharm Sci. 2014;103(7):2157–65.
116. Salunkhe SS, Bhatia NM, Kawade VS, Bhatia MS. Development of lipid based nanoparticulate drug delivery systems and drug carrier complexes for delivery to brain. J Appl Pharm Sci. 2015;5:110–29.
117. Göppert TM, Müller RH. Polysorbate-stabilized solid lipid nanoparticles as colloidal carriers for intravenous targeting of drugs to the brain: comparison of plasma protein adsorption patterns. J Drug Target. 2005;13(3):179–87.
118. Busquets MA, Espargaró A, Sabaté R, Estelrich J. Magnetic nanoparticles cross the blood-brain barrier: when physics rises to a challenge. Nanomaterials. 2015;5(4):2231–48.
119. Chen P-Y, Liu H-L, Hua M-Y, Yang H-W, Huang C-Y, Chu P-C, Lyu L-A, Tseng I-C, Feng L-Y, Tsai H-C. Novel magnetic/ultrasound focusing system enhances nanoparticle drug delivery for glioma treatment. Neuro Oncol. 2010;12:1050–60.
120. Hua M-Y, Liu H-L, Yang H-W, Chen P-Y, Tsai R-Y, Huang C-Y, Tseng I-C, Lyu L-A, Ma C-C, Tang H-J. The effectiveness of a magnetic nanoparticle-based delivery system for BCNU in the treatment of gliomas. Biomaterials. 2011;32(2):516–27.
121. Qiao R, Jia Q, Huwel S, Xia R, Liu T, Gao F, Galla H-J, Gao M. Receptor-mediated delivery of magnetic nanoparticles across the blood–brain barrier. ACS Nano. 2012;6(4):3304–10.
122. Shubayev VI, Pisanic TR, Jin S. Magnetic nanoparticles for theragnostics. Adv Drug Deliv Rev. 2009;61(6):467–77.
123. Umezawa F, Eto Y. Liposome targeting to mouse brain: mannose as a recognition marker. Biochem Biophys Res Commun. 1988;153(3):1038–44.
124. Béduneau A, Saulnier P, Benoit J-P. Active targeting of brain tumors using nanocarriers. Biomaterials. 2007;28(33):4947–67.
125. Lindqvist A, Rip J, van Kregten J, Gaillard PJ, Hammarlund-Udenaes M. In vivo functional evaluation of increased brain delivery of the opioid peptide DAMGO by Glutathione-PEGylated liposomes. Pharm Res. 2016;33(1):177–85.
126. Chen Z-L, Huang M, Wang X-R, Fu J, Han M, Shen Y-Q, Xia Z, Gao J-Q. Transferrin-modified liposome promotes α-mangostin to penetrate the blood–brain barrier. Nanomedicine. 2016;12(2):421–30.
127. van Rooy I, Mastrobattista E, Storm G, Hennink WE, Schiffelers RM. Comparison of five different targeting ligands to enhance accumulation of liposomes into the brain. J Control Release. 2011;150(1):30–6.
128. Fréchet JM. Dendrimers and other dendritic macromolecules: from building blocks to functional assemblies in nanoscience and nanotechnology. J Polym Sci A Polym Chem. 2003;41(23):3713–25.
129. Sarin H, Kanevsky AS, Wu H, Brimacombe KR, Fung SH, Sousa AA, Auh S, Wilson CM, Sharma K, Aronova MA. Effective transvascular delivery of nanoparticles across the blood-brain tumor barrier into malignant glioma cells. J Transl Med. 2008;6(1):80.
130. Yang H. Nanoparticle-mediated brain-specific drug delivery, imaging, and diagnosis. Pharm Res. 2010;27(9):1759–71.
131. Boyd BJ, Kaminskas LM, Karellas P, Krippner G, Lessene R, Porter CJ. Cationic poly-L-lysine dendrimers: pharmacokinetics, biodistribution, and evidence for metabolism and bioresorption after intravenous administration to rats. Mol Pharm. 2006;3(5):614–27.
132. Jannat FT-E. Synthesis of dendrimer molecules based on triazine. Dhaka: Bangladesh University of Engineering and Technology; 2006.
133. Nance E, Zhang F, Mishra MK, Zhang Z, Kambhampati SP, Kannan RM, Kannan S. Nanoscale effects in dendrimer-mediated targeting of neuroinflammation. Biomaterials. 2016;101:96–107.
134. Cabral H, Matsumoto Y, Mizuno K, Chen Q, Murakami M, Kimura M, Terada Y, Kano M, Miyazono K, Uesaka M. Accumulation of sub-100 nm polymeric micelles in poorly permeable tumours depends on size. Nat Nanotechnol. 2011;6(12):815–23.

135. Micheli M-R, Bova R, Magini A, Polidoro M, Emiliani C. Lipid-based nanocarriers for CNS-targeted drug delivery. Recent Pat CNS Drug Discov. 2012;7(1):71–86.
136. Das MK, Sarma A, Chakraborty T. Nano-ART and NeuroAIDS. Drug Deliv Transl Res. 2016;6:452–72.
137. Sonali AP, Singh RP, Rajesh CV, Singh S, Vijayakumar MR, Pandey BL, Muthu MS. Transferrin receptor-targeted vitamin E TPGS micelles for brain cancer therapy: preparation, characterization and brain distribution in rats. Drug Deliv. 2016;23(5):1788–98.
138. Zhang P, Hu L, Yin Q, Feng L, Li Y. Transferrin-modified c [RGDfK]-paclitaxel loaded hybrid micelle for sequential blood-brain barrier penetration and glioma targeting therapy. Mol Pharm. 2012;9(6):1590–8.
139. Yarris L. Nanocarriers may carry new hope for brain cancer therapy. Science Daily, November. 2015;19:2015.
140. Misra A, Ganesh S, Shahiwala A, Shah SP. Drug delivery to the central nervous system: a review. J Pharm Pharm Sci. 2003;6(2):252–73.
141. Masserini M. Nanoparticles for brain drug delivery. ISRN Biochem. 2013;2013:18.
142. Gao H, Pang Z, Jiang X. Targeted delivery of nano-therapeutics for major disorders of the central nervous system. Pharm Res. 2013;30(10):2485–98.
143. Fischer HC, Chan WC. Nanotoxicity: the growing need for in vivo study. Curr Opin Biotechnol. 2007;18(6):565–71.
144. Fazil M, Md S, Haque S, Kumar M, Baboota S, Kaur Sahni J, Ali J. Development and evaluation of rivastigmine loaded chitosan nanoparticles for brain targeting. Eur J Pharm Sci. 2012;47(1):6–15.
145. Costa PM, Cardoso AL, Mendonça LS, Serani A, Custódia C, Conceição M, Simões S, Moreira JN, de Almeida LP, de Lima MCP. Tumor-targeted chlorotoxin-coupled nanoparticles for nucleic acid delivery to glioblastoma cells: a promising system for glioblastoma treatment. Mol Ther Nucleic Acids. 2013;2(6):e100.
146. Hamaguchi T, Doi T, Eguchi-Nakajima T, Kato K, Yamada Y, Shimada Y, Fuse N, Ohtsu A, S-i M, Takanashi M. Phase I study of NK012, a novel SN-38–incorporating micellar nanoparticle, in adult patients with solid tumors. Clin Cancer Res. 2010;16(20):5058–66.
147. Zhan C, Gu B, Xie C, Li J, Liu Y, Lu W. Cyclic RGD conjugated poly (ethylene glycol)-co-poly (lactic acid) micelle enhances paclitaxel anti-glioblastoma effect. J Control Release. 2010;143(1):136–42.

Chapter 11
Bioinformatics for Diseases Management: A Personalized Therapeutics Prospective

Krishna Kanhaiya

Abstract The advancements in omics technologies and the emergence of modern bioinformatics tools transport significant changes in understanding of the mechanism of complex disease and revolutionized the healthcare to better disease diagnostics and management. These developments bring a sea of data for the physician and biological researchers to analyze and overcome the challenges of poor penetration of the available drugs into the diseases. A proper way of data management and integration technology can transform the big biomedical datasets into high-value, cost-effectiveness and rational drug target for effective personalized treatment. Moreover, it can reduce diagnostic costs, improve patient care and help the physician to develop individualistic patient care. Also, it enables the researcher to map disease molecules towards the discovery of distinct biomarkers for effective diagnosis through personalized therapeutics. This chapter provides an overview of available and integrated bioinformatics approaches in the role of effective disease managements, which can further establish safer, accurate, and reliable healthcare for every patient.

Keywords Biomedical data • Personalize therapeutics • Drug-target • Drug discovery • Disease managements • Effective healthcare

1 Introduction

Advancements in bioinformatics have opened up a new realm of possible observations to redefine diseases, develop drugs, and other therapeutical approaches. Developments of high throughput-omics (genomics, transcriptomics, proteomics, transcriptomics, metabolomics etc.) technology have revolutionized the biological research by bringing flood of data on everyday basis. Decoding of these data can

K. Kanhaiya (✉)
Computational Biomodeling Laboratory, Turku Centre for Computer Science, Åbo Akademi University, Agora, Domkyrkotorget 3, 20500 Turku, Finland
e-mail: kkanhaiy@abo.fi

A. Kaushik et al. (eds.), *Advances in Personalized Nanotherapeutics*,
https://doi.org/10.1007/978-3-319-63633-7_11

only be explained by bioinformatics which involve development of methods of storing, retrieving, organizing, and analyzing the large scale robust biomedical data to process and interpret them for translating into better clinical output. All these approaches invite and open new integrated area of bioinformatics research such as, systems biology, network pharmacology, biomedical data management and health informatics. These new approaches are using computational and mathematical approaches to understand the dynamics of complex diseases. It covers both basic experimental molecular validation research as well as explanatory approaches to optimize the shift of increasingly big biomedical data into clinical applications [1]. Amidst all of these approaches the main objective is to develop high-value, cost effective and rational drug targets for effective personalized treatment and it could be achievable by bringing together effective bioinformatics data analytic approaches. Analyze Big data in medical research can bring an effective personalize therapeutics and assemble new insights from patient data for effective treatment like never before [1, 2]. The approach could be critical to understand emerging biomedical problems, deep analysis of disease causality to make valid correlation, and to predict health trends to evaluate effective treatments. This also creates new prospects for researchers to develop bioinformatics based tools to personalize the diagnostic approaches, to identify and validate targets in more accurate way, and to identify novel biomarkers for designing personalized therapies and effective disease management.

One other issue is the management of all these biomedical data from different sources and technologies for effective personalized treatments by gathering them at a single platform Modern data management approach on a single platform can allow physician to mine their own data quickly to find out risk factor in patients and to analyze patients more closely [3]. This can be applied toward better preventive measures for health monitoring and therapeutics and for treating patients with a personalized approach at an early stage of disease development. Moreover, it can reduce diagnostic costs, improve patient care and help physician to develop personalized treatment plans and spot prevention opportunities. In this context, Blue Shield of California is trying to lead the personalize medicine approach by using patient data to develop and integrate technology system that will allow physician to deliver evidence-based care.

Further, all these approaches can be applied to better disease management if we use available bioinformatics resources wisely with its integration of other available approaches. Integrated approaches will help drug developer to design rational target for therapies [4–6] and optimize patient outcomes to build efficient bioinformatics tool for diagnosis [4, 7]. The real challenge is to provide useful bioinformatics and healthcare related tool to use in effective personalized treatments. Some approaches may be able to do these analytics of biomedical (disease) data, but its proper managements could be determined in improving of better disease diagnostic outcomes. From pharmacogenomics standpoint, the ability to manage, integrate, and integrate disease datasets will enable to discover biomarkers to effective diagnosis by personalize medicine. Here, we discuss the impact of new bioinformatics approaches and its integrated fields such as systems biology, network pharmacology, pharmacogenomics, health informatics and their role in personalize therapeutics for effective disease management.

2 Personalize Medicine: Ensuring the Availability of Safe and Effective Therapeutics

Personalized medicine' defines a 'coordinated strategic approach to individualistic patient care with the right drug at the right dose in right time. More broadly personalized medicine is a medical procedure to tailor made medical treatments of individual patients based on family (genetic) history, lifestyle, and environmental factors—to customize healthcare management [8, 9]. By combining the disease data from individual patients and proper use of diagnostic device and a therapeutic product, personalized medicine improve the efficiency of the healthcare system by improving quality, accessibility, affordability and reduce the burden of disease by targeting prevention and more rational therapeutics [9, 10]. The promise of personalized medicine could help patients to avoid harmful side effects, adverse drug interactions, and more effective treatment. Moreover, personalized therapeutics can be used to select optimal solutions and reduce trial and error prescribing.

Personalized medicine has potential to revolutionize the current drug discovery sector by matching to the best therapy based on their genetic makeup and other predictive factors and by the development of new technologies for biomarker testing [11]. It also has the potential of smart drug-delivery systems by matching the drug to the right patient. Developing disease-specific biomarkers for identifying specific stage of disease such as, the levels of certain enzymes, or the presence or absence of drug receptors, and for quantifying disease progression thereafter is an important requirement [12]. Biomarkers can play a crucial role in understanding differences among patients at individual level and help to build computational model of the drug discovery research to move away from conventional drug delivery approach to personalized targeted therapy [13]. Biomarker technologies can be employed in to identify expression signature to disease phenotype and to provide the possibility of identifying disease mechanisms and pathways initiated further to the exposure of pharmaceutical compounds [13].

With the use of personalized medicine, the disease management program focus on prevention, disease classification, and from treating disease to maintaining health. By improving the ability of disease prediction and diagnosis, experience, and targeted therapy response, personalized medicine offers a new paradigm of clinical practice for curtailing the duration and severity of illness, and improving success rates. As with diagnosis, information from personalized medicine and other biomedical technologies is increasing effective diagnosis of patient while reducing costs associated with ineffective treatment and avoidable adverse events. In the future, it will increase the possibility to provide better treatment and management regime that will be optimal for an individual patient. In addition, personalized medicine with use of clinical management could be valuable to match specific individual subgroups that behave similarly clinically and apply particular therapeutic intervention whose average response to a treatment is known [9]. Combination of personalized therapy and information management may also pave the way to improve the ability to track average treatment effects, disease progression and prognostic value in preclinical testing of novel compounds and the management of patients [14]. Moreover, this approach can improve optimal health and disease management and lower health care cost and enables physicians to make appropriate management decisions.

3 Building Omics Data Platform to Transforming Pharmacogenomics

Modern-omics technology has accelerated the pace of medical discovery and at the same time maintained the cost-effectiveness and time constrain required to sequence an entire genome and/or protein and open new way for a better understanding of biological mechanisms to resolve difficult diagnosis [2, 15]. It also drives modern biological and clinical researchers from traditional hypothesis-driven designs toward data-driven optimization studies. At the personal level it includes genomic information to develop a more holistic, and proactive treatment method [2]. This combined information is not only valuable to determine the rare genetic variants of person patients, but it also improves understanding of both human population variation and genetic diseases, and predicts more accurate meaning of disease susceptibility and diagnostic response [3]. As a consequence of cutting-edge omics facilities and analytical tools, data generation is no more a drawback for researchers, at least in the middle and high-income countries [16].

Major scientific investments are required in bioinformatics and data management to accelerate the transition to personalized medicine. Specific investments is also required in computational and collaborative data management infrastructure with cutting-edge, high-throughput omics technology, advanced digital technologies (high performance computing and cloud based multi storage facilities), and highly-qualified multi-disciplinary teams (physicians, biomedical researchers, computer scientists, and data scientists) to manage, integrate, analyze and interpret omics data. Processing of large scale omics data needs sufficient storage and computational resources and extravagant or sub-optimal servers. Cloud computing is an optimum solution to close the gap between data generations and handling of omics data [17, 18]. For instance, BaseSpace Sequence Hub with Amazon web service solves the challenges of biological data management and storage by providing a practically infinite Next Generation Sequencing data storage solution. More recently, Graphics Processing Units (GPUs) are gaining a lot of attention by the scientific community, as they can provide faster computations, general-purpose computing in a cloud environment [19], and allow more intensive research of biological systems [20]. There is need for integration and validation of omics research tools before considering their clinical utility. In this case Galaxy project draws the attention of computational scientists across globe to upload and validate their software tools. Much more collaborative efforts for the integration and interpretation of omics data are still needed for the effective and proper development of personalized therapeutics.

All these recent progress in modern genomic technologies and data analysis and management tools are transforming traditional pharmacogenetics into pharmacogenomics (identifying and studying the genomic contribution to targeted drug therapy and new drug discovery) research into new directions [2, 15]. These genome data approaches demonstrate enormous opportunities for computational pharmacogenomics and drug design. Also, it shows substantial progress in finding new forms of drugs, predicting multiple drug candidates, pharmacological and toxicological properties in chemicals. In recent addition, computational screening of drugs has been greatly facilitated by the advent of connectivity mapping methods to make a meaningful representation of the genome-wide gene expression profiles [4]. Thus, the design and development of data management systems leverage to extract and integrate pharmacogenomics information, retrieval of information regarding diseased genes and their respective variants and drugs [21]. This high-performance computing resources in pharmacogenomics research can automatically reverse-engineer molecular networks from the big omics data, and make its application available for most advanced tools to the real-world pharmacogenomics studies and way the door into a new era of drug discoveries [22].

4 Network Medicine: Introducing New Paradigm in Targeted Therapy

Network medicine is a new evolving research strives from network science and systems pharmacology to investigate disease mechanisms and pharmaco-therapeutics to discover the high efficiency and low side effects of multi-target drugs [22, 23]. With the advent of large sets of genome data, biomedical research enable us to find truly personalized treatment options, and provide the options to explore disease-causing events and understand it through available data integration approach [2, 24]. Doing so requires mining these data sets to translate into specific treatments options by decoding molecular networks within which a disease causing module operates. In disease networks, identifying disease modules defines the disease, where interactions are generated by gene, proteins and other small molecules. The identification of disease module(s) recognizes genes that drive the disease factors within the networks, identify novel relationships among disease specific genes modules, and also predict network blueprint of relationship with potential drugs [23].

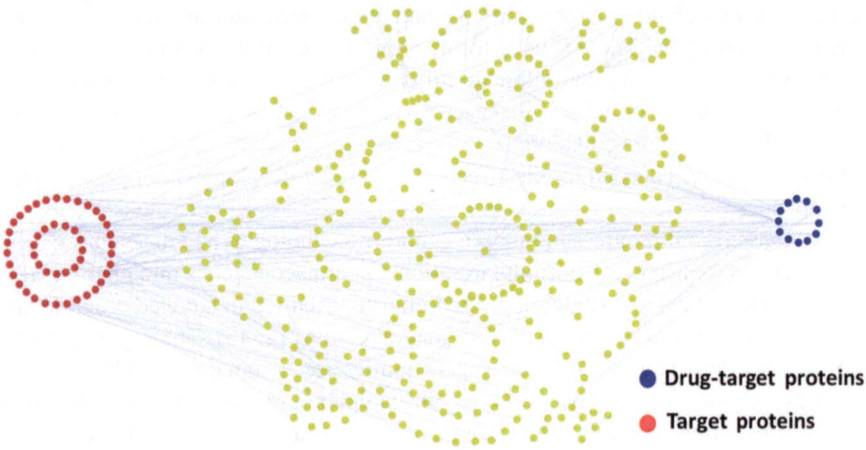

Fig. 11.1 In a protein–protein interaction network of cancer, network medicine approach shows how a single drug target protein can successfully target multiple disease proteins

One valiant prediction of the network approach is to identify potential individual/multi target for the development of targeted therapy. Moreover, the main perspective in shifting from the gene-centric pattern of drug discovery towards a network-centric medicine approach is the efficacy of combination/multi targeted therapy to effectively treat many cancer types and infectious diseases [25, 26]. Such multi-target therapy are being considered as a promising approach to tackle the regulatory mechanisms and robustness in disease (Fig. 11.1), as well as to reduce side effects that often limit the clinical efficacy of many conventional drug therapy [27–29]. In this way, multi-targeted therapies could be essential steps towards the proper and efficient disease management and personalizes treatments. Multi/combinatorial targeted therapies use drug combinations of two or more drugs on the basis of screening results and the patterns of drug resistance. Therefore, discovering effective and safe multi-target drugs or drug combinations need a proper understanding of how and where the essential pathways can be inhibited by drug therapy during a disease progression. However, the efficacy and toxicity of multi-target drugs is not only a priori to complex interactions among drugs and their respective targets, but various others factors (effects of drugs and environmental) that affect the drug response phenotypes in vivo [28]. Efforts of effective molecular target with safer drug combinations could be possible by combining multiple data sources of drug features [30]. Proper implementations of combinatorial drug therapy which can be applied simultaneously at multiple therapeutic targets have the potential to dramatically enhance the effectiveness and the applicability of future therapeutic treatments [31].

More comprehensive information about the drug-target relationships through network approach can be also made by combining data from various publicly available data resources. These data resources contain the detailed information of drugs, targets, pharmacology, bioactive compounds and pathways which can be useful in defining multiple drugs and their targets involved in the regulation of the individual pathways. These important drug-target information can be retrieved from multiple databases, including DrugBank [32], The Therapeutic Target Database [33], PharmGKB [34], STITCH [35], Potential Drug Target Database [36], SuperTarget and Matador [37], etc. Combination of network pharmacology and bioinformatics approaches pave the way towards the development of new network pharmacology strategies in the more efficient manner [38, 39]. It enhances the knowledge of drug-regulated signaling pathways, including experimental drugs as well as approved drugs in complex biological and physiological systems [33]. Therefore, providing the comprehensive information about the drugs, their respective targets and the targeted pathways can reveal and predict drug response most effectively, so that drugs are developed for and targeted to the right patients.

5 From Biomedical Data to Effective Disease Managements

The size and scale of biomedical data (In some cases, it's size in petabyte and exabyte) determines their acquisition, cleansing, administration, curation, storage, retrieval and analysis in a way that is easy to handle for the experimental researchers and it is a necessity of modern research [40, 41]. Therefore, to learn how to handle the complexity of biological information, how to integrate the data from very heterogeneous resources, and what kind of principles or standards to be adopted for analyzing big biomedical data allow us to translate the massive amount of information into a better understanding of various disease mechanisms, which can be further applied to personalized medicine (Fig. 11.2). To apply extensive uses of these big data, it is essential to establish interoperable databases, make state-of-the-art tools available for the research community, and construct resources and infrastructure for high-performance computing systems to provide a straightforward way to researchers [42]. Therefore the primary goal of a modern data management system should combine all types of data obtained from various experimental verification performed on diverse equipment. One integrated database can be used to develop

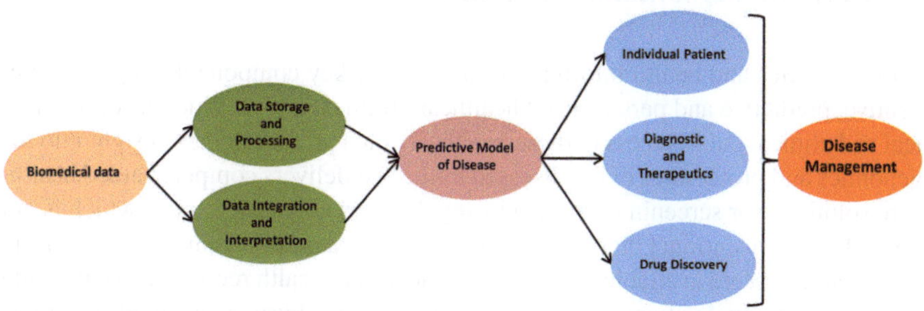

Fig. 11.2 A schematic representation shows step towards the effective disease management from heterogeneous biomedical data

applications to arrange internal and management reports and perform other functions that preserve detailed research data to be performed in a particular laboratory. Hence, to build unifying platforms, gathering, organizing, and storing data by integrating the existing biological data and corresponding mining tools can be ideal for better understanding [4]. Especially for big heterogeneous data computationally optimized efficient algorithms assist to store data in quickest possible access [22] and provide efficient data storage and memory utilization.

The vast amount of data generated by various experiment in modern laboratories enables researchers to propose data management application for extracting more useful and interesting information about various molecules [42]. Therefore a successful data management technique requires a systematic developmental framework. Here, to ensure the reproducibility of the biological analysis and explanation of the experimental findings, data sharing, integration and annotation could be essential. Combination of heterogeneous biomedical data for their management and integration by scientists and researchers from around the world can be a new landmark in the finding of new diseases. For instance, data integration and analysis approach in a milestone study by using omics technology researchers discovered 12 human exomes [43]. Overall, they analyzed nearly 30 GB of data and discovered a single nucleotide polymorphism (SNP) in gene MYH3 straightforward, which is directly contributing for Freeman–Sheldon syndrome (a monogenic human disease called). In another impressive study, scientists analyzed one terabyte (TB) of genome sequencing data to finding causative variations and found a novel missense mutation Y169H on gene SH3TC2 associated with Charcot-Marie-Tooth disease [44]. All such studies indicate that biomedical data management and their analysis are essential parts in biomedical research for discoveries breakthrough [45]. It is need of hour to develop innovative methods for data management, analysis and availability. Also, an integrated and comprehensive genomic information management system in clinical use can be a relevant source of information for related disease in one place and present the information in an easy-to-understand format.

6 Bioinformatics and Health Informatics: Managing Patient Data for Personalized Therapeutics

Bioinformatics and health informatics domains are key components to deliver preventive, predictive and personalized healthcare from enormous amounts of biomedical and clinical data. Access to these healthcare data provides better insight into the genomics and proteomics of the diseased tissue and delivers computer-aided healthcare solutions for screening and diagnosing diseases at an early stage, which is not possible by conventional therapeutics [46]. Also, because of recent developments in healthcare, data are available in the form of electronic health records, and can easily integrate individual patients' data throughout hospitals. Further, availability of electronic health records, which includes all-time accessibility to complete and all-inclusive information available of individual patients (such as symptoms, conventional medical test results and genomic information of an individual) will greatly enhance the ability to improve the quality of traditional public healthcare service [47]. It helps physicians and healthcare providers to identify complex information useful in supporting medical diagnostic to deliver the best standard of care for an individual patient.

The primary goal in health informatics systems is to define a distributed user-centered healthcare system from a centralized hospital-centered healthcare system. For this, there is a need of synergy among different component of information in healthcare informatics (patients, healthcare informatics personnel and providers and management of healthcare facilities). Advancements in information system within the hospital environment also help in improving the quality of healthcare and present disease discovery systems [48]. Smart information management systems detect real time issues such as, those related to clinical care, billing and compliance, biomedical waste, and pop-up reminding messages. Advancement in information management system support to collect individuals' information (from grenomics to system levels) and then use them together for personalized treatments. It also combines information from different sources such that objective measures can be derived to effective disease diagnosis by developing computer-assistive algorithms and classifiers, especially to recognize early stage of disease progression during which treatments are often more progressive [49, 50]. Here good addition could be the development of more efficient algorithms to screen individual patient from a huge population.

The primary purpose of integrating biomedical research and healthcare information is to allow exchange in technology, data, and knowledge. This will in turn provide an integrated database of diseases, their causes and the trigger patterns. This will further improve possible treatment and cure and speed up healthcare access and thus provide a reliable and efficient healthcare system [51]. These all are possible by advancements in acquisition systems to provide smart sensing, personal, and hospital information management systems, and modern platform for managing health care ubiquitously and inconspicuously, as well as smart systems to maintain the practice of healthcare information. The advanced healthcare systems generate insights into the lower costs to reduce diagnostic inefficiencies, improvement in identification of risk populations, and enable physicians' to personalized treatments through health care. It might be beneficial to develop a nationwide information system infrastructure with specific standards implemented to guide physicians in the use of personalized medicine in their practices [52, 53].

7 Conclusion

Advances in high throughput technologies and the development and implementation of bioinformatics and integrated tools provide an opportunity to analyze large amount of diverse biomedical datasets to extract more useful information. This further build accurate complex disease model and allow effective personalized

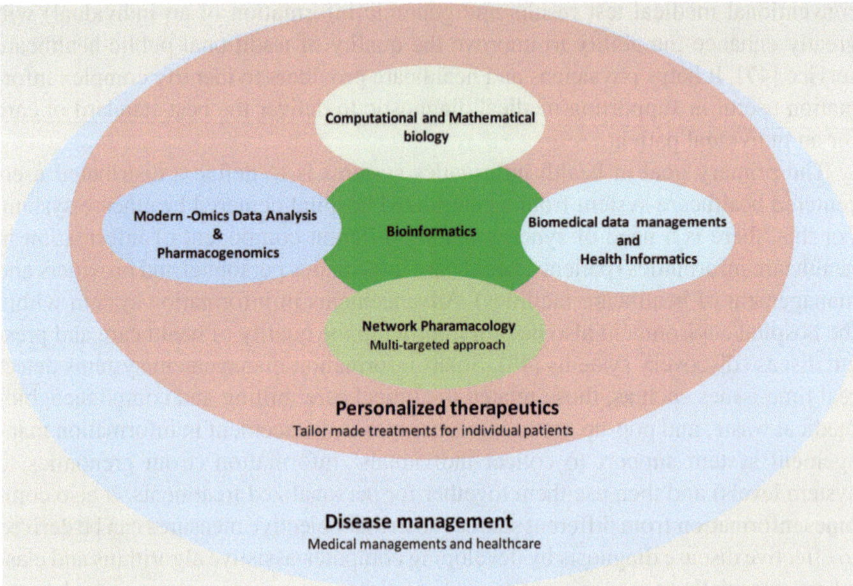

Fig. 11.3 Bioinformatics relation with modern-omics data analysis and pharmacogenomics, Network Pharmacology, computational and mathematical biology, and biomedical data managements and health informatics for effective personalized therapeutics and proper disease management

therapeutics for better disease management (Fig. 11.3). In addition, for better data management in biomedical research, modern research laboratories require to integrate and mine the paramount amount of multiple sets of heterogeneous dataset from different experiments. For successful disease management within a systematic development framework, experimental researchers should get more involved with bioinformaticians and computational scientists to achieve uniformity of their data.

Developments in this direction may also indeed revolutionize biological research, and personalized medicine towards cost effective. For example, incorporating computational systems biology (network pharmacology) and available dataset resources will be useful in understanding the effect of individual/multi drug target in specific disease locations and environmental factors that are more effective in drug discovery applications [4, 25]. In another addition to successful disease management, health information systems must collect and analyze personal genomic information for physicians to make decisions based on the vast genetic patterns of diseases for genomics-based personalized medicine [47]. In order to sustain continued advances in bioinformatics' role in personalized therapeutics and disease managements, use of experimental research and technological innovation combined with new advances must be made toward building the infrastructure for personalized medicine. This will further establish safer, accurate, reliable healthcare for each and every patient.

Acknowledgements This work was supported by the Academy of Finland through grant 272451, by the Finnish Funding Agency for Innovation through grant 1758/31/2016, and by Center of International Mobility through grant TM-15-9933.

References

1. Fernald HG, Capriotti E, Daneshjou R, et al. Bioinformatics challenges for personalized medicine. Bioinformatics. 2011;27:1741–8.
2. Alyass A, Turcotte M, Meyer D, et al. From big data analysis to personalized medicine for all: challenges and opportunities. BMC Med Genomics. 2015;8:33.
3. Chen R, Snyder M. Promise of personalized omics to precision medicine. Rev Syst Biol Med. 2013;5(1):73–82.
4. Musa A, Ghoraie SL, Zhang S-D. A review of connectivity map and computational approaches in pharmacogenomics. Brief Bioinform. 2017:1–18. doi:10.1093/bib/bbw112.
5. Mullen J, Cockell SJ, Woollard P, et al. An integrated data driven approach to drug repositioning using gene-disease associations. PLoS Onc. 2016;11(5):e0155811.
6. Lu J-J, Pan W, Hu Y-J, et al. Multi-target drugs: the trend of drug research and development. PLoS One. 2012;7(6):e40262.
7. Mooney SD, Krishnan VG, Evani US. Bioinformatic tools for identifying disease gene and SNP candidates. Methods Mol Biol. 2010;628:307–19.
8. Zhang P, Wang F, Hu J, et al. Towards personalized medicine: leveraging patient similarity and drug similarity analytics. AMIA Summ Transl Sci Proc. 2014;2014:132–6.
9. Cornetta K, Brown CG. Perspective: balancing personalized medicine and personalized care. Acad Med. 2013;88(3):309–13.
10. Brown C. Targeted therapy: an elusive cancer target. Nature. 2016;537:S106–8.
11. Bibault JE, Fumagalli I, Ferté C, et al. Personalized radiation therapy and biomarker-driven treatment strategies: a systematic review. Cancer Metastasis Rev. 2013;32(3-4):479–92.
12. Scherer HU, Dörner T, Burmester GR. Patient-tailored therapy in rheumatoid arthritis: an editorial review. Curr Opin Rheumatol. 2010;22(3):237–45.
13. Aslani A-A, Mangematin V. The future of drug discovery and development: shifting emphasis towards personalized medicine. Technol Forecast Soc Change. 2010;77(2):203–17.
14. Downing JG, Boyle NS, Brinner NK. Information management to enable personalized medicine: stakeholder roles in building clinical decision support. BMC Med Inform Decis Mak. 2009;9:44.
15. Egea RR, Puchalt NG, Escrivá MM, et al. OMICS: current and future perspectives in reproductive medicine and technology. J Hum Reprod Sci. 2014;7(2):73–92.
16. Shendure J, Lieberman AE. The expanding scope of DNA sequencing. Nat Biotechnol. 2012;30(11):1084–94.
17. Schatz MC, Langmead B, Salzberg SL. Cloud computing and the DNA data race. Nat Biotechnol. 2010;28(7):691–3.
18. Schadt EE, Linderman MD, Sorenson J, et al. Cloud and heterogeneous computing solutions exist today for the emerging big data problems in biology. Nat Rev Genet. 2011;12(3):224.
19. Jo H, Jeong J, Lee M, et al. Exploiting GPUs in virtual machine for BioCloud. Biomed Res Int. 2013;2013:1–11.
20. Nobile SM, Cazzaniga P, Tangherloni A. Graphics processing units in bioinformatics, computational biology and systems biology. Brief Bioinform. 2016:bbw058. doi:10.1093/bib/bbw058.
21. Potamias G, Lakiotaki K, Katsila T, et al. Deciphering next-generation pharmacogenomics: an information technology perspective. Open Biol. 2014;4(7) doi:10.1098/rsob.140071.

22. Rodin AS, Gogoshin G, Boerwinkle E. Systems biology data analysis methodology in pharmacogenomics. Pharmacogenomics. 2011;12(9):1349–60.
23. Barabási A-L, Gulbahce N, Loscalzo J. Network medicine: a network-based approach to human disease. Nat Rev Genet. 2011;12:56–68.
24. Loscalzo J. Systems biology and personalized medicine: a network approach to human disease. Proc Am Thorac Soc. 2011;8(2):196–8.
25. Tang J, Aittokallio T. Network pharmacology strategies toward multi-target anticancer therapies: from computational models to experimental design principles. Curr Pharm Des. 2014;20(1):23–36.
26. Talevi A. Multi-target pharmacology: possibilities and limitations of the "skeleton key approach" from a medicinal chemist perspective. Front Pharmacol. 2015;6:205.
27. Kitano H. A robustness-based approach to systems-oriented drug design. Nat Rev Drug Discov. 2007;6:202–10.
28. Xie L, Xie L, Kinnings SL, et al. Novel computational approaches to polypharmacology as a means to define responses to individual drugs. Annu Rev Pharmacol Toxicol. 2012;52:361–79.
29. Lehár J, Krueger AS, Avery W, et al. Synergistic drug combinations tend to improve therapeutically relevant selectivity. Nat Biotechnol. 2009;27:659–66.
30. Zhao X-M, Iskar M, Zeller G, et al. Prediction of drug combinations by integrating molecular and pharmacological data. PLoS Comput Biol. 2011;7:e1002323.
31. Zanzonia A, Soler-Lópeza M, Aloy P. A network medicine approach to human disease. FEBS Lett. 2009;583(11):1759–65.
32. Wishart DS, Knox C, Guo AC, et al. DrugBank: a knowledgebase for drugs, drug actions and drug targets. Nucleic Acids Res. 2008;36:D901–6.
33. Yang H, Qin C, Li HY, et al. Therapeutic target database update 2016: enriched resource for bench to clinical drug target and targeted pathway information. Nucleic Acids Res. 2016;44:D1069–74.
34. Thorn CF, Klein TE, Altman RB. PharmGKB: the pharmacogenomics knowledge base. Methods Mol Biol. 2013;1015:311–20.
35. Kuhn M, von Mering C, Campillos M, et al. STITCH: interaction networks of chemicals and proteins. Nucleic Acids Res. 2008;36:D684–8.
36. Gao Z, Li H, Zhang H, et al. PDTD: a web-accessible protein database for drug target identification. BMC Bioinform. 2008;9:104.
37. Günther S, Kuhn M, Dunkel M, et al. SuperTarget and Matador: resources for exploring drug-target relationships. Nucleic Acids Res. 2008;36:D919–22.
38. Berg JM, Rogers ME, Lyster PM. Systems biology and pharmacology. Clin Pharmacol Ther. 2010;88:1719.
39. Hopkins AL. Network pharmacology: the next paradigm in drug discovery. Nat Chem Biol. 2008;4:682690.
40. Luo J, Wu M, Gopukumar D, et al. Big data application in biomedical research and health care: a literature review. Biomed Inform Insights. 2016;8:1–10.
41. Schadt EE, Linderman MD, Sorenson J, et al. Computational solutions to large-scale data management and analysis. Nat Rev Genet. 2010;11(9):647–57.
42. Li Y, Chen J. Big biological data: challenges and opportunities. Genomics Proteomics Bioinformatics. 2014;12(5):187–9.
43. Ng SB, Turner EH, Robertson PD, et al. Targeted capture and massively parallel sequencing of 12 human exomes. Nature. 2009;461:272–6.
44. Lupski JR, Reid JG, Gonzaga-Jauregui C, et al. Whole-genome sequencing in a patient with Charcot–Marie–Tooth neuropathy. N Engl J Med. 2010;362:1181–91.
45. Howe D, Costanzo M, Fey P, et al. Big data: the future of biocuration. Nature. 2008;455:47–50.
46. Belle A, Thiagarajan R, Soroushmehr SMR, et al. Big data analytics in healthcare. Biomed Res Int. 2015;2015:370194.

47. Alzu'bi A, Zhou L, Watzlaf V. Personal genomic information management and personalized medicine: challenges, current solutions, and roles of HIM professionals. Perspect Health Inf Manag. 2014;11:1c.
48. Nair BG, Newman SF, Peterson GN, et al. Smart Anesthesia Manager™ (SAM)—a real-time decision support system for anesthesia care during surgery. IEEE Trans Biomed Eng. 2013;60(1):207–10.
49. Gomez-Cabrero D, Abugessaisa I, Maier D, et al. Data integration in the era of omics: current and future challenges. BMC Syst Biol. 2014;8(Suppl 2):I1.
50. Buil-Bruna N, Sahota T, López-Picazo JM, et al. Early prediction of disease progression in small cell lung cancer: toward model-based personalized medicine in oncology. Cancer Res. 2015;175(12):2416–25.
51. Oyelade J, Soyemi J, Isewon I, et al. Bioinformatics, healthcare informatics and analytics: an imperative for improved healthcare system. Int J Appl Inform Syst. 2015;8(5):1–6.
52. Kawamoto K, Lobach DF, Willard HF, et al. A national clinical decision support infrastructure to enable the widespread and consistent practice of genomic and personalized medicine. BMC Med Inform Decis Mak. 2009;9:17.
53. Buntin MB, Jain SH, Blumenthal D. Health information technology: laying the infrastructure for national health reform. Health Aff. 2010;29(6):1214–9.

Chapter 12
Nano-Enabled Sensing Platforms for Personalized Care

Renu Singh, Chandini C. Mohan, and Abhijit Chandra Roy

Abstract The present book chapter concerns the recent developments of nano-sensing platforms based biosensors for personalized health care as is currently of prime interest to circumvent the delay in diagnosis. Exploiting the burgeoning traits of nanomaterials i.e. high surface to volume ratio, enhanced electron transfer, extraordinary optical, magnetic, and electrical properties improve chances for the transduction technology development. Efforts have been made to discuss the importance of various nanomaterials used with several techniques i.e. electrochemical, magneto-resistance, localized surface plasmon resonance, surface-enhanced Raman scattering, fluorescence; for biosensor construction towards their applications to personalized health care, which is still a "Holy Grail".

Keywords Personalized care • Nanomaterials • Electrochemical • SERS • Fluorescence • GMR • LSPR

1 Introduction

The last few decades showed drastic advancements in the field of biosensors. In short, biosensors are miniaturized portable devices that detect and generate quantifiable signals corresponding to the amount of a target bio-analyte, upon interaction with a biologically responsive platform [1–5]. Biological recognition components of a biosensor could be an antibody, enzymes, nucleic acid, aptamers, polysaccharides

R. Singh (✉)
Department of Bioproducts and Biosystems Engineering, University of Minnesota, Minneapolis, MN, USA
e-mail: singhr@umn.edu; raina1785@gmail.com

C.C. Mohan
Center for Soft and Living Matter, Institute for Basic Science (IBS), Ulsan National Institute of Science and Technology, UNIST-gil 50, Ulsan, Republic of Korea

A.C. Roy
Soft Matter Laboratory, Department of Chemical Engineering, Indian Institute of Technology, Kanpur, UP 208016, India

A. Kaushik et al. (eds.), *Advances in Personalized Nanotherapeutics*,
https://doi.org/10.1007/978-3-319-63633-7_12

or even biological tissues/organelles [6–9]. Upon binding of a specific target molecule, a transducer, converts the gathered information into measurable signals such as an electronic signal. A signal processing unit then performs as a user-friendly interphase to digitally display the test result. Extensive multidisciplinary research has led to the development of a wide variety of biosensor prototypes for different applications, more significantly for health care [10–14]. Currently, there is an increasing demand for biosensors in public health/clinics as rapid diagnostic tools for early detection or to study the disease progression [15, 16]. Despite the tremendous progress made by researches all over the world, there is still immense room for highly sensitive, multiplexing, low cost biosensing devices for successful clinical translation (from bench to bed side). Many of the commercially available kits are hampered by a lower sensitivity (false results) upon multiplexing, need for a large test sample volume and moreover its time consuming [17, 18].

Emergence of nanotechnology could possibly address the current limitations pertaining to biosensors such as low sensitivity, multiplexing wherein, various nanomaterials are used as sensing platforms to generate faster and accurate results [19–23]. Nanomaterials offer many exceptional physico-chemical properties such as optical and electrical conductivity, higher surface energy, catalytic activity and stronger adsorption ability in comparison with its bulk counterpart. In addition, owing to their higher surface area, nanomaterials offer more no. of molecular reactive sites and more stable solid matrix for biomolecules [19, 20]. Nanomaterials of varied morphologies and dimensions that differ in their functionality, reactivity and loading efficiency have contributed for various biosensors components, as both recognition elements and transducers. Nano biosensors have exhibited remarkable sensitivity, accuracy and reproducibility compared to traditional sensors possibly due to the maximum biological activity exhibited by the biomolecule, minimal interference by the nano form of matrices and/or high performance of transducers.

Extensive research works are done to develop low cost future bioelectronics devices without compromising its effectiveness for high selectivity. Integration of unique nanomaterial properties with the emerging new fabrication technologies viz., bioinspired materials, biomimetics, synthetic ligands, plastic antibodies etc., may help to design commercially viable and affordable nanobiosensors for various biomedical applications in near future [24–28]. Integration of microfluidics/nanofluidics technologies with nanosensors have enabled to develop easy-to-use chips for analysis with further improvement in sensitivity even at lower concentration [29–31].

In healthcare, nanobiosensors could be beneficial for both therapy as well as diagnosis which is often referred to as "theragnostics". Regardless of all these advancements in R&D, very few biosensors have been commercially successful and being used in clinics like Glucose sensors. Several researchers are involved in the developing nanobiosensors to detect several pathogens and early clinical diagnosis and highlighting the advancements in developing smart sensing strategies for point-of-care diagnostics [32–37]. These sensor prototypes are proposed to have great chance for the future point of care diagnostic market as an easy, quick, accurate and affordable means of disease detection by patients. In the new era of personalized

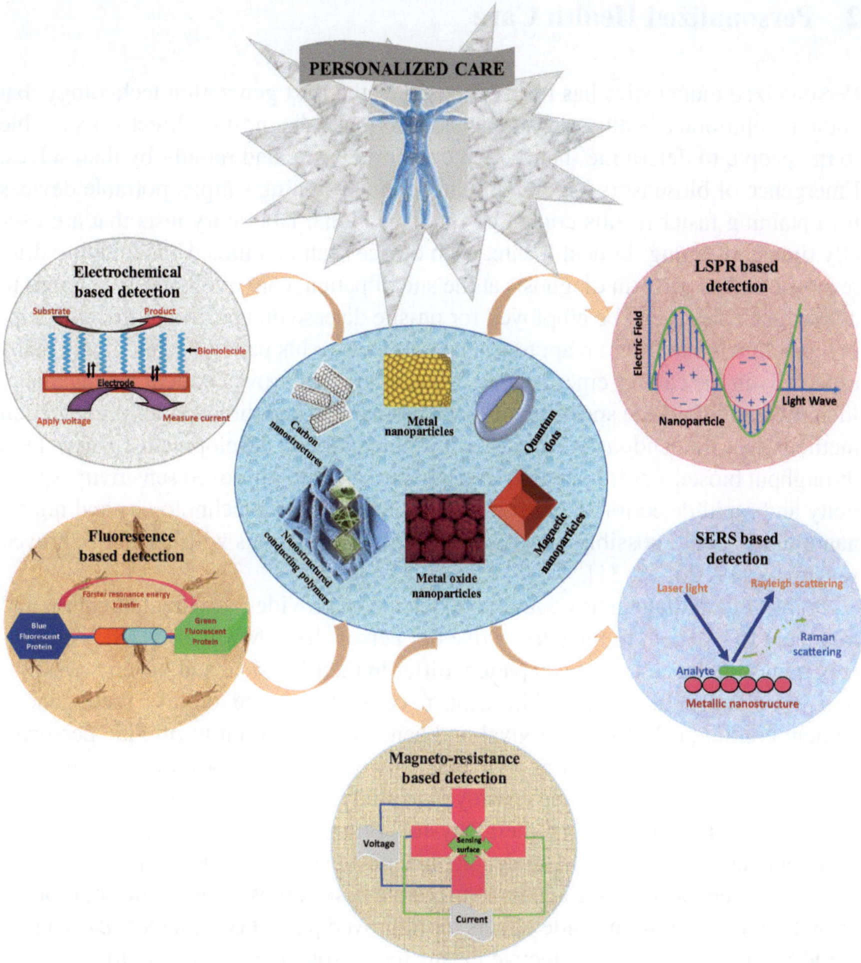

Fig. 12.1 Schematic showing the usage of various nanomaterials for different type of biosensors construction for personalized health care

health care, the relevance for a faster, high performance nanosensor is of utmost importance in clinics where the patient sample volumes are limited.

The aim of this chapter is to cover in detail, a comprehensive over view of the fast-growing aspects of the various nanomaterials used with several techniques i.e. electrochemical, magneto-resistance, localized surface plasmon resonance, surface-enhanced Raman scattering, fluorescence; for biosensor construction towards health care applications (Fig. 12.1). It surveys on the current prospects and the need for the further research on nano based sensing platforms for effective personalized care in future.

2 Personalized Health Care

Personalized diagnostics has been proposed as the next generation technology that could revolutionize health care. It envisions to make diagnostics directly accessible to the people to determine their health condition easily and rapidly by themselves. Emergence of biosensors have highly helped in designing simple, portable devices for obtaining faster results compared to conventional laboratory tests that are usually time consuming. In health care, such devices can be utilized for studying drug response and to perform diagnosis at the site of patient care or otherwise referred to as point-of-care. It can be employed for on-site disease diagnosis and thereby helping the clinician to offer an appropriate treatment for his patients without any delay especially in the case of emergencies. This aid in an improved patient management. Identification of a wide spectrum of novel biomarkers and direct pathogen detection methods has tremendously increased the potential for development of new high throughput biosensors for personalized assessment with improved sensitivity, specificity and rapid detection. Convergence of newer bio-nanotechnologies and micro/nanofluidics could possibly give rise to diagnostic platforms with further improved sensitivity [19, 20, 30, 31].

Point of care diagnostics could be extended to provide individual attention and treatment to a patient in the perspective of "Personalized Medicine" [38]. The concept relays on the fact that each patient differ in their physical and genetic constitution. Essentially, in diseases like cancer, cells evolve and modify their genetic content continuously for its survival and hence it's important to provide personalized treatment for their faster as well as speedy recovery [39, 40]. In addition, point of care diagnostics can also be employed to study the drug response in individuals for a better understanding of their physiology and/or mutations and to offer an excellent personalized care. A molecular diagnostic test that simply requires a blood sample can replace invasive and uncomfortable tissue biopsies or fix the appropriate drug dosage with minimal side effects for improved patient compliance. Biosensors could be utilized as a cost-effective means for personalized medicine: to detect the disease or continuous monitoring of its progression in each individual patient and thereby providing a higher standard of medical care.

3 Importance of Nanomaterials in Constructing Biosensors

Even though the existence and the application of the nanomaterials existed back to many civilizations, it gained tremendous popularity during last few decades due to their unique properties such as high surface to volume ratio, enhanced electron transfer, extraordinary optical, magnetic, and electrical properties with respect to its bulk counterpart [41]. In this context, it is found that some of these special properties of the nanomaterials can fit perfectly to the requirement of fabricating biosensors to provide superior performance. Therefore, a huge response is seen in the field of biosensor for these nanomaterials to get better performance. In terms of material, there are different kinds of nanoparticles, such as metal nanoparticles, conducting polymeric nanoparticles, quantum dots, magnetic nanoparticles, metal oxide nanoparticles, carbon nanostructures [42] and even composite nanoparticles, have been widely used to construct nanobiosensors. Although these nanoparticles play different roles in different biosensor systems based on their unique properties but immobilization of biomolecules; catalysis of electrochemical reactions; enhancement of electron transfer are the basic functions of nanoparticles based matrices for electrochemical biosensors [43].

4 Electrochemical Based Detection

Easy electron transfer between the redox-active centers of biomolecules with an electrode surface is an essential requirement for a successful functioning of an electrochemical biosensor. In most of the cases, the electron transfer between electrodes and the active centers are hindered by considerably thick insulating biomolecular shells. In addition to that, the substrate to fabricate biosensors are usually coated with monolayer of molecules to immobilize biomolecules. Due to this insulating coating of monolayer, no appreciable Faradaic electrochemistry between the electrode and a redox species in solution is occurred (Fig. 12.2).

Graber et al. disclosed that presence of nanoparticles on the surface of the insulating layer, mediate charge transfer across the insulating layer, where silver or gold nanoparticles (AuNPs) were attached to a polymerized silane modified platinum electrode [44]. In another significant report authors used AuNPs (~1.4 nm) coated with N6-(2-aminoethyl)-flavin adenine dinucleotide to enhance the electron transfer capability between the gold electrode coated with thiolated monolayer and the enzyme apo-glucose oxidase and exhibited fast electron transfer between the enzyme redox center and the substrate with a rate constant about 5000 s^{-1} which is about seven times faster than that between glucose oxidase and its natural substrate, oxygen [45]. Another mandatory requirement for the successful biosensor fabrication is oriented/proper immobilization of biomolecules on the nanomaterials coated platforms as nanometer sized particle naturally acquire large surface to volume ratio, higher surface energy, which strongly helps in strongly absorbing biomolecules on the surface of the nanomaterials. Raj et al., developed an electrochemical sensor for selective detection of dopamine in the presence of ascorbic acid, which was based on the catalytic effect of AuNPs on the ascorbic acid oxidation. This resulted in decrease of the oxidation overpotential of ascorbic acid and effective separation of the oxidation potentials of ascorbic acid and dopamine, thus allowing the selective electrochemical detection and showed nanoforms of material facilitates the catalytic activity [46]. Singh et al., fabricated DNA sensor based on nucleic acid probe (from *Opa*, a multi-copy gene of *N. gonorrhoeae*) functionalized using avidin–biotin as cross-linking agent onto nanostructured conducting polymer polyaniline having 0.5×10^{-15} M of complementary target by differential pulse

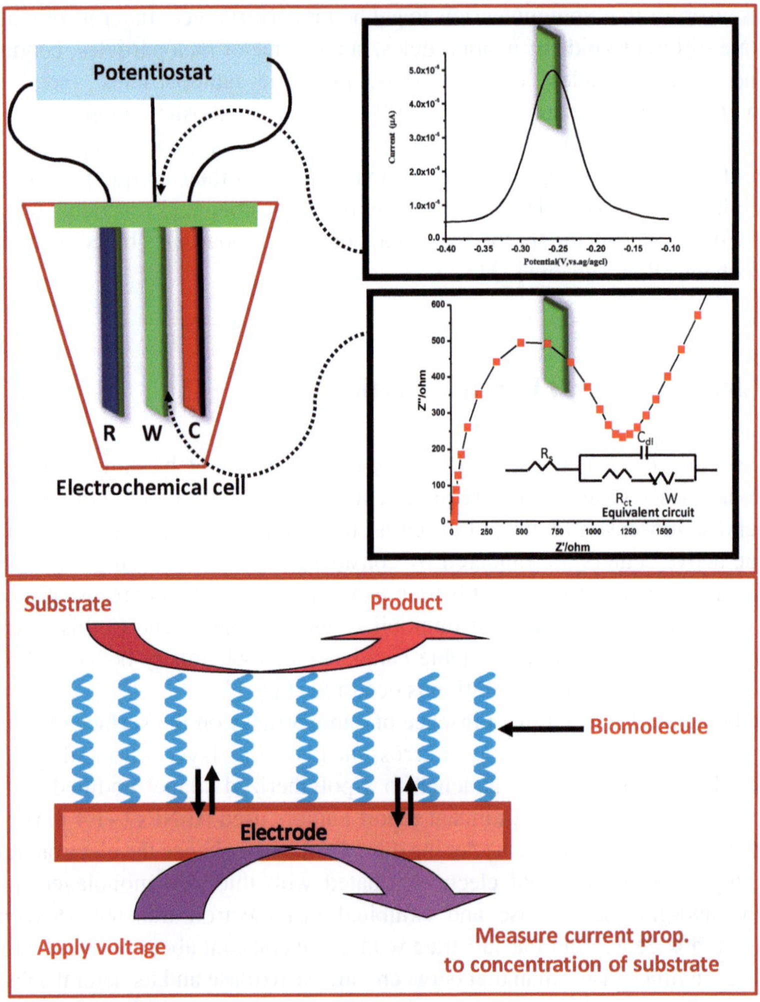

Fig. 12.2 Schematic representation of electrochemical cell and process

voltammetry using redox indicator methylene blue [47]. Multiwalled carbon nanotubes were introduced with the nanostructured polyaniline to enhance the electron transfer which provided conducting channels via electrons from one polymer chain to other and finally to the electrode surface and improved LOD 1.2×10^{-17} M [7]. Nanostructured polyaniline-iron oxide nanocomposite based biosensor was developed for the clinical diagnosis and exhibited good specificity for *N. gonorrhoeae* species against nongonorrhoeae type of *Neisseria* species and other gram-negative bacterias and showed the improved affinity constant $3.39 \times 108\ M^{-1}$ for hybridization calculated via Langmuir adsorption isotherm model [48].Trio matrix having nanostructured conducting polymer, Fe_3O_4 nanoparticles, carbon nanotubes was also explored for sexually transmitted infection detection and tested with positive and negative PCR amplicons of gonorrhoea affected patient samples [49]. In another study, Fe_3O_4 nanoparticles (~18 nm) increased the electro-active surface area of chitosan–Fe_3O_4 nanobiocomposite that provided desirable environment for loading of DNA with better conformation leading to increased electron transfer kinetics between the medium and electrode for detection of gonorrhea having LOD of 1×10^{-15} M within range of 1×10^{-16} M–1×10^{-6} M [50]. An impedimetric DNA sensor for HIV and HBV detection was fabricated by functionalizing magnetic nanoparticles onto bare Au electrode using a magnet via non-faradic impedance spectroscopy to detect 50 pmol of HBV DNA and 160 pmol of HIV DNA on sample of 20 μL [51]. Bacillus anthracis spores in food samples was detected by utilizing electrically-active polyaniline-coated magnetic NPs [52]. Nanostructured zinc oxide was exploited for clinical diagnosis [53]. Fe_3O_4 nanoparticles contributed to enhance electron transfer and enhanced sensitivity by providing suitable microenvironment and low LOD (50 pM) in nanobiocomposite form with chitosan matrix for electrochemical detection of HIV-1 [54]. Polydimethylsiloxane/glass based flow injection microfluidic chip was fabricated for cholesterol detection using carbon nanotubes via flow injection chronoamperometric measurement and very rapid and sensitive real-time detection was achieved with high throughput of more than 60 samples per hour with small sample volume of 15 μl having the linear detection range between 50 and 400 mg/dl. Additionally, they also verified the low cross reactivity with glucose, ascorbic acid, acetaminophen and uric acid and confirmed that the CNT based microfluidic system was promising for clinical diagnostics of cholesterol with high speed real-time detection capability, very low sample consumption, high sensitivity, low interference and good stability (Fig. 12.3) [55].

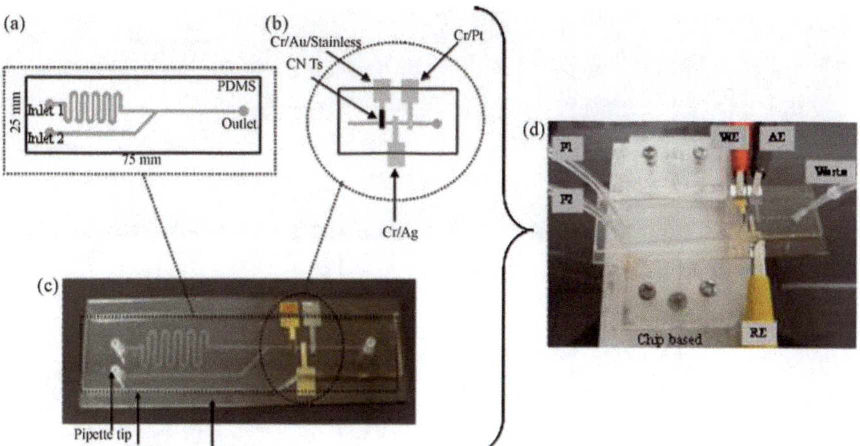

Fig. 12.3 Diagram and photographs of the carbon nanotubes based flow injection microfluidic device for rapid cholesterol detection [55]

5 Magneto-Resistance Based Detection

Despite of significant work in the field of biosensor, many biosensor detection technique fails to universally detect biomolecule in diverse clinical samples i.e. serum, urine, cell lysates etc. concisely, these techniques are not matrix insensitive. A major problem of these technologies is the signal distortion that occurs in various matrices due to heterogeneity in ionic strength, pH, and temperature and auto fluorescence etc. On the other hand, magneto-resistance is a promising technology that can be utilized to fabricate biosensor for the detection of biomolecules tagged with magnetic nanoparticles and it is matrix insensitive in nature [56]. The magneto resistive materials are multilayer thin-film structures that has quantum mechanical effect and it induces a change in electric resistance of the sensor [57]. Hall et al., reported a matrix-insensitive detection of protein assay in which an array of giant magnetoresistance (GMR) sensors was used to detect binding events of protein by immobilizing antibody molecule on the array of GMR sensor head and the protein molecules selectively bound to the receptor molecules and finally real time detection was performed by binding of magnetic nanoparticle tags to the protein (Fig. 12.4) [56].

Application of external magnetic field induced magnetization on the nanoparticles, and their presence or absence could be detected by underlying GMR sensor. In a recent report, GMR biosensor was used to detect influenza A virus using the novel magnetoresistance effect by engaging magnetic nanoparticles with monoclonal antibodies to viral nucleoprotein. This binding phenomenon of magnetic NPs caused changes in the resistance and the output changed in resistance real time through electrical devices and could detect as low as 1.5×10^{-2} TCID 50/ml virus [58]. GMR based sensor technology incorporated with a three-layer competition

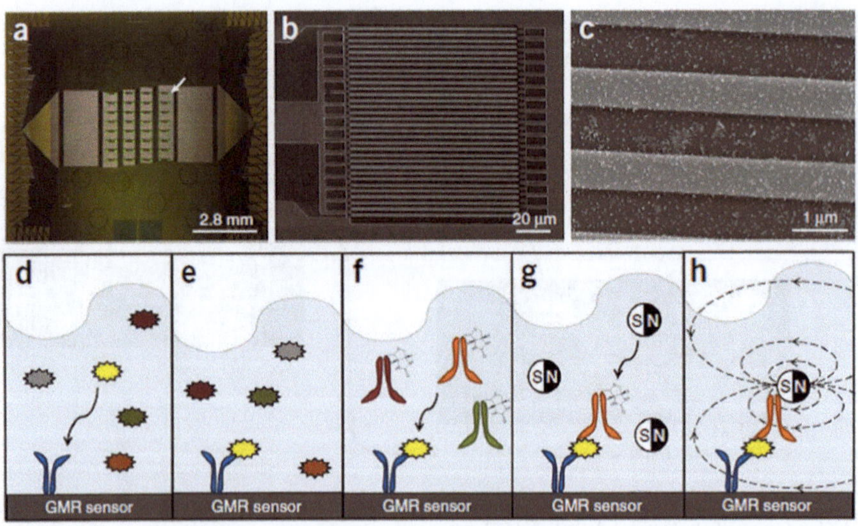

Fig. 12.4 Sensor architecture of magnetonanosensing chip for protein assay [56]

based assay was reported to detect endoglin from unprocessed human urine samples and could detect the endoglin concentration as low as 83 fM with sufficiently high selectivity [59].

6 Localized Surface Plasmon Resonance Based Detection

Localized surface plasmon resonance (LSPR) is an optical phenomenon that occur due to the interaction of electromagnetic wave with the conduction electrons in metal nanoparticles and yields an intense absorbance peak of visible range, and very sensitive to its minute change in the local refractive index of its surrounding. Metal nanoparticles show a strong absorption and scattering with applied electromagnetic field due to the collective oscillation of the conduction electrons located at the surface of the nanoparticles [60]. The resonance frequency and absorption bands depends on the type of the materials i.e. silver, gold, and platinum as well as on the shape and size distribution of the nanostructured material. Therefore, it gives a huge choice to selectively tune the sensitivity of the device by changing the shape, size, and material composition depending on the requirement and the types of applications [61]. Therefore, many researchers have been adapting this phenomenon to make biosensor to detect different biomolecules [62, 63]. Although, most of the nanoparticles exhibit LSPR, gold and silver NPs show higher LSPR due to exceptionally high absorption coefficients and scattering properties within the UV/visible wavelength range compared to conventional organic dyes [64]. One of the pioneering work, in which Englebienne observed that antibody coated AuNPs show longer SPR wavelength than that of the uncoated colloidal AuNPs due to the red shift in SPR wavelength which originated neither from agglutination nor aggregation, but was due to a change in the refractive index of the individual particles subsequent [65]. After that the same idea has been utilized to further explore the application to detect different biomolecules. Hence, it gained tremendous popularity among the researcher. He et al., conjugated oligonucleotide probes with colloidal AuNPs as amplification tags and used to selectively recognize surface-confined target DNA via sequence-specific hybridization and revealed >1000-fold improvement of sensitivity over the unamplified methods [66]. Yang et al. reported the enhanced sensitivity for detected DNA using catalytically grown AuNPs based LSPR biosensor [67]. Li and his co-worker developed a LSPR immunosensor using AuNPs to detect an ischemia modified albumin (IMA), a biomarker representing myocardial ischemia condition and resulted into high sensitivity thus providing an effective new approach for a direct assay of IMA [68]. Choi et al. developed a highly sensitive affinity based SPR based immunosensor using AuNPs attached with polyclonal antibody for the detection of prostate-specific antigen (PSA) in the blood and claimed 1000-fold increase in the sensitivity of PSA detection than existing biosensor for the detection of PSA [69]. In addition, some of the biomolecules detected by using SPR technique are human epididymis secretory protein 4 (HE4) biomarker [70], HIV-1 virus [71], Ischemia-Modified Albumin [72] etc.

7 Surface-Enhanced Raman Scattering Based Detection

When light photon inelastically scatter after hitting from a molecule due to gain or loss of energy from molecule's vibrational and rotation motion, then the scattering is called Raman scattering. This scattering generates a very narrow band of spectrum which is unique for each molecules or analyte. Studies showed that the Raman Scattering signal can be drastically enhanced by incorporation of metal particles, the enhanced scattering is known as surface-enhanced Raman scattering (SERS). Fleishmann and co-workers first demonstrated that molecules absorbed to the silver surface can significantly enhance the Raman signals from different types of molecules and subsequently the same phenomenon was observed on other metals also [73]. Otto et al. reported that the introduction of nanoparticles can enhance the scattering in the order of 10^5–10^6 times more than Raman signals [74]. There are several reports on the utilization of AuNPs using SERS for single molecule detection [75–78].

8 Fluorescence Based Detection

The phenomenon Förster resonance energy transfer (FRET) shows, when an incident light excites a donor fluorophore come very close to a noble metal NPs or a chromophore, it transfers the excitation energy, thus intensity of the donor fluorescence decreases significantly. FRET based biosensors show superior sensitivity due to exceptional quenching of fluorescent dyes facilitated by noble metal NPs [79]. FRET is extremely sensitive to small changes in the distance, since the efficiency of the energy transfer is inversely proportional to the sixth power of the distance between donor and acceptor. Mainly two approaches have been used: (1) molecular beacons that rely on NPs functionalized with fluorescent-labeled ssDNA strands that form a hairpin structure, and (2) noble metal nanoparticles functionalized with single strand nucleic acid strands and allowed to hybridize with another fluorescent-labeled single strand nucleic acid probe [60] (Fig. 12.5). When hairpin fluorescent

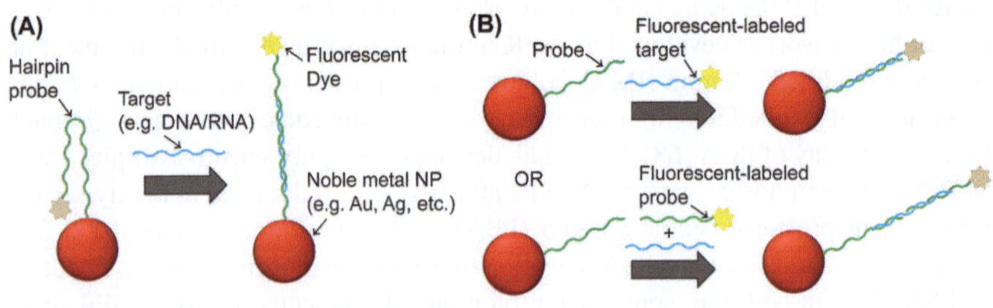

Fig. 12.5 Different approaches for fluorescent-based noble metal NPs biosensing. (**a**) molecular nanobeacons and (**b**) other nanoprobes [60]

molecular beacons come close to nanoparticles, the quenching of the beacon occurs due to nanoparticle surface energy transfer takes place between donor molecule and nanoparticle surface (acceptor) [80, 81].

The introduction of a complementary target DNA/RNA, the hairpin structure is unfolded by target hybridization and fluorescence is restored. Several methods are reported using this method to monitor specific nucleic acid hybridizations and cleavage processes induced by nucleases, mainly using gold nanoparticles [82]. AuNPs size ranging from 1.4 to 13 nm to detect single-base mismatches in DNA, proteins, mutation associated to cystic fibrosis with enhanced sensitivity compared to conventional molecular beacons [83–85]. Guirgis et al., proposed an immunoassay to diagnose malaria based on the fluorescence quenching of Cy3B-labeled recombinant *Plasmodium falciparum* heat shock protein 70 (PfHsp70) using AuNPs functionalized with an anti-Hsp70 monoclonal antibody [86]. Noble metal nanoparticles and semiconductor quantum dots as a FRET donor–acceptor couple has been used to develop fluorescence competition assays for nucleic acid, protein and antibody/antigen detection, by replacing the dye with quantum dots [87–89]. Multiplexing approach was also developed using more than one dye simultaneously by tracking the specific emission of each dye because of widespread quenching effectiveness across UV to visible spectra of the electromagnetic spectrum for DNA detection [90, 91], small analyte detection (adenosine, potassium and cocaine) using aptamers [92], and endonucleases monitoring [93].

9 Challenges and Future Prospects

One of the main challenges associated with biosensors for personalized care is the development of a simple user friendly and cost effective device, ideally suitable for faster and equipment-free sample testing. Despite the massive development of biosensor technologies for health care and a handful of testing devices to make commercial success, its adoption in clinics remains sluggish. Currently, the major challenge faced by the biosensor technologies that hampers its clinical success is often associated with its limited sensitivity and specificity while working with patient samples. Newer fabrication technologies viz, micro, nano-bio strategies seem to offer great platform for sensing, however their usefulness still needs to be demonstrated in clinics. Nano/micro fluidics are projected to be advantageous in designing new generation biosensing platforms with high sensitivity capable of sensing even at a very low concentration of analyte. Multiplexing will be an indispensable feature for future high performance diagnostics, as it simultaneously gives more information about a particular health condition from a single test. Although, over the few years R&D works have attempted to develop several proofs of concept multiplexing biosensor platforms, its efficiency and sensitivity is considered to be very low for clinical samples. Future diagnostics should do more focused research on novel biomarkers for development of effective personalized drugs to improve the patient non-compliance with a treatment. This could considerably reduce the overall expenses for health care as the patient receives a more streamlined medical care by eliminating unwanted tests and determination of appropriate drug dosage with minimal side effects.

10 Conclusions

As highlighted in the chapter, nano-enabled sensing platforms possess an overall advantage above routinely used techniques although, it needs further improvement to establish its effectiveness in a clinical scenario. Clinical access to the multiple established biomarkers via detection using high throughput multiplex devices could contribute to better information regarding a disease condition. Considering an increasing trend in future diagnostics towards personalized medicine, nanosensor technologies offer tremendous potential in terms of its high sensitivity, cost effectiveness and rapid multiplex detection ability for the development of point of care devices. Future personalized medicine aims to make use of such newer technologies for gauging the biological information (genetics, proteomic or metabolic) to foresee a new era of improved health care by making its practice more predictive and preventive. Through these emerging fields of nanotechnology approaches to diagnosis, drug development and ultimately personalized health care and point-of-care testing for real-world applicability are possible and must ultimately provide a cost-effective alternative to expensive and time-consuming laboratory tests.

Acknowledgments The authors acknowledge Department of Bioproducts and Biosystems Engineering, University of Minnesota, United States.

References

1. Arnold MA, Meyerhoff ME. Recent advances in the development and analytical applications of biosensing probes. Crit Rev Anal Chem. 1988;20(3):149–96.
2. Belkin S. Microbial whole-cell sensing systems of environmental pollutants. Curr Opin Microbiol. 2003;6:206–12.
3. Eggins BR. Chemical sensors and biosensors. Chichester: Wiley; 2002.
4. Wilson GS, Gifford R. Biosensors for real-time in vivo measurements. Biosens Bioelectron. 2005;20(12):2388–403.
5. Wilson JS. Sensor technology handbook. Amsterdam/Boston: Elsevier; 2005.
6. Conroy PJ, Hearty S, Leonard P, O'Kennedy RJ. Antibody production, design and use for biosensor-based applications. Semin Cell Dev Biol. 2009;20(1):10–26.
7. Singh R, Dhand C, Sumana G, Verma R, Sood S, Gupta RK, Malhotra BD. Polyaniline/carbon nanotubes platform for sexually transmitted disease detection. J Mol Recognit. 2010;23(5):472–9.
8. Malhotra BD, Chaubey A, Singh SP. Prospects of conducting polymers in biosensors. Anal Chim Acta. 2006;578(1):59–74.
9. Singh R, Mukherjee MD, Sumana D, Gupta RK, Sood S, Malhotra BD. Biosensors for pathogen detection: a smart approach towards clinical diagnosis. Sens Actuators B. 2014;197:385–404.

10. Belluzo MS, Ribone ME, Lagier CM. Assembling amperometric biosensors for clinical diagnostics. Sensors. 2008;8(3):1366–99.
11. Mozaza SR, López de Aldaa MJ, Marcob MP, Barcelóa D. Biosensors for environmental monitoring: a global perspective. Talanta. 2005;65(2):291–7.
12. Mulchandani A, Bassi AS. Principles and applications of biosensors for bio-process monitoring and control. Crit Rev Biotechnol. 1995;15(2):105–24.
13. Leonard P, Hearty S, Brennan J, Dunne L, Quinn J, Chakraborty T, O'Kennedy R. Advances in biosensors for detection of pathogens in food and water. Enzyme Microb Technol. 2003;32(1):3–13.
14. Li S, Simonian A, Chin BA, et al. Electrochem Soc Interface. 2010;19(4):41–6.
15. Sharmat A, Rogers RK. Meas Sci Technol. 1994;5:461.
16. Orazio PD. Biosensors in clinical chemistry. Clin Chim Acta. 2003;334(1-2):41–69.
17. Hunsperger EA, Yoksan S, Buchy P, et al. Evaluation of commercially available diagnostic tests for the detection of dengue virus NS1 antigen and anti-dengue virus IgM antibody. Morrison AC, ed. PLoS Negl Trop Dis. 2014;8(10):e3171.
18. Salez N, Vabret A, Leruez-Ville M, Andreoletti L, Carrat F, Renois F, et al. Evaluation of four commercial multiplex molecular tests for the diagnosis of acute respiratory infections. PLoS One. 2015;10(6):e0130378.
19. Huang L, Peng Z, Guo Y, Porter AL. Characterizing a technology development at the stage of early emerging applications: nanomaterial-enhanced biosensors. Technol Anal Strat Manag. 2011;23(5):527–44.
20. Bellan LM, Wu D, Langer RS. Current trends in nanobiosensor technology. Nanomed Nanobiotechnol. 2011;3(3):229–46.
21. Gooding JJ. Biosensor technology for detecting biological warfare agents: recent progress and future trends. Anal Chim Acta. 2006;559(2):137–51.
22. Syam R, Davis KJ, Pratheesh MD, Joseph BS. Biosensors: a novel approach for pathogen detection. VETSCAN. 2012;7(1):14–8.
23. Belluzo MS, Ribone ME, Lagier CM. Assembling amperometric biosensors for clinical diagnostics. Sensors. 2008;8(3):1366–99.
24. Kashefi-Kheyrabadi L, Mehrgardi MA, Wiechec E, Turner AP, Tiwari A. Ultrasensitive detection of human liver hepatocellular carcinoma cells using a label-free aptasensor. Anal Chem. 2014;86(10):4956–60.
25. Chandra P, Noh HB, Pallela R, Shim YB. Ultrasensitive detection of drug resistant cancer cells in biological matrixes using an amperometric nanobiosensor. Biosens Bioelectron. 2015;70:418–25.
26. Blind M, Blank M. Aptamer selection technology and recent advances. Mol Ther Nucleic Acids. 2015;4:e223.
27. Noh HB, Chandra P, Moon JO, Shim YB. In vivo detection of glutathione disulfide and oxidative stress monitoring using a biosensor. Biomaterials. 2012;33(9):2600–7.
28. Yadav SK, Agrawal B, Chandra P, Goyal RN. In vitro chloramphenicol detection in a Haemophilus influenza model using an aptamer-polymer. Biosens Bioelectron. 2014;15(55):337–42.
29. Godin M, Delgado FF, Son S, Grover WH, Bryan AK, Tzur A, Jorgensen P, Payer K, Grossman AD, Kirschner MW, et al. Using buoyant mass to measure the growth of single cells. NatMeth. 2010;7(5):387–90.
30. von Muhlen MG, Brault ND, Knudsen SM, Jiang S, Manalis SR. Label-free biomarker sensing in undiluted serum with suspended microchannel resonators. Anal Chem. 2010;82(5):1905–10.
31. Barton RA, Ilic B, Verbridge SS, Cipriany BR, Parpia JM, Craighead HG. Fabrication of a nanomechanical mass sensor containing a nanofluidic channel. Nano Lett. 2010;10(6):2058–63.
32. Kaushik A, Vabbina PK, Atluri V, Shah PK, Vashist A, Jayant RD, Yandart A, Nair M. Electrochemical monitoring-on-chip (E-MoC) of HIV-infection in presence of cocaine and therapeutics. Biosens Bioelectron. 2016;86:426–31.
33. Kaushik A, Tiwari S, Jayant RD, Vashist A, Nikkhah-Moshaie R, El-Hage N, Nair M. Electrochemical biosensors for early stage Zika diagnostics. Trends Biotechnol. 2016;35:308–17. doi:10.1016/j.tibtech.2016.10.001.

34. Kaushik A, Jayant RD, Tiwari S, Vashist A, Nair M. Nano-biosensors to detect beta-amyloid for Alzheimer's disease management. Biosens Bioelectron. 2016;80:273–87.
35. Kaushik A, Tiwari S, Jayant RD, Marty A, Nair M. Towards detection and diagnosis of Ebola virus disease at point-of-care. Biosens Bioelectron. 2016;75:254–72.
36. Kaushik A, Dixit CK. Nanobiotechnology for sensing applications: from lab to field, vol. 104. Oakville, ON: Apple Academic; 2016. Isbn:9781771883283., CAT# N11612.
37. Dixit CK, Kaushik A. Microfluidics for biologists—fundamentals and applications. Berlin: Springer; 2016.
38. Luppa PB, Muller C, Schlichtiger A. Point-of-care testing (POCT): current techniques and future perspectives. TrAC Trends Anal Chem. 2011;30(6):887–98.
39. Brock G, Castellanos-Rizaldos E, Hu L, Coticchia C, Skog J. Liquid biopsy for cancer screening, patient stratification and monitoring. Transl Cancer Res. 2015;4(3):280–90.
40. Girotti MR, Gremel G, Lee R, Galvani E, Rothwell D, Viros A, Mandal AK, et al. Application of sequencing, liquid biopsies, and patient-derived xenografts for personalized medicine in melanoma. Cancer Discov. 2016;6(3):286–99.
41. Sattler IN, Klaus D. History nanoparticles: Reiss, Gunter; Hutten, Andreas. "Magnetic nanoparticles". In: Handbook of nanophysics: nanoparticles and quantum dots. Boca Raton: CRC; 2010. p. 2–1. Isbn:9781420075458.
42. Singh R, Sharma A, Hong S, Jang J. Electrical immunosensor based on dielectrophoretically-deposited carbon nanotubes for detection of influenza virus H1N1. Analyst. 2014;139(21):5415–21.
43. Luo X, Morrin A, Killard A, Smyth M. Application of nanoparticles in electrochemical sensors and biosensors. Electroanalysis. 2006;18:319–26.
44. Grabar KC, Allison KJ, Baker BE, Bright RM, Brown KR, Freeman RG, Fox P, Keating CD, Musick MD, Natan MJ. Two-dimensional arrays of colloidal gold particles: a flexible approach to macroscopic metal surfaces. Langmuir. 1996;12:2353–61.
45. Xiao Y, Patolsky F, Katz E, Hainfeld JF, Willner I. Plugging into enzymes: nanowiring of redox enzymes by a gold nanoparticle. Science. 2003;299:1877–81.
46. Raj CR, Okajima T, Ohsaka T. Gold nanoparticle arrays for the voltammetric sensing of dopamine. J Electroanal Chem. 2003;543:127–33.
47. Singh R, Prasad R, Sumana G, Arora K, Sood S, Gupta RK, et al. STD sensor based on nucleic acid functionalized nanostructured polyaniline. Biosens Bioelectron. 2009;24(7):2232–8.
48. Singh R, Matharu Z, Srivastava AK, Sood S, Gupta RK, Malhotra BD. Nanostructured platform for the detection of *Neisseria gonorrhoeae* using electrochemical impedance spectroscopy and differential pulse voltammetry. Microchim Acta. 2012;177(1-2):201–10.
49. Singh R, Verma R, Sumana G, Srivastava AK, Sood S, Gupta RK, et al. Nanobiocomposite platform based on polyaniline-iron oxide-carbon nanotubes for bacterial detection. Bioelectrochemistry. 2012;86:30–7.
50. Singh R, Verma R, Kaushik A, Sumana G, Sood S, Gupta RK, et al. Chitosan-iron oxide nanocomposite platform for mismatch-discriminating DNA hybridization for detection of *Neisseria gonorrhoeae* causing sexually transmitted disease. Biosens Bioelectron. 2011;26(6):2967–74.
51. Hassen WM, Chaix C, Abdelghani A, Bessueille F, Leonard D, Jaffrezic-Renault N. An impedimetric DNA sensor based on functionalized magnetic nanoparticles for HIV and HBV detection. Sens Actuators B. 2008;134(2):755–60.
52. Pal S, Alocilja EC. Electrically active polyaniline coated magnetic (EAPM) nanoparticle as novel transducer in biosensor for detection of Bacillus anthracis spores in food samples. Biosens Bioelectron. 2009;24(5):1437–44.
53. Ansari AA, Singh R, Sumana G, Malhotra BD. Sol–gel derived nano-structured zinc oxide film for sexually transmitted disease sensor. Analyst. 2009;134(5):997–1002.
54. Tran LD, Nguyen BH, Hieu NV, Tran HV, Nguyen HL, Nguyen PX. Electrochemical detection of short HIV sequences on chitosan/Fe_3O_4 nanoparticle based screen printed electrodes. Mater Sci Eng C. 2011;31(2):477–85.
55. Wisitsoraat A, Sritongkham P, Karuwan C, Phokharatkul D, Maturos T, Tuantranont A. Fast cholesterol detection using flow injection microfluidic device with functionalized carbon nanotubes based electrochemical sensor. Biosens Bioelectron. 2010;26(4):1514–20.

56. Hall DA, Nielsen CH, Osterfeld SJ, Yu H, Mach KE, Wilson RJ, Murmann B, Liao JC, Gambhir SS, Wang SX, Gaster RS. Matrix-insensitive protein assays push the limits of biosensors in medicine. Nat Med. 2009;15:1327–32.
57. Barnas J, Fuss A, CamleyGrunberg REP, Zinn W. Novel magnetoresistance effect in layered magnetic structures: theory and experiment. Phys Rev B. 1990;42:8110–20.
58. Krishna VD, Wu K, Perez AM, Wang J-P, Giant VD. Magnetoresistance-based biosensor for detection of influenza a virus'. Front Microbiol. 2016;7:400.
59. Srinivasan B, Li Y, Jing Y, Xing C, Slaton J, Wang J-P. A three-layer competition-based giant magnetoresistive assay for direct quantification of endoglin from human urine. Anal Chem. 2011;83:2996–3002.
60. Doria G, Conde J, Veigas B, Giestas L, Almeida C, Assunção M, Rosa J, Baptista PV. Noble metal nanoparticles for biosensing applications. Sensors. 2012;12:1657–87.
61. Mayer KM, Lee S, Liao H, Rostro BC, Fuentes A, Scully PT, Nehl CL, Hafner JH. A label-free immunoassay based upon localized surface plasmon resonance of gold nanorods. ACS Nano. 2008;2(4):687–92.
62. Wilcoxon J. Optical absorption properties of dispersed gold and silver alloy nanoparticles. J Phys Chem B. 2009;113:2647–56.
63. Jain PK, Lee KS, El-Sayed IH, El-Sayed MA. Calculated absorption and scattering properties of gold nanoparticles of different size, shape, and composition: applications in biological imaging and biomedicine. J Phys Chem B. 2006;110:7238–48.
64. Barnes WL, Dereux A, Ebbesen TW. Surface plasmon subwavelength optics. Nature. 2003;424:824–30.
65. Englebienne P. Use of colloidal gold surface plasmon resonance peak shift to infer affinity constants from the interactions between protein antigens and antibodies specific for single or multiple epitopes. Analyst. 1998;123:1599–603.
66. He L, Musick MD, Nicewarner SR, Salinas FG, Benkovic SJ, Natan MJ, Keating CD. Colloidal Au-enhanced surface plasmon resonance for ultrasensitive detection of DNA hybridization. J Am Chem Soc. 2000;122:9071–7.
67. Yang X, Wang Q, Wang K, Tan W, Li H. Enhanced surface plasmon resonance with the modified catalytic growth of Au nanoparticles. Biosens Bioelectron. 2007;22:1106–10.
68. Li G, Li X, Yang M, Chen M-M, Chen L-C, Xiong X-L. A gold nanoparticles enhanced surface plasmon resonance immunosensor for highly sensitive detection of ischemiamodified albumin. Sensors. 2013;13(10):12794–803.
69. Choi JW, Kang DY, Jang YH, Kim HH, Min J, Oh BK. Ultra-sensitive surface plasmon resonance based immunosensor for prostate-specific antigen using gold nanoparticle-antibody complex. Colloids Surf A Physicochem Eng Asp. 2008;313–314:655–9.
70. Yuan J, Duan R, Yang H, Luo X, Xi M. Detection of serum human epididymis secretory protein 4 in patients with ovarian cancer using a label-free biosensor based on localized surface plasmon resonance. Int J Nanomedicine. 2012;7:2921–8.
71. Lee JH, Kim BC, Oh BK, Choi JW. Highly sensitive localized surface plasmon resonance immunosensor for labelfree detection of HIV-1. Nanomedicine. 2013;9(7):1018–26.
72. Li G, Li X, Yang M, Chen MM, Chen LC, Xiong XL. A gold nanoparticles enhanced surface plasmon resonance immunosensor for highly sensitive detection of ischemia-modified albumin. Sensors. 2013;13:12794–803.
73. Fleishmann M, Hendra PJ, Mcquillan AJ. Chem Phys Lett. 1974;26:163.
74. Otto A, Mrozek I, Grabhorn H, Akemann W. Surface-enhanced Raman scattering. J Phys Condens Matter. 1992;4:1143–212.
75. Nie SM, Emory SR. Probing single molecules and single nanoparticles by surface-enhanced Raman scattering. Science. 1997;275(5303):1102–6.
76. Kneipp K, Wang Y, Kneipp H, Perelman LT, Itzkan I, Dasari R, Feld MS. Single molecule detection using surface enhanced Raman scattering (SERS). Phys Rev Lett. 1997;78(9):1667–70.
77. Krug JT, Wang GD, Emory SR, Nie SM. Efficient Raman enhancement and intermittent light emission observed in single gold nanocrystals. J Am Chem Soc. 1999;121(39):9208–14.

78. Emory SR, Nie S. Screening and enrichment of metal nanoparticles with novel optical properties. J Phys Chem B. 1998;102(3):493–7.
79. Harris DC. Applications of spectrophotometry. In: Quantitative Chemical Analysis. 8th ed. New York: W. H. Freeman and Co; 2010. p. 419–44.
80. Jennings TL, Singh MP, Strouse GF. Fluorescent lifetime quenching near d = 1.5 nm gold nanoparticles: probing NSET validity. J Am Chem Soc. 2006;128:5462–7.
81. Griffin J, Singh AK, Senapati D, Rhodes P, Mitchell K, Robinson B, Yu E, Ray PC. Size- and distance-dependent nanoparticle surface-energy transfer (NSET) method for selective sensing of hepatitis C virus RNA. Chemistry. 2009;15:342–51.
82. Ray PC, Darbha GK, Ray A, Walker J, Hardy W. Gold nanoparticle based FRET for DNA detection. Plasmonics. 2007;2:173–83.
83. Dubertret B, Calame M, Libchaber AJ. Single-mismatch detection using gold-quenched fluorescent oligonucleotides. Nat Biotechnol. 2001;19:365–70.
84. Benia V, Hayesa K, Lerga MT, O'Sullivan CK. Development of a gold nano-particle-based fluorescent molecular beacon for detection of cystic fibrosis associated mutation. Biosens Bioelectron. 2010;26:307–13.
85. Wang W, Chen C, Qian M, Zhao XS. Aptamer biosensor for protein detection using gold nanoparticles. Anal Biochem. 2008;373:213–9.
86. Guirgis BS, Cunha CS, Gomes I, Cavadas M, Silva I, Doria G, Blatch GL, Baptista PV, Pereira E, Azzazy HM, Mota MM, Prudêncio M, Franco R. Gold nanoparticle-based fluorescence immunoassay for malaria antigen detection. Anal Bioanal Chem. 2012;402(3):1019–27.
87. Dyadyusha L, Yin H, Jaiswal S, Brown T, Baumberg JJ, Booye FP, Melvin T. Quenching of CdSe quantum dot emission, a new approach for biosensing. Chem Commun. 2005;(25):3201–3.
88. Oh E, Hong MY, Lee D, Nam SH, Yoon HC, Kim HS. Inhibition assay of biomolecules based on fluorescence resonance energy transfer (FRET) between quantum dots and gold nanoparticles. J Am Chem Soc. 2005;127:3270–1.
89. Chang E, Miller JS, Sun J, Yu WW, Colvin VL, Drezek R, West JL. Protease-activated quantum dot probes. Biochem Biophys Res Commun. 2005;334:1317–21.
90. Song S, Liang Z, Zhang J, Wang L, Li G, Fan C. Gold-nanoparticle-based multicolor nanobeacons for sequence-specific DNA analysis. Angew Chem Int Ed. 2009;48:8670–4.
91. Zhang Y, Wang L, Tian J, Li H, Luo Y, Sun X. Ag@poly(m-phenylenediamine) core-shell nanoparticles for highly selective, multiplex nucleic acid detection. Langmuir. 2011;27:2170–5.
92. Zhang J, Wang L, Zang H, Boey F, Song S, Fan C. Aptamer-based multicolor fluorescent gold nanoprobes for multiplex detection in homogeneous solution. Small. 2010;6:201–4.
93. Huang Y, Zhao S, Liang H, Chen ZF, Liu YM. Multiplex detection of endonucleases by using a multicolor gold nanobeacon. Chem A Eur J. 2011;17:7313–9.

Chapter 13
Biosensing Devices for Personalized Healthcare

Zimple Matharu and Ozge Polat

Abstract Application of novel biosensor platforms has acquired paramount importance in the field of point-of-care diagnostics, environmental monitoring and drug delivery. The recent biosensor device development using submicron-sized dimensions has opened new horizons for detection of clinical analytes as well as signaling molecules secreted from living cells. Taking advantage of the unique properties of the micro/nano fabrication technology integrated with electrochemical techniques, faster and sensitive biosensors can be developed. The present chapter discusses the basic concepts of electrochemical biosensors, recent developments in the biosensor industry and importance of biosensors in personalized drug delivery.

Keywords Electrochemical biosensors • Microfabrication • Biological fluids • Drug delivery

1 Introduction

The capability to identify diseases and pathogens via detection of associated proteins, nucleic acids, or other biomarkers can provide biomedical researchers and healthcare professionals with highly detailed knowledge of patients' conditions, disease pathways, or the presence of infection. However, many of the tests commercially available and universally used such as ELISA, PCR and other bioassays have disadvantages of being time consuming and requiring large amounts of sample materials. The results of these problems range from misdiagnosis of disease, inability to timely detect lethal pathogens, and difficulty in understanding the development and progression of the disease. Thus, there is a strong push towards developing improved diagnostic technologies that would allow for rapid, trustworthy, low-cost and multiplexed screening to detect a wide range of biomolecules. To this end,

Z. Matharu (✉) • O. Polat
Department of Electrical and Computer Engineering, University of California–Davis, Davis, California, USA
e-mail: zmatharu@ucdavis.edu

A. Kaushik et al. (eds.), *Advances in Personalized Nanotherapeutics*,
https://doi.org/10.1007/978-3-319-63633-7_13

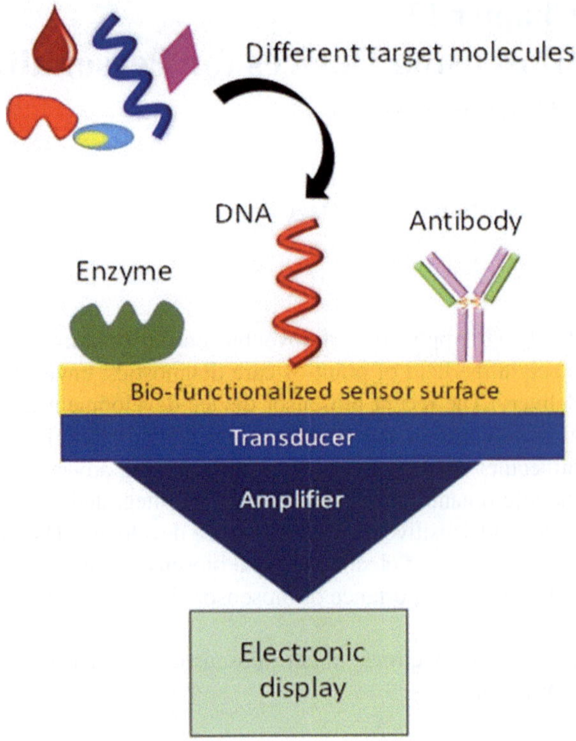

Fig. 13.1 Schematic of biosensor

chemical sensors and biosensors have been widely used as attractive alternatives to the bulky, expensive and complex analytical instruments being used in the healthcare sector [1, 2]. A biosensor is a biomolecular electronic device that translates a biological response into an equivalent recordable signal. This device utilizes sensing biomolecules (enzymes, nucleic acid, antibody, virons), support matrix for biomolecule functionalization (planar metal electrodes, carbon electrodes, nanostructured metal electrodes, metal nanoparticles, organic/inorganic nanoparticles) and signal transducers (electrical, optical, mechanical, magnetic) (Fig. 13.1).

In order to overcome the issues associated with current convention diagnostic techniques, a wide range of new biosensors are being developed. Several of these biosensors rely on nanotechnological platforms [3, 4]. The development of analytical tools and processes utilized to fabricate, measure, and image nanoscale objects has paved a path for fabrication of nanosensors which can interact with extremely small number (or an extremely small concentration) of analyte molecules. These advancements are particularly exciting in the field of point-of-care biosensing, where the demands for low concentration detection with high specificity are enormous. Nanoscale biosensors provide researchers with an unparalleled level of sensitivity and detection, often to the single molecule level. In addition, nanostructuring of the sensor surface have enabled significant enhancement in sensor performance surpassing what has been possible with smooth electrode surfaces [5, 6]. These enhancements include increased surface coverage and favorable orientation of biomolecules, and higher electro-catalytic activity at the electrode surface (in case of electrochemical detection systems) [7]. However, after so many advantages proposed by nanobiosensor platforms, biocompatibility and biofouling (specificity/selectivity in complex media) are the two key issues hindering the progress of these sensor for diseases detection in clinical samples. The non-specific accumulation of biological components such as proteins, lipids, and polysaccharides on the sensor surface is referred to as biofouling [8, 9]. Biofouling is disadvantageous to the sensor performance as it impedes the transport of target analyte onto the electrode surface and hinders the reaction between redox moieties and the electrode. This has been a persistent challenge particularly for electrochemical biosensing devices and has motivated the development of several methods to sustain sensor performance in biofouling conditions. To this end, the fouling of sensor can be addresses by controlling the nanostructure geometry which can act as a bio-filter and only allow analyte of interest to pass through it. As nanobiosensor technology becomes more refined and reliable, it is likely that it will eventually make its way from the lab to the clinic, where future lab-on-a-chip devices incorporating arrays of nanobiosensors could be used for rapid screening of a wide variety of analytes at low-cost using small patient samples. The current biosensor market can be broadly divided in two major branches of sensing platforms (1) electrochemical and (2) optical. However, major thrust is shifting towards the use of electrochemical techniques because of their low-cost, portability and multiplexing capabilities. This chapter gives a brief overview of the progress in the field of portable electrochemical biosensing devices, their working principles and further connection of sensors to personalized drug delivery.

2 Electrochemical Methods for Disease Detection

Electrochemical biosensors interface biochemical recognition events (e.g., enzymatic reactions, DNA hybridization, antibody/antigen binding) at molecular level and the corresponding conversion of biochemical event into an electronic signal (Fig. 13.2) in the form of current and voltage [10]. Electrochemical sensors have gained a dominating role in clinical diagnostics owing to their high performance, portability, simplicity, and low-cost [10]. The major benefit of electroanalytical techniques over other detection methods (e.g,. optical, magnetic, and mechanical) is their multiplexing via seamless connectivity with signal processing electronics, without sacrificing the sensor performance [11]. Amperometric, potentiometric and voltammetric techniques are the most widely applied methods in electrochemical analysis and have been utilized for qualitative and quantitative biosensing. Each technique has its own strengths and weaknesses in terms of sensor performance characteristics, however, all of them depend strongly on the electrode, electrolyte, and biomolecule properties. Particularly, the size and morphology of the sensing electrode and the fabrication method used can be very influential on the voltammetric response of the system. The electrolyte containing the redox marker, buffer

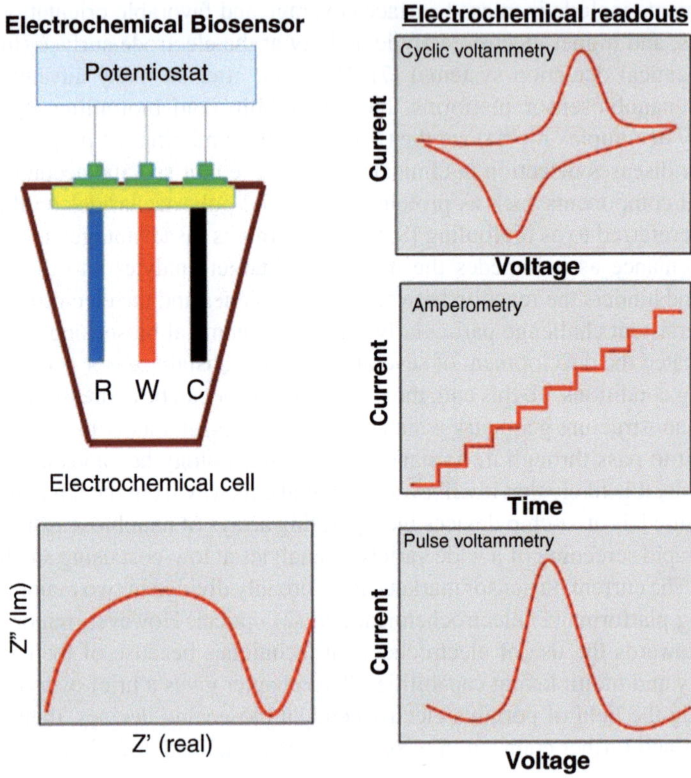

Fig. 13.2 Electrochemical cell-setup and possible signal readouts

constituents and supporting salts, has an important effect on the sensor performance. Many redox markers or mediators (MED) have been utilized for signal amplification in case of enzymatic biosensors and hybridization indicators in case of DNA biosensors due to their fast electron transfer properties [10]. The first generation of enzymatic biosensors did not rely on the use of external redox markers and thus suffered problem due to non-specific interfering species (Fig. 13.3a). The added redox markers in the second generation of enzymatic biosensors showed faster electron transfer and allowed low potential device operations (Fig. 13.3b). However, they indeed added extra steps to the sensor preparation process. This issue was addressed in the third generation of enzymatic biosensors where the electrode used for biomolecule immobilization itself shows excellent redox properties and provided a way to achieve direct electron transfer from biomolecule to the electrode (Fig. 13.3c). Most of the current commercial electrochemical biosensor belong to either second or third generation schemes. $K_4[Fe(CN)_6]$ and $Ru(NH_3)_6$ are the most widely used redox markers to study enzyme reactions [12–14], while, methylene blue (MB), has been extensively used as DNA hybridization indicator [15].

The electrochemical reactions of these redox markers at nanostructured electrodes are also dictated by their reaction rates and transport properties. For example,

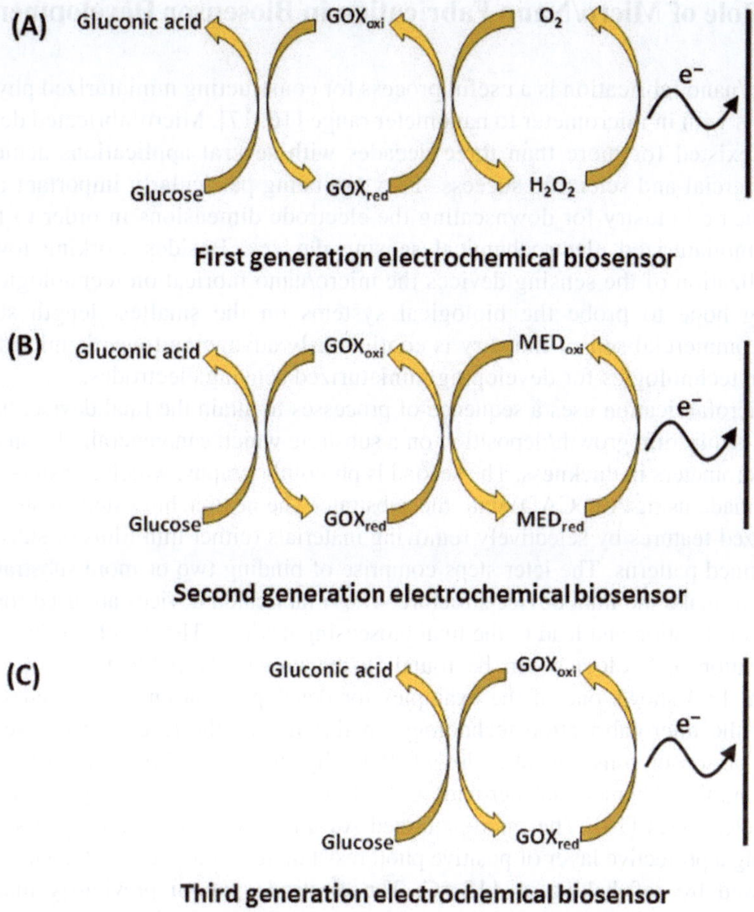

Fig. 13.3 Various mechanisms of electrochemical enzyme biosensors

some redox markers such as MB undergo sluggish reactions with the electrode surface; thereby they can travel further into high aspect ratio structures (such as nanopores, nano-wells) without losing their activity by fast reactions at the electrode surface [5]. On the other hand, other redox markers for example $K_4[Fe(CN)_6]$ and $Ru(NH_3)_6$ react rapidly with the electrode surface because of their high reaction kinetics and therefore lose their activity on the surface without accessing the entire surface areas in case of nanostructured electrodes [12]. The former examples of redox markers are reaction-limited while the latter are diffusion-limited. Thus, the right selection of the redox markers is important for the fabrication of sensitive electrochemical sensor. Taken together, nanostructured electrochemical sensors in combination with biorecognition elements and accompanying selection of redox markers offer a rich design space for developing tunable sensor platforms for disease detection.

3 Role of Micro/Nano Fabrication in Biosensor Development

Micro/nano fabrication is a useful process for constructing miniaturized physical objects with in micrometer to nanometer range [16, 17]. Microfabricated devices have existed for more than three decades with several applications achieving commercial and scientific success. They are being particularly important in the healthcare industry for downscaling the electrode dimensions in order to fabricate miniaturized electrochemical sensing devices. Besides working towards mobilization of the sensing devices the micro/nano fabrication technologies are giving hope to probe the biological systems on the smallest length scales. The commercial sensor industry is continuously advancing towards microfabrication technologies for developing miniaturized sensing electrodes.

Microfabrication uses a sequence of processes to attain the final device, first of them is thin-film growth/deposition on a substrate which can generally be an order of micrometers in thickness. The second is photolithography, which transfers a pattern (made using AutoCAD) onto the substrate. The next etching step creates miniaturized features by selectively removing materials (either thin films or substrate) in defined patterns. The later steps comprise of binding two or more substrates in order to make the final device structure. These fabricated devices are used for biofunctionalization and lead to the final biosensing devices. The details of the microfabrication technologies can be found in the previously published books [18]. Figure 13.4 shows one of the examples for developing an enzyme based sensor using the microfabrication technology. In this report, the researchers developed H_2O_2 biosensor consisted of hydrogel (Polyethylene glycol, PEG) microstructures with entrapped horseradish peroxidase (HRP) immobilized on an array of miniature gold electrodes [19]. The micropatterned Au arrays were prepared by first spin-coating a protective layer of positive photoresist on top of the Au-coated glass slide followed by soft-baking at 115 °C. The electrode pattern previously made in AutoCAD was formed on the slides using a mask aligner. During the UV exposure process the exposed regions of photoresist in the mask gets dissolved in the developer solution and leaves the underneath gold region unprotected which can be removed by wet metal etching process and results in the final array pattern. In this way the photolithography can be used to pattern largely varying geometries and dimensions on a single chip.

The microfabrication technology confers various advantages on the biomedical industry as it provides scalability to the devices by reducing the system size for point-of-care use. One such example is microfabricated flow cytometers [20–22], which employ microfabrication techniques to make miniature flow chambers. Although these devices cannot compete with the power and versatility of conventional flow cytometers, their portability may be useful for point-of-care hematological tests. Additionally, the close relationship between microfabrication and conventional semiconductor fabrication allows one to integrate electronics or electrical components with microfabricated biochips. The microfabricated biochips also offer detection of various analytes simultaneously and thus are useful for multiplexed

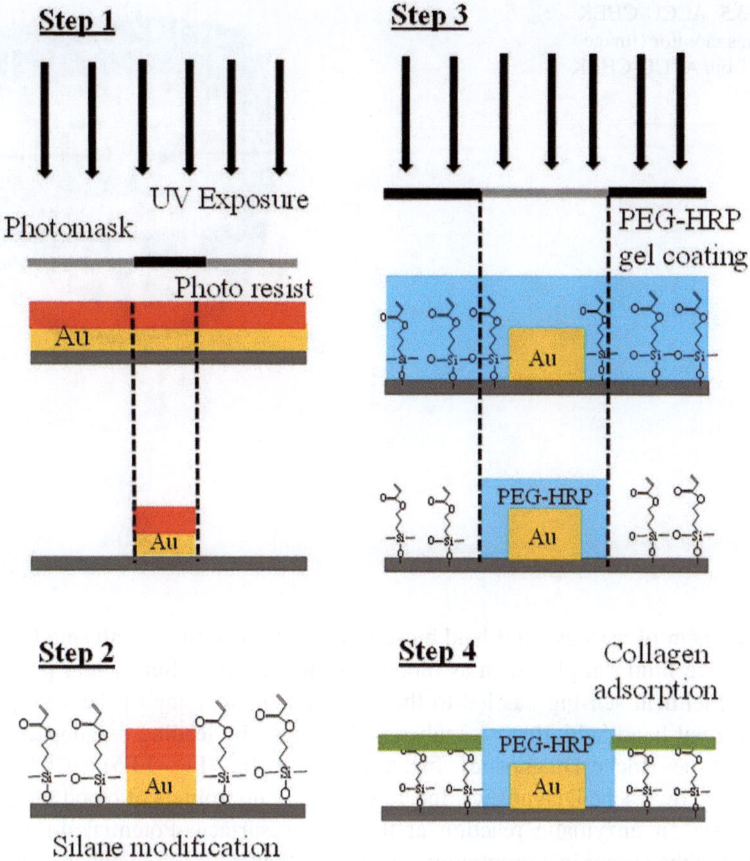

Fig. 13.4 Fabrication of Au patterned electrodes using photolithography and deposition of enzyme hydrogels for biosensing (Figure taken from Matharu et al. Anal Chem 2013, 85, 932–93) [19]

detection. The microfabrication and microfluidics are also giving a new direction and level to single-cell analysis which is essential for fatal diseases like cancer [23].

4 Commercially Available Electrochemical Biosensing Devices

4.1 Detection from Invasive Fluids

Electrochemical methods have played a pivotal role in detecting the changes that occur during a biorecognition event (e.g. enzymatic reaction, DNA hybridization) and the merging of microfabrication with electrochemical detection has enabled

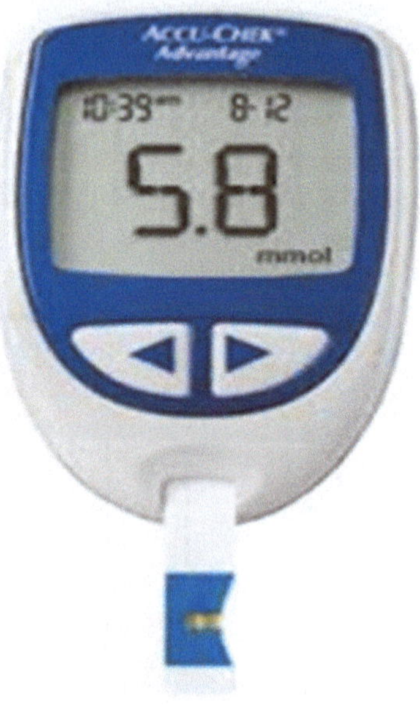

Fig. 13.5 ACCU-CHEK diabetes monitor (Image taken from ACCU-CHEK website)

development of various hand-held biosensor devices operating with small amounts of invasive fluid samples such as blood, serum and urine. Substantial progress in electrochemical sensing has led to the development of some of the very popular commercial hand-held glucose analyzers developed by leading diagnostic companies such as Roche Diagnostics, Inc. and Abbott, Inc (Fig. 13.5). ACCU-CHEK developed by Roche Diagnostics, Inc. relies on few microliters of blood sample that undergoes an enzymatic reaction at the sensor surface. Potential difference is applied to the sensor in a programmed sequence that results in biamperometric current data [24].

Abbott, Inc. has developed the world's first hand-held device (i-STAT) for point-of-care clinical assay of blood analysis employing several electrochemical-based transduction methods (i.e., potentiometric, amperometric, conductometric) (Fig. 13.6). The i-STAT Portable Clinical Analyser™ is a hand-held silicon-based multiple-analyte sensor array which is used to monitor various blood electrolytes (i.e., sodium, potassium, chloride, calcium, pH), gases (i.e., carbon dioxide, oxygen) and molecules (i.e., urea, glucose, hematocrit) [25]. The individual test for blood gases, electrolytes, creatinine and urea are performed using a miniaturized version of traditional electrode technology contained within a single use test cartridge. To switch from one test group to another, one has to simply change to a different cartridge type. Cardiac troponin I and brain natriuretic peptide tests utilize a sandwich type immunoassay with electrochemical detection. Coagulation tests uses a recombinant human thromboplastin with electrochemical detection of the clotting. Measurement of hematocrit (the ratio of the volume of red blood cells to the total volume of blood) is based on plasma conductivity; the electrical resistance of a whole blood sample is proportional to its hematocrit. Further the hemoglobin

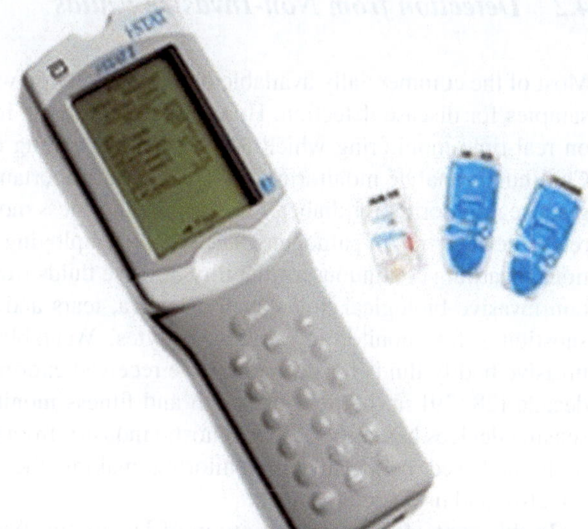

Fig. 13.6 i-STAT handheld analyzer (Image taken from Abbott website)

concentration is calculated from the hematocrit using a standard equation. Likewise, SenDx Medical Inc. (acquired by Radiometer) has developed a compact and portable potentiometric sensor array for determining various ions in the blood.

Oxford Biosensors have also developed a portable hand-held device (Multisense™) for cholesterol quantification. The biosensor consists of disposable test strips (microelectrodes) and utilizes the electrochemical detection strategy. Chemel (Lund, Sweden) have developed a portable biosensor (SIRE biosensor) that is based on enzymatic/amperometric measuring principles. This technology has been used to measure various sugars and alcohols. Sensor Tech. Ltd. (Cambridge, UK) developed an immunosensor (Universal Transducer System) for in vitro diagnostic and biosensor applications. The Universal Transducer System (UTS™) employs a potentiometric detection platform. The multi-array biosensor was fabricated using screen-printing technology based on glassy carbon powder containing tyrosinase and transduction chemistry [26]. The University of Ulster has developed a biosensor for the determination of flavanols using either plant tissue material (polyphenoloxidases or commercial tyrosinase) [27] immobilized on either a carbon paste electrode or screen-printed electrode with modified polypyrrole.

Various organizations have directed their efforts towards developing DNA biochips. For example, GeneOhm Sciences have developed a DNA chip that employs the electrochemical detection platform. Established in San Diego, California, in 2001, GeneOhm Sciences is a company that focuses on molecular diagnostics for a wide range of diseases. Likewise, Motorola Life Sciences Inc. is another organization that has produced an electrochemical-based DNA chip. A number of small companies appear to be making some progress in the development of various handheld devices.

4.2 Detection from Non-Invasive Fluids

Most of the commercially available biosensors rely on whole blood, urine or serum samples for disease detection. However, reliability on invasive fluids pose hurdles on real-time monitoring which is essential for finding out the onset of a disease. Continuous analyte monitoring is of particular importance in areas such as regular glucose monitoring for diabetes management, fitness monitoring, drug efficacy and real-time detection of pathogens. The sensors employing invasive fluids have limitations regarding continuous availability of these fluids from the patients. To this end, non-invasive biological fluids such as saliva, tears and sweat can serve as useful substitutes for monitoring various analytes. Wearable sensors employing non-invasive bodily fluids for detection have received enormous attention over the past decade [28, 29] in the field of health and fitness monitoring. These miniaturized sensing devices has shifted the diagnostic industry from hospital-based patient care to home-based personalized monitoring making the field of diagnostics cost-effective and user friendly.

In this context, the research group of Dr. Joseph Wang have taken the electrochemical biosensors to a whole new level by using non-invasive biological fluids for detection of several analytes of clinical interest. Recently, Wang and co-workers reported a wearable non-invasive mouthguard amperometric biosensor for detecting lactate [30] (Fig. 13.7). This was acknowledged as the first example of a wearable electrochemical salivary sensor for continuous metabolite sensing. The sensor was fabricated on polyethylene terephthalate (PET) substrate and subsequently affixed to a mouthguard for continuous saliva lactate detection in undiluted human saliva samples. The group is continuously working on non-invasive monitoring of salivary lactate with integrated electronic backbone and wireless transmitter within the mouthguard.

In an another report, Parviz's and co-workers have demonstrated biofunctionalization of a PET-based contact lens with glucose oxidase enzyme (GOx) within a Titania sol–gel matrix for continuous glucose monitoring [31].They further developed an amperometric glucose sensors in the form of contact lens with in-built wireless electronics for continuous data transmission [32, 33]. The authors later modified their wireless sensor to incorporate a dual sensor consisting of activated and deactivated GOx for minimizing interference effect [33].The authors studied the effects of ageing, temperature, and biofouling on the sensor response and demonstrated excellent sensitivity, detection limit, and linear range using a polymer eye model.

NovioSense glucose sensor (NovioSense BV) is another potential commercial product being developed in the Netherlands. It consists of a number of parallel coiled wires forming a micro-electrochemical cell with immobilized GOx. The coiled device can be easily placed within the eyelid for continuous wireless glucose sensing. Efforts have also been made in developing tear-based sensors for detecting lactate as an indicator of hypoxic conditions.

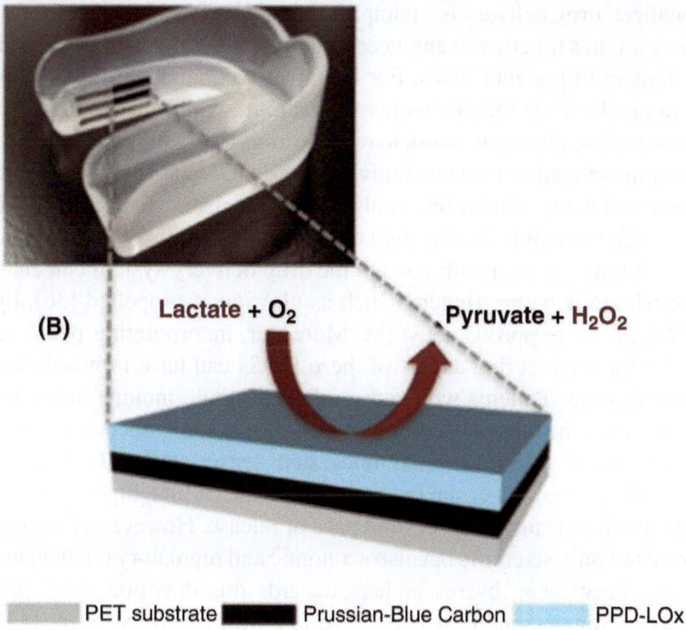

Fig. 13.7 Mouth guard sensor and its sensing mechanism (Figure taken from Kim et al. Analyst, 139, 1632–1639) [30]

5 Importance of Sensing Devices in the Field of Personalized Drug Delivery

Advances in molecular biology have altered the pace of research on the therapeutic development to increase the efficiency of the treatment by the concept of personalized medicine [34, 35]. While there is an increasing trend for personalized medicine, the role of personalized drug delivery systems (DDSs) have not been explored to its full potential. The development of new pharmaceutics for personalized therapy has been increasing rapidly, yet the optimum outcome for the patient requires testing the medicine via a drug delivery system tailored for each patient. Patient-tailored drug delivery becomes more challenging with an increased number of complexities, however; it may improve the overall efficacy of the treatment. The overarching goal of a drug delivery system is to improve the efficacy by releasing the needed amount of pharmaceuticals at a specific rate, time and location over a long period of time while maintaining the therapeutic dose. This can only be achieved by real-time monitoring of the diseases which will further decide the amount of pharmaceutics for a particular patient. To this end, the integration of the drug delivery systems with biosensing devices can be highly advantageous. The incorporation of sensors to track the delivered drug dose volume and flow rate and report to the user in real-time may offer active control over the delivery profile.

Personalized drug delivery is a complex therapy and requires incorporating interdisciplinary studies together to engineer smart multifunctional materials tailored for a bigger impact in patients' lives. For example, a conductive implant surface can function as an electrode for electrochemical studies, which can be used as a biosensor and also as drug eluting network to release drugs iontophoretically. Incorporating these modalities together into one individualized DDS can enable patient to get the best accuracy in the treatment since only one device itself can detect what is needed and deliver that amount in a controlled manner with a feedback loop. Or integrating materials with imaging capabilities into the drug delivery system can enable imaging while release is being triggered such as ultrasound-propelled [36], light-gated [37, 38] release from porous networks. Moreover, incorporating pharmacogenetic factors with the engineering aspect of these DDSs can have tremendous advancement in the designs. Patients with cancer, diabetes, autoimmune disorders, cardiovascular diseases, infectious diseases or neurological disorders can get immense benefit from these multifunctional integrated sensor and tailored drug eluting devices with improved target site recognition, enhanced imaging during the delivery of multiple loads of drugs at the desired pace of release. However, all these improvements bring not only scientific but also economic and regulatory challenges together. Efforts have been taken by researchers towards this direction show promise for future devices for personalized healthcare [39–41]. Sheybani et al. [42] demonstrated a fully integrated implantable electrolysis-based micropump with incorporated electrochemical impedance dosing sensors. This system provided successful delivery, infusion rate control, and dose sensing in simulated brain tissue.

As the degree of personalization increases, development, production, storage and transportation of the system will change and be reflected in the price. Depending on the choice of DDS, a medical procedure may be required for the placement or additional procedures depending on the depletion rate of the drug. From patient's perspective, the health and economic outcome will be positive in the long run, however it will create a specific and small market for a certain type of devices, which can be a significant barrier to overcome in the industry. Nevertheless, combining both sensor and drug delivery system in one device can reduce the production costs. However, the interdisciplinary approach in today's fast-paced technology and growing interest in personalized medicine will pave the way to the development of novel individualized drug delivery systems.

6 Conclusions

The progress made in commercialization of electrochemical biosensors for diabetes, pathogens and several other important biomarkers shows significant promise for future biosensor development specially for early and sensitive detection of life threating diseases such as cancer and HIV. Although, the current available sensors mostly utilize invasive body fluids for disease detection, the on-going research by several pioneer groups are showing potential for designing biosensors detecting target molecules in non-invasive biological fluids such as sweat and tears. The next steps in this field would be fabrication of robust sensing platforms that can show detection in smallest possible quantities of the disease related target molecules in biological samples. This can only be achieved by sensor chips on which isolation, amplification and detection of the target can be performed simultaneously. These tools further integrated with the personalized drug delivery systems can be a boon for healthcare industry.

Acknowledgements The authors acknowledge financial support from University of California-Davis, Research Investments in the Sciences & Engineering, and National Science Foundation (Awards CBET-1512745 and CBET&DMR-1454426).

References

1. Banica F-G. Chemical sensors and biosensors: fundamentals and applications. Chichester: Wiley; 2012.
2. Malhotra BD, Chaubey A. Biosensors for clinical diagnostics industry. Sens Actuators B. 2003;91:117–27.
3. Bellan LM, Wu D, Langer RS. Current trends in nanobiosensor technology. Wiley Interdiscip Rev Nanomed Nanobiotechnol. 2011;3:229–46.
4. Jain KK. Applications of nanobiotechnology in clinical diagnostics. Clin Chem. 2007;53:2002–9.
5. Daggumati P, Matharu Z, Seker E. Effect of nanoporous gold thin film morphology on electrochemical DNA sensing. Anal Chem. 2015;87:8149–56.
6. Gasparac R, Taft BJ, Lapierre-Devlin MA, Lazareck AD, Xu JM, Kelley SO. Ultrasensitive electrocatalytic DNA detection at two-and three-dimensional nanoelectrodes. J Am Chem Soc. 2004;126:12270–1.
7. Lapierre-Devlin MA, Asher CL, Taft BJ, Gasparac R, Roberts MA, Kelley SO. Amplified electrocatalysis at DNA-modified nanowires. Nano Lett. 2005;5:1051–5.
8. Barfidokht A, Gooding JJ. Approaches toward allowing electroanalytical devices to be used in biological fluids. Electroanalysis. 2014;26:1182–96.
9. Shin DS, Matharu Z, You J, Siltanen C, Vu T, Raghunathan VK, Stybayeva G, Hill AE, Revzin A. Sensing conductive hydrogels for rapid detection of cytokines in blood. Adv Healthc Mater. 2016;
10. Wang J. Electrochemical biosensors: towards point-of-care cancer diagnostics. Biosens Bioelectron. 1887-1892;2006:21.
11. Ronkainen NJ, Halsall HB, Heineman WR. Electrochemical biosensors. Chem Soc Rev. 2010;39:1747–63.
12. Anandan V, Yang X, Kim E, Rao YL, Zhang G. Role of reaction kinetics and mass transport in glucose sensing with nanopillar array electrodes. J Biol Eng. 2007;1:5–5.
13. Chaubey A, Malhotra BD. Mediated biosensors. Biosens Bioelectron. 2002;17:441–56.
14. Cui G, Yoo JH, Woo BW, Kim SS, Cha GS, Nam H. Disposable amperometric glucose sensor electrode with enzyme-immobilized nitrocellulose strip. Talanta. 2001;54:1105–11.
15. Kelley SO, Boon EM, Barton JK, Jackson NM, Hill MG. Single-base mismatch detection based on charge transduction through DNA. Nucleic Acids Res. 1999;27:4830–7.
16. Singh PS. From sensors to systems: CMOS-integrated electrochemical biosensors. IEEE Access. 2015;3:249–59.
17. Voldman J, Gray ML, Schmidt MA. Microfabrication in biology and medicine. Annu Rev Biomed Eng. 1999;1:401–25.
18. Franssila S. Introduction to microfabrication. Chichester: Wiley; 2010.
19. Matharu Z, Enomoto J, Revzin A. Miniature enzyme-based electrodes for detection of hydrogen peroxide release from alcohol-injured hepatocytes. Anal Chem. 2012;85:932–9.

20. Sung-Yi Y, Suz-Kai H, Yung-Ching H, Chen-Min C, Teh-Lu L, Gwo-Bin L. A cell counting/sorting system incorporated with a microfabricated flow cytometer chip. Meas Sci Technol. 2001;2006:17.
21. Voldman J, Gray ML, Toner M, Schmidt MA. A microfabrication-based dynamic array cytometer. Anal Chem. 2002;74:3984–90.
22. Chung TD, Kim HC. Recent advances in miniaturized microfluidic flow cytometry for clinical use. Electrophoresis. 2007;28:4511–20.
23. Son KJ, Rahimian A, Shin D-S, Siltanen C, Patel T, Revzin A. Microfluidic compartments with sensing microbeads for dynamic monitoring of cytokine and exosome release from single cells. Analyst. 2016;141:679–88.
24. Hill B, Accu-Chek Advantage. Electrochemistry for diabetes management. Curr Sep. 2005;21:45–8.
25. Lauks IR. Microfabricated biosensors and microanalytical systems for blood analysis. Acc Chem Res. 1998;31:317–24.
26. Hedenmo M, Narváez A, Domínguez E, Katakis I. Improved mediated tyrosinase amperometric enzyme electrodes. J Electroanal Chem. 1997;425:1–11.
27. Cummings EA, Mailley P, Linquette-Mailley S, Eggins BR, McAdams ET, McFadden S. Amperometric carbon paste biosensor based on plant tissue for the determination of total flavanol content in beers. Analyst. 1998;123:1975–80.
28. Windmiller JR, Wang J. Wearable electrochemical sensors and biosensors: a review. Electroanalysis. 2013;25:29–46.
29. Bandodkar AJ, Wang J. Non-invasive wearable electrochemical sensors: a review. Trends Biotechnol. 2014;32:363–71.
30. Kim J, Valdés-Ramírez G, Bandodkar AJ, Jia W, Martinez AG, Ramírez J, Mercier P, Wang J. Non-invasive mouthguard biosensor for continuous salivary monitoring of metabolites. Analyst. 2014;139:1632–6.
31. Yao H, Shum AJ, Cowan M, Lähdesmäki I, Parviz BA. A contact lens with embedded sensor for monitoring tear glucose level. Biosens Bioelectron. 2011;26:3290–6.
32. Liao Y-T, Yao H, Lingley A, Parviz B, Otis BP. A 3-CMOS glucose sensor for wireless contact-lens tear glucose monitoring. IEEE J Solid-State Circuits. 2012;47:335–44.
33. Yao H, Liao Y, Lingley A, Afanasiev A, Lähdesmäki I, Otis B, Parviz B. A contact lens with integrated telecommunication circuit and sensors for wireless and continuous tear glucose monitoring. J Micromech Microeng. 2012;22:075007.
34. Colombo P, Sonvico F, Colombo G, Bettini R. Novel platforms for oral drug delivery. Pharm Res. 2009;26:601–11.
35. Crommelin DJ, Storm G, Luijten P. 'Personalised medicine' through 'personalised medicines': time to integrate advanced, non-invasive imaging approaches and smart drug delivery systems. Int J Pharm. 2011;415:5–8.
36. Garcia-Gradilla V, Sattayasamitsathit S, Soto F, Kuralay F, Yardımcı C, Wiitala D, Galarnyk M, Wang J. Ultrasound-propelled nanoporous gold wire for efficient drug loading and release. Small. 2014;10:4154–9.
37. Yang X, Liu X, Liu Z, Pu F, Ren J, Qu X. Near-infrared light-triggered, targeted drug delivery to cancer cells by Aptamer gated nanovehicles. Adv Mater. 2012;24:2890–5.
38. Santos GM, Zhao F, Zeng J, Shih W-C. Characterization of nanoporous gold disks for photothermal light harvesting and light-gated molecular release. Nanoscale. 2014;6:5718–24.
39. Razzacki SZ, Thwar PK, Yang M, Ugaz VM, Burns MA. Integrated microsystems for controlled drug delivery. Adv Drug Deliv Rev. 2004;56:185–98.
40. Staples M, Daniel K, Cima MJ, Langer R. Application of micro-and nano-electromechanical devices to drug delivery. Pharm Res. 2006;23:847–63.
41. Santini JT Jr, Cima MJ, Langer RS. Microchip drug delivery devices. Google Patents. 1998.
42. Sheybani R, Cobo A, Meng E. Wireless programmable electrochemical drug delivery micropump with fully integrated electrochemical dosing sensors. Biomed Microdevices. 2015;17:1–13.

Chapter 14
Future Prospects and Vision

Ajeet Kaushik, Babak Kateb, and Madhavan Nair

Abstract The nano-enabled approached for (a) nanoparticles surface chemistry needed to bind with therapeutic agent, (b) stimuli-response drug delivery and site-specific release of therapeutic agent, (c) development of pharmacological relevant personalized nanomedicine, (d) bioinformatics to manage a targeted diseases, (e) smart sensing system for diagnostics have been described in this book. However, there is a considerable explore novel biocompatible nanostructures to develop nano-formulation (NF) exhibit targeted delivery, better efficacy, and no adverse effects. Beside this novel nanostructures are also in demand to fabricate miniaturized sensing system for POC applications for rapid diagnostics needed for timey therapeutics. The state-of-the-art of advancements in personalized nanotherapeutics, related challenged, potentials future vision are discussed briefly in this chapter

Keywords Personalized nanomedicine • POC systems • Health care management

1 Discussion

Nanoscience and nanotechnology have emerged with ability to tune or investigate a materials within nano-dimension criteria to exhibit desired optical, electrical, magnetic, and molecular properties. In the beginning, these nanostructures were a subject of a debate to conclude their practicality in biomedical research. This was due to various reports confirmed toxicity produced by nanostructures, mainly metal and metal

A. Kaushik (✉) • M. Nair
Center for Personalized Nanomedicine, Institute of Neuro immune Pharmacology, Department of Immunology, Herbert Wertheim College of Medicine, Florida International University, Miami, FL, USA
e-mail: akaushik@fiu.edu

B. Kateb
National Center for NanoBioElectronics, West, Hollywood, CA, USA

California Neurosurgical Institute, Los Angeles, CA, USA

Brain Mapping Foundation, West, Hollywood, CA, USA

Society for Brain Mapping and Therapeutics, West, Hollywood, CA, USA

A. Kaushik et al. (eds.), *Advances in Personalized Nanotherapeutics*,
https://doi.org/10.1007/978-3-319-63633-7_14

oxides. Beside this, nanostructures due to the fantastic tunable properties were in demand and thus raised the demand to modify synthesize methods aiming to use in biomedical applications. Keeping this in view, various methodologies such as sol-gel, biomolecule/biomaterials-assisted synthesis have been investigated to improve the bio-computability and processing of nanomaterials for better use. Such novel nanostructures facilitated research toward development of specific nanomedicine and smart sensing systems required for the management of personalized health care [1].

Smart and intelligent nanostructures of metals (Au), metal oxides (ZnO, Fe_3O_4 etc.), conducting polymers (polyaniline, polypyrrole etc.), biopolymers (hydrogels, chitosan etc.,), and hybrids (organic-inorganic composite systems) have been synthesis in desired shape, size and surface properties resulting in advanced state-of-the-art of personalized nano-therapeutics, mainly including nanomedicine and diseases detection along with monitoring [2]. Significant efforts are continuously being made to investigate new synthesis methods and modification in state-of-the-art of nanostructures fabrication using biomolecules such DNA, biopolymers etc., to introduce biocompatibility and therapeutic feature for example Fe_3O_4 useful for sensor development, drug delivery, and chemotherapy [3, 4].

Numerous reports describe the significance the nano-enabled personalized nanomedicine capable to eradicate targeted diseases [4–7]. In spite of significant advancements in nano-pharmacology to treat peripheral diseases, a considerable efforts are required to developed nanomedicine for the treatment of brain diseases mainly brain cancer, and neurocognitive disorders including neuro-HIV/AIDS, Parkinson's, Alzheimer's etc. The CNS diseases related specific drug are unable to cross the blood-brain-barriers (BBB) which make complete treatment of these diseases very challenging [8, 9]. To develop an efficient treatment for the brain diseases, safe and non-invasive novel methods of brain delivery and stimuli responsive drug release are urgently required. Recently demonstrated one such approach is magnetically guided delivery of a nanoformulation across the BBB. For example, a nanomedicine based layer-by-layer methodology using Fe_3O_4 nanoparticles was developed to deliver across the BBB using static magnetic field (0.8 T) to eradicate neuro-HIV. This magnetic nanoformulation may contains single or multiple drugs and exhibited drug release until 5 days in sustained manner across the BBB [9]. Another approach showed magnetically guided drug delivery of magneto-electro nanoparticles (MENPs) and controlled on-demand release of anti-HIV drug for the eradication of neuroHIV/AIDS [4]. The MENPs are biocompatible act as nanovehicles to deliver therapeutic agents/anti-HIV drugs across the BBB. The 100% drug release was achieved on applying ac-magnetic magnetic field of 60 Oe through electromagnetic coils [10]. For clinical utility, MENPs have also been delivered to the mice brain under static magnetic field influence [8]. The MENPs presence in the brain did not cause any toxicity and adverse motor coordination function of mice. In spite of these significant advancements, the targeted delivery of diseases specific drug in desired concentration is focus of nano-pharmacologists and nano-bio-engineers. We believe that a significant attention must be paid towards investigating safe delivery methods developing personalized nanomedicine for efficient treatment.

Nanomaterials have proven significant contribution in disease monitoring through assisting image enhancement and processing for diseases and treatment monitoring [11]. Such image guided nanomaterial are biocompatible optically sensitive materials and can be deliver or administrated in animal showed high cell-uptake. The well proven high cell uptake of nanomaterial can considerable be useful for the nano-therapeutics and nano-neurosurgery strategies. For example, various strategies have been utilized for delivery of a therapeutic agent inside the cells through electroporation. Electroporation has been useful to facilitate treatment. However, this stimuli dependent cell membrane pore opening procedure may affect cell health and need to be controlled manner. Recently, ac-magnetically stimulated electroporation of MENPs has been demonstrated inside the brain cells [12]. This methods is very rapid, can safely be performed, and with the tunable feature-according to diseases.

Beside this, the biopolymers, for example hydrogels, have emerged as multifunctional biocompatible materials for various biomedical applications with a significant contribution in nano-therapeutics. Multi-institutional based considerable efforts have been made to scale-up gels of desired properties for translation research including mainly easy surgery, organ transplantation, tissues engineering, would healing, drug delivery and imaging guided process [13]. These smart hydrogels need lot of attention to promote them toward FDA approval and then further for clinical trials.

Nanostructures have contributed significantly in developing very efficient sensing systems to detect target biomarkers in physiological range [14]. The nano-enabled sensors are being integrated BioMEMS and interfacing with miniaturized analyzed to develop smart portable sensing systems [15–17]. The operation and management such systems is very user friendly and can be performed at POC application. Significant effort can be made in the direction to improve the performance of BioMEMS and miniaturized analyzer to make them cost-effective and precise. Beside this, sincere efforts also need toward investigating nanostructures useful for to fabrication flexible and implantable sensors capable to detect diseases at early stage.

2 Summary

Nanomaterials have contributed significantly in order to develop personalized nano-therapeutics including nanomedicine and smart sensing systems for health care management. Along with valuable on-going research projects, there is a considerable scope to promote research and funding in the area of investigating novel nano-structures for nano-neurosurgery, nano-enabled diseases monitoring, fabrication of pharmacological nanomedicine, and developing miniaturized sensing system for POC application. The outcomes of such efforts will greatly be useful to establish efficient therapies to combat against targeted diseases, to generate bioinformatics needed for diseases management, and for rapid diagnostics needed to make timely therapeutic decision.

Acknowledgments Authors acknowledge NIH grants namely RO1DA027049, RO1DA034547, RO1-DA037838, RO1DA040537, and RO1DA042706.

Competing Financial Interests Authors declare no conflict of interest

References

1. Fadeel B, Garcia-Bennett AE. Better safe than sorry: understanding the toxicological properties of inorganic nanoparticles manufactured for biomedical applications. Adv Drug Deliv Rev. 2010;62(3):362–74.
2. Thanh NTK, Green LAW. Functionalisation of nanoparticles for biomedical applications. Nano Today. 2010;5(3):213–30.
3. Gupta AK, Gupta M. Synthesis and surface engineering of iron oxide nanoparticles for biomedical applications. Biomaterials. 2005;26(18):3995–4021.
4. Kaushik A, Jayant RD, Sagar V, Nair M. The potential of magneto-electric nanocarriers for drug delivery. Expert Opin Drug Deliv. 2014;11(10):1635–46.
5. Ruiz A, Nair M, Kaushik A. Recent update in NanoCure of NeuroAIDS. Sci Lett J. 2015;4:172.
6. Nair M, Jayant RD, Kaushik A, Sagar V. Getting into the brain: potential of nanotechnology in the management of NeuroAIDS. Adv Drug Deliv Rev. 2016;103:202–17.
7. Kaushik A, Jayant RD, Nair M. Advancements in nano-enabled therapeutics for neuroHIV management. Int J Nanomedicine. 2016;11:4317.
8. Kaushik A, Jayant RD, Nikkhah-Moshaie R, Bhardwaj V, Roy U, Huang Z, Ruiz A, Yndart A, Atluri V, El-Hage N, Khalili K. Magnetically guided central nervous system delivery and toxicity evaluation of magneto-electric nanocarriers. Scientific Rep. 2016;6:25309.
9. Jayant RD, Atluri VS, Agudelo M, Sagar V, Kaushik A, Nair M. Sustained-release nanoART formulation for the treatment of neuroAIDS. Int J Nanomedicine. 2015;10:1077.
10. Nair M, Guduru R, Liang P, Hong J, Sagar V, Khizroev S. Externally controlled on-demand release of anti-HIV drug using magneto-electric nanoparticles as carriers. Nat Commun. 2013;4:1707.
11. Tomitaka A, Arami H, Raymond A, Yndart A, Kaushik A, Jayant RD, Takemura Y, Cai Y, Toborek M, Nair M. Development of magneto-plasmonic nanoparticles for multimodal image-guided therapy to the brain. Nanoscale. 2017;9(2):764–73.
12. Kaushik A, Nikkhah-Moshaie R, Sinha R, Bhardwaj V, Atluri V, Jayant RD, Yndart A, Kateb B, Pala N, Nair M. Investigation of ac-magnetic field stimulated nanoelectroporation of magneto-electric nano-drug-carrier inside CNS cells. Scientific Rep. 2017;7:45663.
13. Vashist A, Kaushik A, Vashist A, Jayant RD, Tomitaka A, Ahmad S, Gupta YK, Nair M. Recent trends on hydrogel based drug delivery systems for infectious diseases. Biomater Sci. 2016;4(11):1535–53.
14. Kaushik A, Tiwari S, Jayant RD, Vashist A, Nikkhah-Moshaie R, El-Hage N, Nair M. Electrochemical biosensors for early stage Zika diagnostics. Trends Biotechnol. 2017;35(4):308–17.
15. Kaushik A, Tiwari S, Jayant RD, Marty A, Nair M. Towards detection and diagnosis of Ebola virus disease at point-of-care. Biosens Bioelectron. 2016;75:254–72.
16. Kaushik A, Jayant RD, Tiwari S, Vashist A, Nair M. Nano-biosensors to detect beta-amyloid for Alzheimer's disease management. Biosens Bioelectron. 2016;80:273–87.
17. Kaushik A, Dixit CK. Nanobiotechnology for sensing applications: from lab to field. Oakville, Canada: Apple Academic; 2016.

Index

A. Kaushik et al. (eds.), *Advances in Personalized Nanotherapeutics*,
https://doi.org/10.1007/978-3-319-63633-7

MIX
Papier aus verantwortungsvollen Quellen
Paper from responsible sources
FSC® C105338

If you have any concerns about our products,
you can contact us on
ProductSafety@springernature.com

In case Publisher is established outside the EU,
the EU authorized representative is:
Springer Nature Customer Service Center GmbH
Europaplatz 3, 69115 Heidelberg, Germany

Printed by Libri Plureos GmbH
in Hamburg, Germany